Peptides Across The Pacific

Proceedings of the Twenty-Third American
and the Sixth International Peptide Symposium

Peptides Across The Pacific

Proceedings of the Twenty-Third American
and the Sixth International Peptide Symposium

Peptides Across The Pacific

Proceedings of the Twenty-Third American
and the Sixth International Peptide Symposium
June 22-27, 2013, Hilton Waikoloa Village, HI, U.S.A.

Edited by
Michal Lebl
Prompt Scientific Publishing
San Diego, CA
michallebl@gmail.com

American Peptide Society

Sold and distributed by www.promptpublishing.com
ISBN 978-0-9839741-3-0

Produced by Prompt Scientific Publishing, San Diego, U.S.A.
www.promptpublishing.com

Introduction - ALOHA!

We extend to you a warm and sunny Aloha in celebration of the 23rd American Peptide Symposium and the 6th International Symposium. The meeting theme, ***Peptides Across the Pacific***, embraced the spirit of the scientific and social program. ***Peptides Across the Pacific*** encompassed the important role that peptide science currently plays in so many disciplines and explored the potential impact peptides can make in scientific fields that have yet to realize the utility of these wonderful molecules.

The scientific program for 2013 was framed by distinguished lectures delivered by two renowned peptide chemists. Professor Chris Walsh of Harvard opened the Symposium with the first Distinguished Lecture, followed by a reception overlooking the Pacific Ocean. The Symposium wrapped up on with Philippine native Baldomero "Toto" Olivera of the University of Utah.

Within the framework set by the Distinguished Lectures, a series of talks were delivered by Merrifield Award Winner, James Tam of Nanyang Technological University of Singapore, Makineni Award winner, Samuel Gellman of the University of Wisconsin, the two du Vigneaud Award lecturers Kit Lam (UC Davis) and Michael Chorev (Harvard), and the Goodman Award winner, Robert Hodges of the University of Colorado at Denver.

Interspersed between Distinguished and Award talks were presentations encompassing a broad spectrum of peptide science: Bioactive Peptides, Post Translational Modification of Peptides and Proteins, Peptide Natural Products and Biosynthesis, Peptide Libraries and Arrays, Peptidomimetics, Structure and Function, Peptide Based Materials, Peptides in the Clinic, Peptide Delivery, Peptide Based Bioprobes, and Peptides and Proteins in Imaging

Augmenting the scientific program were poster presentations, including a remarkable number of young scientists who elected to be a part of the Young Investigator Poster Competition. In addition, following Symposium tradition, a series of social events were scheduled to facilitate networking, conversation and camaraderie including the Opening Reception, the closing Gala Banquet and, for the first time ever, a coffee and pastries get-together for the spouses and families of attendees.

As always, but particularly this year with the challenging economic environment, we are especially appreciative of our many sponsors and exhibitors who, without their generous support, the Symposium would not have been possible. The significant contributions of these organizations enabled us to put together a program of outstanding scientific depth and richness and a social program of delightful food, conversation and fun.

Mahalo for joining us in Hawaii. It was our pleasure and honor to assemble a week of peptide chemistry, biochemistry, and biology at the forefront of science.

David Lawrence, Co-Chair   Marcey Waters, Co-Chair

Editor's Remarks

The American Peptide Society used the same production team (scientific and technical editor) as in 2011. And again, manuscripts were accepted through the internet page, entered into the database which allowed creating indexes on the fly and simplified communication with authors. Processed manuscripts were available for proofing – helping us to correct blunders made by us and authors in the rush to deliver the book as soon as possible. The finalized book was available on the website for downloading 8 weeks after the symposium ended. The "real book" is available from "just in time" printing process at www.lulu.com. Due to the fact that the American Peptide Society is a nonprofit organization, there is no margin charged for the book, and all members can order it for production cost. Our thanks go to all authors who delivered manuscripts of a high technical quality before the deadline.

We believe that the quality of the figures and schemes is a part of the presentation skills of the authors and we let them show it. In the past we were insisting on the delivery of only monochromatic graphics with large enough details (fonts). However, since only less than 5% of all distributed books were produced in paper form (95% of proceeding books from the last symposium were downloaded free by the public as a pdf file), we accepted color graphics as well, which, of course, is showing only in pdf version of the book. The downloaded file can be displayed in "zoomed" version showing details which are not distinguishable in the printed version due to its color or size. We hope to be able to produce the proceedings of the next symposia even faster than this volume. We asked during the submission process whether it would be desirable to have the proceedings available at the time of the symposium. The answers were not uniform "yes" – only 48% scientists wanted this early availability, 28% did not want this option and the rest did not care one way or the other. The production of the proceedings during the symposium is actually achievable – the crude, unedited, version can be produced from manuscripts delivered (electronically) at the time of the symposium by our semi-automated process. The final, edited, version would be available several weeks after your return home. The requirement of delivery of your manuscript at the time of the symposium is not that draconic – it was done like that in the past when all presenters had to deliver the hard copy of their paper during the meeting. We would like to hear your opinion.

If you were downloading the pdf file from our web site, you were asked several questions. The majority of scientists (65%) want the proceedings to be produced in pdf form, about one third wants both pdf and printed book, and 3% want only the printed book. And an overwhelming majority of you expressed the wish that everything presented at the symposium should be available through the internet. Actually only 4.6% of the responders did not want everything to be available electronically – we will try to find out what was their reason. Every presentation actually exists in electronic form and therefore it would be extremely simple to make it available. We will explore this idea in the coming months and hopefully we will have an answer for you before the next symposium.

Roseann Story-Lebl
Technical Editor
rpstory@gmail.com

Michal Lebl
Scientific Editor
m@5z.com

A Message from the President of the American Peptide Society

On behalf of my fellow officers and councilors of the American Peptide Society, I was very pleased to welcome you to the 23nd American Peptide Symposium which was held jointly with the 6th International Peptide Symposium. The American Peptide Society was founded in 1990 as the result of the rapid growth of peptide-related research. The Society is a nonprofit organization dedicated to advancing and promoting knowledge of the chemistry and biology of peptides and proteins. A major component of this charter is our biennial Symposium. David Lawrence and Marcey Waters and their associates have organized a truly outstanding program for us in Kona on "The Big Island" with the theme of the meeting "Peptides Across the Pacific." This was the continuation of the theme of the 2011 APS in San Diego, which had as its theme "Building Bridges". This time we were literally building a much longer bridge.

Our Society's official journal, Biopolymers (Peptide Science) publishes both original articles and reviews covering all aspects of peptide science. The Editor-in-Chief is Joel Schneider (NCl-Frederick). He welcomes your manuscript submissions. Members also have free access to the Society's continually evolving web site (www.americanpeptidesociety.org) where the latest information on American Peptide Society activities and developments in peptide science may be found. Free professional position and resume posting is offered at the site. Through the membership of the American Peptide Society in the Federation of American Societies for Experimental Biology (FASEB), our members have a strong voice advocating for support of biomedical research. During these times of budgetary pressure on all federal discretionary funding, our advocacy efforts are critical. The American Peptide Society maintains relationships with other key scientific societies including all of the other regional peptide societies and the Protein Society. Representatives of all the peptide societies met during the Symposium to discuss ongoing activities. Due to many changes in the world of peptide science, this meeting was coming at a critical time. Many companies in the United States have closed their doors or shifted operations to the Pan-Pacific arena. Therefore, we have planned this meeting to bring scientists from that region to meet jointly with scientists from the US, Europe, Australia, and all other areas to discuss common interests and new opportunities.

The American Peptide Society strongly believes in supporting the young scientists entering our field. Travel grants to the Symposium have been provided to qualified students and postdoctoral fellows. The Society is proud to award at the Symposium the prestigious R. Bruce Merrifield, Vincent du Vigneaud, and the Rao Makineni Lecture Awards. We were pleased to announce that the recipient of this year's Merrifield Award was Dr. James Tam (Singapore), the recipients of this year's two du Vigneaud Awards were Michael Chorev (Harvard Medical School) and Kit Lam (University of California, Davis). It was also a pleasure to announce that Sam Gellman (University of Michigan) has been chosen to present this year's Rao Makineni Lecture and Robert S. Hodges for Murray Goodman Scientific Excellence & Mentorship Award.

I thank all of you who were involved in planning, organizing, and providing support for the Symposium and all those who made scientific contributions and provided exhibits. Your hard work and dedication made this a very stimulating and rewarding week. I hope you enjoyed the meeting and your stay in beautiful Hawai'i.

Ben M. Dunn
University of Florida
Gainesville

23rd AMERICAN PEPTIDE SYMPOSIUM
6th INTERNATIONAL PEPTIDE SYMPOSIUM
June 22-27, 2013
Waikoloa, Hawaii

Co-Chairs

David Lawrence
Department of Chemistry
The University of North Carolina at Chapel Hill
NC 27599

Marcey Waters
Department of Bioorganic Chemistry
The University of North Carolina at Chapel Hill
NC 27599

The Scientific Committee

Jean Chmielewski, Purdue University, Lafayette
Jianmin Gao, Boston College, Chestnut Hill
Tomi Sawyer, Aileron Therapeutics Inc., Cambridge
Helma Wennemers, ETH Zürich
Les Miranda, Amgen, Thousand Oaks
Jon Lai, Albert Einstein College of Medicine, Bronx
Jon-Paul Bingham, University of Hawaii at Manoa
Lei Liu, Tsinghua University, Beijing
Wendy Hartsock, Biotage
Kaiulani Houston, University of North Carolina
Bikash Manandhar, University of Texas, Dallas
Sujeewa Ranatunga, Moffit Cancer Center, Tampa
Johnny Pham, University of California, Irvine
Sha "Lisa" Li, Emory University, Atlanta
Melanie Blevins, University of Colorado Denver

Student Travel Award Committee

Bradley Nilsson, University of Rochester, Rochester

Student Activities Committee

Audrey Kelleman, Grace Materials Technologies, Albany
Juan Del Valle, Moffitt Cancer Center, Tampa

List of 23rd American Peptide Symposium Sponsors

The 23rd American Peptide Symposium was made possible through the generous support of the following organizations:

GOLD SPONSORS
aapptec
IPSEN

SILVER SPONSORS
Aileron Therapeutics
CEM Corporation
CovX
Ferring Pharmaceuticals
PolyPeptide Group
Protein Technologies, Inc.
Senn Chemicals

BRONZE SPONSORS
American Peptide Company, Inc.
Bachem Americas, Inc.
Bentham Science Publishers
CPC Scientific
Eli Lilly and Company
Peptisyntha S.A.

ALOHA SPONSORS
Amgen Inc.
Biotage
Escom Science Foundation
GlaxoSmith Kline
Hoffman-La Roche
Novo Nordisk

TRAVEL AWARD SPONSORS
aapptec
Bentham Science Publishers
CPC Scientific
GL Biochem (Shanghai) Ltd.
New England Biolabs
New England Peptide
Peptides International
Peptisyntha S.A.
Sigma-Aldrich

List of 23rd American Peptide Symposium Exhibitors

aapptec
Advanced ChemTech
Altoris, Inc.
American Peptide Company, Inc.
American Peptide Society
Bachem Americas, Inc.
BCN Peptides S.A.
Bentham Science Publishers
Biopeptek, Inc.
Biotage
C.A.T. GmbH&Co
CEM Corporation
CPC Scientific Inc.
CS Bio Co.
EMD Millipore
FASEB
GL Biochem (Shanghai) Ltd.
Grace/Synthetech
Intavis, Inc.
International Peptide Societies
IRIS Biotech GmbH
Peptide International, Inc.
Peptide Scientific, Inc.
Peptisyntha SA
PolyPeptide Group
Protein Technologies, Inc.
Rapp Polymere GmbH
Suzhou Tianma Pharma Group Tianji Biopharma Co., Ltd
Tianjinj Nankai Hecheng Science & Technology Co. Ltd
Wiley-Blackwell

The American Peptide Society

The American Peptide Society (APS), a nonprofit scientific and educational organization founded in 1990, provides a forum for advancing and promoting knowledge of the chemistry and biology of peptides. The approximately 450 members of the Society come from North America and from more than thirty other countries throughout the world. Establishment of the American Peptide Society was a result of the rapid worldwide growth that has occurred in peptide-related research, and of the increasing interaction of peptide scientists with virtually all fields of science. A major function of the Society is the biennial American Peptide Symposium. The Society recommends awards to outstanding peptide scientists, works to foster the professional development of its student members, interacts and coordinates activities with other national and international scientific societies, sponsors travel awards to the American Peptide Symposium, and maintains a website at www.ampepsoc.org.

The American Peptide Society is administered by Officers and Councilors who are nominated and elected by members of the Society. The Officers are: President: Robin E. Offord, Mintaka Foundation for Medical Research, President Elect: Philip Dawson, Scripps Research Institute, Secretary: DeAnna Wiegandt-Long, Bachem Americas Inc., Treasurer: Pravin T.P. Kaumaya, The Ohio State University, Past President: Ben M. Dunn, University of Florida. The councilors are: Jung-Mo Ahn, University of Texas at Dallas, Maria Bednarek, MedImmune Ltd, Michael Carrasco, Santa Clara University, Waleed Danho, Retired, Charles Deber, Hospital for Sick Kids, Emanuel Escher, Institute de Pharmacology, University of Sherbrooke, Robert P. Hammer, New England Peptide, Carrie Haskell-Luevano, University of Minnesota, Michal Lebl, Spyder Institute Praha, John Mayer, Eli Lilly and Co., Tom Muir, Rockefeller University, Laszlo Otvos, Temple University, Joel Schneider, University of Delaware, Andrei Yudin, University of Toronto.

Membership in the American Peptide Society is open to scientists throughout the world who are engaged or interested in the chemistry or biology of peptides and small proteins. For application forms or further information on the American Peptide Society, please visit the Society web site at www.americanpeptidesociety.org or contact Becci Totzke, Association Manager, P.O.Box 13796, Albuquerque, NM 87192, U.S.A., tel (505) 459-4808; fax (775) 667-5332; e-mail "APSmanager@americanpeptidesociety.org".

American Peptide Symposia

1st	1968	Saul Lande & Boris Weinstein	New Haven	CT
2nd	1970	F. Merlin Bumpus	Cleveland	OH
3rd	1972	Johannes Meienhofer	Boston	MA
4th	1975	Roderich Walter	New York	NY
5th	1977	Murray Goodman	San Diego	CA
6th	1979	Erhard Gross	Washington	DC
7th	1981	Daniel H. Rich	Madison	WI
8th	1983	Victor J. Hruby	Tucson	AZ
9th	1985	Kenneth D. Kopple & Charles M. Deber	Ontario	Canada
10th	1987	Garland R. Marshall	St. Louis	MO
11th	1989	Jean E. Rivier	San Diego	CA
12th	1991	John A. Smith	Cambridge	MA
13th	1993	Robert S. Hodges	Alberta	Canada
14th	1995	Pravin T.P. Kaumaya	Columbus	OH
15th	1997	James P. Tam	Nashville	TN
16th	1999	George Barany & Gregg B. Fields	Minneapolis	MN
17th	2001	Richard A. Houghten & Michal Lebl	San Diego	CA
18th	2003	Michael Chorev & Tomi K. Sawyer	Boston	MA
19th	2005	Jeffery W. Kelly & Tom W. Muir	San Diego	CA
20th	2007	Emanuel Escher & William D. Lubell	Quebec	Canada
21st	2009	Richard DiMarchi & Hank Mosberg	Bloomington	IN
22nd	2011	Philip Dawson & Joel Schneider	San Diego	CA
23rd	2013	David Lawrence & Marcey Waters	Waikoloa	HI

2013 R. Bruce Merrifield Award

(previously the Alan E. Pierce Award) The Merrifield Award was endowed by Dr. Rao Makineni in 1997, in honor of R. Bruce Merrifield (1984 Nobel Prize in Chemistry), inventor of solid phase peptide synthesis. Previously, it was called the Alan E. Pierce Award and was sponsored by the Pierce Chemical Company from 1977-1995.

James P. Tam

James P. Tam is the Director of the Drug Discovery Laboratory at Nanyang Technological University, Singapore. He was the founding dean of the School of Biological Sciences, founding director of the double-degree program in Biomedical Science and Chinese Medicine, and the founding director of NTU Biological Research Center.

His research focuses on synthetic methodology, drug design of metabolic-stable peptidyl biologics and intracellular delivery of peptides. He developed peptide dendrimers as synthetic vaccines. His current research also includes herbalomics in traditional medicines to discover novel peptides as potential therapeutics.

He received his Ph.D. in Medicinal Chemistry from the University of Wisconsin, Madison, USA and held appointments as Associate Professor at The Rockefeller University, USA (1982-1991), Professor at Vanderbilt University, USA (1991-2004) and The Scripps Research Institute, USA (2004-2008). Professor Tam has published more than 330 papers in these areas of research. He received the Vincent du Vigneaud Award in 1986, the Rao Makineni Award by American Peptide Society in 2003, the Ralph F. Hirschmann Award by the American Chemical Society (ACS) in 2005, and the Merrifield Award by American Peptide Society in 2013 for his outstanding contributions to peptide and protein sciences. In addition to his scientific research, he has also been active in the peptide community. Besides serving on many editorial boards, he organized international peptide and protein symposia and was co-founder of the past ten International Chinese Peptide Symposia. He received the Cathay Award from the Chinese Peptide Society, China in 1996.

R. Bruce Merrifield Award winners

2013 - James P. Tam, Nanyang Technical University, Singapore
2011 - Richard DiMarchi, Indiana University
2009 - Stephen Kent, University of Chicago
2007 - Isabella Karle, Naval Research Laboratory, D.C.
2005 - Richard A. Houghten Torrey Pines Institute for Molecular Studies
2003 - William F. DeGrado, University of Pennsylvania
2001 - Garland R. Marshall, Washington University Medical School
1999 - Daniel H. Rich, University of Wisconsin-Madison
1997 - Shumpei Sakakibara, Peptide Institute, Inc.
1995 - John M. Stewart, University of Colorado-Denver
1993 - Victor J. Hruby, University of Arizona
1991 - Daniel F. Veber, Merck Sharp & Dohme
1989 - Murray Goodman, University of California-San Diego
1987 - Choh Hao Li, University of California-San Francisco
1985 - Robert Schwyzer, Swiss Federal Institute of Technology
1983 - Ralph F. Hirschmann, Merck Sharp & Dohme
1981 - Klaus Hofmann, University of Pittsburgh, School of Medicine
1979 - Bruce Merrifield, The Rockefeller University
1977 - Miklos Bodansky, Case Western Reserve University

2013 Vincent du Vigneaud Award

Sponsored by BACHEM Inc.

Kit S. Lam

Professor Kit Lam is a physician-scientist and an expert in combinatorial chemistry, peptide chemistry, chemical biology, drug discovery and development, molecular imaging, nanotherapeutics and medical oncology. He obtained his B.A. in Microbiology in 1975 at the University of Texas at Austin, his Ph.D. in Oncology in 1980 from McArdle Laboratory for Cancer Research, University of Wisconsin, and his M.D. in 1984 from Stanford University School of Medicine. He completed his Internal Medicine residency training and Medical Oncology Fellowship training at the University of Arizona. He is board certified in both Internal Medicine and Medical Oncology. He is currently Chair of the Department of Biochemistry and Molecular Medicine, University of California Davis School of Medicine, Professor of Hematology and Oncology, a leader of the UC Davis Comprehensive Cancer Center, and a Fellow of the American College of Physicians. He has made a seminal contribution to the peptide field through the development of the one-bead-one-compound (OBOC) approach to combinatorial chemistry.

He is a founding scientist of the Selectide Corporation, one of the first start-up companies to specialize in combinatorial chemistry. He has published over 290 peer-reviewed scientific publications and holds 15 patents on inventions.

Over the last two decades, Professor Lam has made a number of advances in the chemistry and screening of OBOC combinatorial methods. He successfully applied these methods for B-cell epitope mapping, discovery of cancer targeting ligands for cancer imaging and therapy, and the development of on-demand protease cleavable linker for radioimmunotherapy. More recently, he and his colleagues reported the successful development of LLP2A-bisphosphonate conjugate that facilitates the homing of mesenchymal stem cells to the bone matrix as a potential treatment for osteoporosis. LLP2A is a high-affinity and high-specificity peptidomimetic ligand against activated $\alpha 4\beta 1$ integrin discovered by the OBOC technology. In addition to screening methods, he has developed novel encoding strategies with topographically segregated bilayer beads, such that library compounds reside on the bead surface and the coding tags reside in the bead interior. He invented the one-bead-two-compound library method for the efficient discovery of pro-apoptotic ligands against cancer. In the last few years, Professor Lam has expanded his research to the development of targeting nanocarriers for cancer imaging and therapy. Using Fmoc-chemistry, multi-gram quantities of amphiphilic telodendrimers comprised of PEG-dendritic lysine/cholic acid can be prepared and used as targeting reversible micellar nanocarriers for drug delivery and cancer imaging.

2013 Vincent du Vigneaud Award

Sponsored by BACHEM Inc.

Michael Chorev

Michael Chorev's contributions to biomedical sciences in general and peptide science in particular are characterized by an integrated approach that combines basic and translational research that is inter- and multi-disciplinary in nature. This research covers subjects in organic-, medicinal-, and bioorganic-chemistries and involves small molecules, peptides and proteins. Through the years, the topics of his research projects changed but were always aimed at furthering understanding of the mechanism of action of bioactive molecules, studying structure-conformation-activity relationships, designing and synthesizing molecular tools and probes, and developing methodologies to accomplish the above. Prof. Chorev success in establishing expansive, extensive and meaningful collaborations with colleagues was always based on bringing together complementing expertise and overlapping interests. Prof. Chorev has published extensively in top peer-reviewed journals, authored several authoritative reviews and book chapters, and was recognized as a co-inventor on more than forty patents. Prof. Chorev co-chaired the 18th American Peptide Symposium and served during 2003-2009 as a council member of the American Peptide Society.

Initially, as a postdoctoral fellow in Prof. Murray Goodman's laboratory and later as a junior faculty, Michael Chorev has been a major formulator of concepts in the retro-inverso peptide chemistry and pioneered the development of the partial retro-inverso modification and the end-group modified retro-inverso peptides. These contributions invigorated a neglected topochemical approach and stimulated a new surge of very fruitful and diverse range of synthetic, structural, and biological investigations.

Prof. Chorev's long lasting interest in G-protein coupled receptors (GPCRs) led to the study of ligands such as enkephalins, substance P, CCK, osteogenic groMh peptide (OGP), parathyroid hormone (PTH), PTH-related protein (PTHTP), and more recently melanotropins. Highlights include preparation of highly potent partial retro-inverso enkephalins, development of peptide bond surrogate containing highly selective agonists of the neurokinin receptor subtypes that become standard tools in neurobiology, structure-activity-conformation relationship studies of antagonists and agonists derived from PTH and PTHTP, and the application of an integrated approach that combines photoaffinity crosslinking studies, conformational analysis and molecular simulations leading to the development of an experimentally-based model for PTH interaction with its cognate GPCR. Prof Chorev's recent introduction of the intramolecular side chain-to-side chain copper(I)-catalyzed Huisgen's azide-alkyne 1,3-cycloaddition (CuAAC) click reaction as a method to stabilize secondary structures of othervise unstructured linear peptides included extensive development of enabling synthetic methodology. This novel heterodetic modification was applied to generate highly potent MTII antagonists as well as potent inhibitors of elF4E/elF4G protein-protein interaction.

More recently, Prof. Chorev's contributions enabled the successful development of an ELISA able to identify glycated-CD59 (gCD59), a novel pathologically relevant biomarker for glycemic control in diabetes. This effort required the development of methods to synthesize this post-translationally modified antigen, immunogen and gCD59-surrogate. This approach offers a general strategy to overcome the inherent difficulty associated with the scarcity and inherent heterogeneity of post- translationally modified molecular tools.

Prof. Chorev's translational research activities included small molecules - a marketed one for treating patients with moderate to mild Alzheimer's disease and one that is currently in clinical trials for treating mild cognition impairment (MCI). In addition, he recently cofounded a startup targeting the development and commercialization of CD59-based diagnostics.

Vincent du Vigneaud Award winners

2013 - Kit Lam, University of California, Davis
2013 - Michael Chorev, University of Colorado Denver
2011 - Fernando Albericio, University of Barcelona
2011 - Morten Meldal Carlsberg Laboratory, Copenhagen
2010 - Phil Dawson, Scripps Research Institute
2010 - Reza Ghadiri, Scripps Research Institute
2008 - Jeffery W. Kelly, Scripps Research Institute
2008 - Tom W. Muir, Rockefeller University
2006 - Samuel H. Gellman, University of Wisconsin
2006 - Barbara Imperiali, Massachusetts Institute of Technology
2004 - Stephen B. H. Kent, University of Chicago
2004 - Dieter Seebach, Swiss Federal Institute of Technology, Zurich
2002 - Horst Kessler, Technical University of Munich
2002 - Robert Hodges, School of Medicine, University of Colorado
2000 - Charles M. Deber, University of Toronto
2000 - Richard A. Houghten, Torrey Pines Institute for Molecular Studies
1998 - Peter W. Schiller, Clinical Research Institute of Montreal
1998 - James A. Wells, Genentech, Inc
1996 - Arthur M. Felix, Hoffmann-La Roche, Inc.
1996 - Richard G. Hiskey, University of North Carolina
1994 - George Barany, University of Minnesota, Minneapolis
1994 - Garland R. Marshall, Washington University Medical School, St. Louis
1992 - Isabella L. Karle, Naval Research Laboratory
1992 - Wylie W. Vale, The Salk Institute for Biological Studies
1990 - Daniel H. Rich, University of Wisconsin-Madison
1990 - Jean E. Rivier, The Salk Institute for Biological Studies
1988 - William F. De Grado, DuPont Central Research
1988 - Tomi K. Sawyer, The Upjohn Company
1986 - Roger M. Freidinger, Merck Sharpe & Dohme
1986 - Michael Rosenblatt, Massachusetts General Hospital
1986 - James P. Tam, The Rockefeller University
1984 - Betty Sue Eipper, The Johns Hopkins University
1984 - Lila M. Gierasch, University of Delaware
1984 - Richard E. Mains, The Johns Hopkins University

2013 The Rao Makineni Lectureship

Endowed by PolyPeptide Laboratories and Murray and Zelda Goodman (2003)

Samuel H. Gellman

Sam Gellman is the Ralph F. Hirschmann Professor of Chemistry at the University of Wisconsin-Madison. He earned his A.B. from Harvard University in 1981 and his Ph.D. from Columbia University, under Ronald Breslow, in 1986. After an NIH post-doctoral fellowship at the California Institute of Technology, with Peter Dervan, Gellman joined the faculty at the University of Wisconsin-Madison in 1987.

The work from Gellman's laboratory has been recognized by the Ralph F. Hirschmann Award in Peptide Chemistry from the American Chemical Society in 2007, the Vincent du Vigneaud Award from the American Peptide Society in 2006 and the Arthur C. Cope Scholar Award from the American Chemical Society in 1997. Gellman was elected to the American Academy of Arts and Sciences in 2010. He has served on the National Institutes of Health Medicinal Chemistry Study Section (1999-2002) and several editorial advisory boards (the Journal of Organic Chemistry, the European Journal of Organic Chemistry, Biopolymers-Peptide Science, Chemical Society Reviews and Organic and Biomolecular Chemistry). Major interests in Gellman's research program have included fundamental studies of non-covalent interactions, elucidation of the origins of peptide and protein folding preferences, development and application of unnatural oligomers that display protein-like conformational behavior ("foldamers"), creation of new amphiphiles for membrane protein manipulation, and development of new biologically active polymers.

The Rao Makineni Lectureship winners

2013 - Sam Gellman, University of Wisconsin, Madison
2011 - Jeffery W. Kelly, The Scripps Research Institute
2009 - William DeGrado, University of Pennyslvania
2007 - Ronald T. Raines, University of Wisconsin - Madison
2005 - Robin E. Offord, Centre Medical Universitaire, Switzerland
2003 - James P. Tam, Vanderbilt University

The 2013 Murray Goodman Scientific Excellence & Mentorship Award

The Goodman Award recognizes an individual who has demonstrated career-long research excellence in the field of peptide science. In addition, the selected individual should have been responsible for significant mentorship and training of students, post-doctoral fellows, and/or other co-workers. The Awards Committee may also take into account any important contributions to the peptide science community made by the candidate, for example through leadership in the American Peptide Society and/or its journals. Endowed by Zelda Goodman (2007).

Robert S. Hodges

Bob Hodges graduated with his Ph.D. in Biochemistry from the University of Alberta in 1971. He then joined the laboratory of Dr. Bruce Merrifield at Rockefeller University from 1971-1974 where he used solid-phase peptide synthesis to study the enzyme, Ribonuclease. He left the Merrifield lab to accept a position as Assistant Professor of Biochemistry at the University of Alberta and became a founding member of the famous Medical Research Council Group in Protein Structure and Function where he remained for more than 25 years. In 1990 Dr. Hodges also joined two Networks of Centers of Excellence which involved outstanding researchers from across Canada to work together on research projects that bridged the gap between academia and industry. These Networks included the Canadian Bacterial Diseases Network (CBDN) and the Protein Engineering Network of Centers of Excellence (PENCE). In 1994 Dr. Hodges took over the leadership of PENCE from Michael Smith, Nobel laureate, and headed this network for 6 years. In 2000 he moved to the University of Colorado Denver, School of Medicine to accept the position as Director of the Program in Biomolecular Structure, Professor of Biochemistry and Molecular Genetics and holder of the John Stewart Endowed Chair in Peptide Chemistry. In the U.S. he was successful at obtaining NIH R01 grants in diverse areas of science all applying peptide chemistry to solve questions about peptides and proteins of biological interest. Bob has won many outstanding awards in both Canada and the U.S. among which are the Distinguished Medical Research Council of Canada Scientist Award (the MRC career awards were considered the most prestigious of such awards in Canada (1995-2000). In 1995 he won the Boehringer-Mannheim Award from the Canadian Society of Biochemistry and Molecular Biology (in recognition of a record of outstanding achievements in research in the field of biochemistry undertaken in Canada by a Canadian Scientist). The Alberta Science and Technology Award for outstanding leadership in Alberta Science in 1995. In 2002 he won the Vincent Du Vigneaud Award from the American Peptide Society for outstanding achievements in peptide research.

Bob has served the peptide science community in a variety of roles. He chaired the 1993 American Peptide Symposium in Edmonton, Alberta. Interestingly, this meeting had the largest number of attendees of any American Peptide Symposium to date. From 1995-1999 he served as President-elect and President of the American Peptide Society. He was Co-chair of the Gordon Research Conference on the Chemistry and Biology of Peptides in 2006. Inventor of the year award for 2009 at University of Colorado Denver.

Dr. Hodges has used peptide chemistry in an exceptionally creative and innovative manner to investigate major challenges in biomedical research. He has published over 500 publications in his career. He is presently focusing his research in the following areas:
(i) Development of a "universal" synthetic peptide influenza A vaccine which is based on the hypothesis that immunization with conformation-constrained peptides derived from highly conserved helical sequences in the stem region of influenza A hemagglutinin (HA) will generate antipeptide antibodies that recognize conformation-dependent epitopes in HA to prevent influenza infection.

(ii) Understanding the relationship of sequence to protein folding, stability and function. We recently made the first identification of a stability control region (SCR) in a protein. This SCR 97-118 in tropomyosin (a 284-residue regulatory protein involved in muscle contraction) transmits information along the two-stranded α-helical coiled-coil to control protein stability (J. Structural Biology, 2010).

(iii) The rational design of α-helical antimicrobial peptides to target Gram-negative pathogens, Acinetobacter baumannii and Pseudomonas aeruginosa (Chem. Biol. and Drug Design, 2011) and Gram-positive pathogens, Methicillin-resistant Staphylococcus aureus to solve the rapidly growing problem of increased resistance to traditional antibiotics, a major issue in human health.

(iv) Development of new HPLC methodology. Our design of peptide standards of the same composition and minimal sequence variation to monitor performance and selectivity of reversed-phase matrices represents a breakthrough in assessing new column packing materials (J. Chrom. A, 2012). We recently introduced a new dimension to traditional hydrophilic interaction chromatography (HILIC) of peptides, termed HILIC/SALT. This approach promises to enable HILIC packings to enter the mainstream of approaches for peptide separations.

(v) Developing anticancer peptidomimetics (500-700 daltons) with nanomolar broad spectrum activity against 9 cancer types (non-small cell lung, prostate, breast, colon, ovarian, renal cancer, melanoma, leukemia and CNS).

Murray Goodman Scientific Excellence & Mentorship Award winners

2013 - Robert Hodges, University of Colorado, Denver
2011 - Victor J. Hruby, University of Arizona
2009 - Charles M. Deber, University of Toronto, Hospital for Sick Children

Young Investigators' Competition

Poster Competition

Grand Prize	Alexander Spokoyny	MIT, Cambridge
First	Christopher Lohans	University of Alberta, Edmonton
First	Jason Arsenault	MRC Laboratory of Molecular Biology, Cambridge
First	Jerrin Kuriakose	Purdue University, West Lafayette
Second	Matteo De Poli	University of Manchester, Manchester
Second	Simon Gregersen	University of Copenhagen, Frederiksberg
Second	Conan Wang	The University of Queensland, Brisbane

Speaker Competition

Outstanding	Chenrui Chen	Emory University, Atlanta
Outstanding	Tobias Postma	Institute for Research in Biomedicine, Barcelona
Outstanding	Yannan Zhao	The Scripps Research Institute, San Diego

Contents

Proceedings of the 23rd American Peptide Symposium
Michal Lebl (Editor)
American Peptide Society, 2013

Design and Synthesis of Peptide Biologics by Deconstruction of Proteins

James P. Tam

School of Biological Sciences, Nanyang Technological University,
60 Nanyang Drive, Singapore, 637551

Introduction

It is indeed a great honor to receive the Merrifield Award and a privilege to join the list of luminous winners. I would like to thank the Award Committee and my nominators for this honor, and Dr. Rao Makineni for his generous donation to make this award possible.

This year marks the 50th anniversary of Bruce's first paper describing the concept of solid-phase synthesis, a groundbreaking paper published in 1963 in the Journal of American Chemical Society. His discovery was recognized by the Nobel Committee for the 1984 Nobel Prize in Chemistry and the American Chemical Society as one of the Top-ten inventions.

I began my peptide research with Professor Daniel Rich at the University of Wisconsin-Madison to work on a fungal dehydro-tetrapeptide, tentoxin and its related analogs. My synthesis of tentoxin by solution synthesis took up a large part of my thesis years, and I believe my thesis work would have been accelerated using the solid-phase method. But synthetic problems associated with tentoxin such as the acid-catalyzed racemization of N-methylated amino acids was not completely understood, and I would revisit this issue many years later.

I met Bruce Merrifield in 1971 in a small peptide conference in Madison, Wisconsin and then again, in the Fourth American Peptide Symposium held at New York City in 1975. I was impressed not only by his science but also his comforting manner. I joined his laboratory in 1976.

The goals of the Merrifield laboratory at the Rockefeller University were methodology development of peptides and proteins as well as their applications pertaining to biology. For five years, I worked on various aspects of solid-phase methodology development together with a very talented group of young scientists. Among them were Alex Mitchell, Bruce Erickson, Stephen Kent, George Barany, and Richard DiMarchi. In 1981, Bruce asked me to set up my own laboratory, and I decided to work on synthetic vaccines. In today's talk, I will present some of our work on the methodology development of chemically defined peptide vaccines, and our earlier work on peptide synthesis.

Sequence-specific antibodies

In the late 1970s and the early 1980s, several laboratories including Micheal Sela and Ruth Arnon showed that synthetic peptides conjugated to a protein carrier can induce anti-peptide antibodies reactive with their cognate sequences in the native proteins [1,2]. This finding was timely because it enhanced the application and acceptance of solid-phase synthesis for preparing anti-peptide antibodies as a useful tool in the emerging field of recombinant DNA methodology and molecular biology. Also during 1980s, old and emerging infectious diseases such as malaria and HIV were unresolved public health issues and working on a peptide-based approach to develop a synthetic vaccine based on anti-peptide antibodies, appeared to be a promising research area for my laboratory. In the next 30 years, our laboratory was engaged in developing synthetic vaccines, first against malaria and then against HIV.

Tetanus toxoid-conjugated vaccine for malaria parasites

It was my good fortune to find two excellent collaborators in the malaria field, Ruth and Victor Nussenzweig who worked at the New York University Medical School. NYU is located about 2 miles from the Rockefeller University, and such a short distance between our laboratories allowed my frequent visits to NYU to move the project along. We had a productive collaboration until I moved from Rockefeller University to Vanderbilt University in 1991.

The lead in developing a sporozoite-based malaria vaccine was derived from the work by Ruth and her coworkers who showed that X-ray irradiated sporozoites induce protective immunity in animal models [3]. The sporozoite stage of malaria parasites carries a protein on its outer surface, which expresses a unique immunodominant epitope recognized by immunized or repeatedly infected hosts. Sera from mice immunized with *Plasmodium berghei* sporozoites immunoprecipitate a single 44,000 Mr protein, the circumsporozoite (CS) protein, from extracts of surface-labeled sporozoite. The CS protein contains a central region encoding two types of tandemly repeated amino acid units, flanked by nonrepeated regions encoding amino- and carboxyl-terminal signal and anchor-like sequences, respectively. One of the central repeated amino acid unit types contains the immunodominant epitopes. Interestingly, all CS proteins from different strains contain a region consisting of tandemly repeating peptides, which are an immunodominant B cell epitopes. These findings provided an essential clue for designing a peptide-based vaccine against malaria parasites [4].

Table 1. Protection from malaria infection by P. berghei sporozoites after immunization with a synthetic peptide vaccine.

Immunogen	Titers x 10 $^{-5}$ (IFA)	Protected/ Challenged	Protect (%)
Peptide 17.1-TT	16-32	7/8	87
X-irradiated sporozoites	4-16	6/7	85
TT	Negative	0/8	0
None	Negative	0/5	0

17.1 B epitope of P. berghei (DPPPPNPN)$_2$D, TT: tetanus toxoid.

Our first-generation design of malaria vaccine was based on the conventional method of conjugating a peptide with the repeating sequences (the immunodominant epitope) from the CS protein to a protein carrier (tetanus toxoid) using a crosslinking reagent [5]. In comparing different peptide epitopes (17.1 DPPPPNPNDPPPPNPND vs 17.2 DPAPPNANDPAPPNAND) and crosslinking reagents (glutaraldehyde vs bisdiazo-benzidine, BDB), we found that tetanus toxoid as the protein carrier conjugated to peptide 17.1 and BDB as the crosslinking reagent provided the best result with 87% protection against malaria parasite-challenged rodent model (Table 1). Importantly, our results achieved the same level of protection from the parasite as the immunization with the X-irradiated sporozoites.

Peptide dendrimers

To advance to the next step in designing a synthetic malaria vaccine for humans, we had to overcome several hurdles. They include producing an immunogen with reproducible chemical composition, the choice of a suitable adjuvant to replace tetanus toxoid as a carrier, which provides T-helper epitopes. My priority was to replace the protein carrier by a peptide carrier. The requirements for a peptide carrier would be diametrical opposite to a protein carrier, which is generally large and highly immunogenic whereas a peptide carrier would be small and immunologically silent (Table 2). Personally, I found the control polymerization to prepare dendrimers has a certain conceptual similarity to solid-phase peptide synthesis in which peptides are growing stepwise from a polystyrene bead. It is, however, important to point out the large size differences in the core used in producing dendrimers and the polystyrene bead in the solid-phase synthesis. From a chemical point of view, a disadvantage of peptide dendrimers prepared directly by solid-phase method is the difficulty of obtaining products of high homogeneity.

Table 2. Comparison of protein carrier and peptide dendrimer as an antigen-carrier.

	Protein carrier (Keyhole limpet hemocyanin)	Peptide dendrimer (Octavalent MAP)
Carrier		
Size	Large, MW >300 kDa	Small, MW <2 kDa
Immunogenicity	Active	Silent
Multivalency	Generally unknown	Known
Copies of antigens conjugated to a carrier	Generally not controllable	Controllable
Chemical composition	Ambiguous	Defined
Spatial geometry of antigens conjugated to a carrier	Ambiguous	Defined
Carrier/conjugated vaccine (% weight) with peptide antigen about 2 kDa	>90%	<10 %

In the early 1980s, R.G. Denkewalter at Allied Chemicals described the use of poly-lysine to form dispersed polymers which could have potential application for drug delivery [6]. This idea seemed promising for designing chemically defined vaccine. In 1984, my laboratory began to experiment with branched polyamino acids such as Asp, Glu, Lys and Orn as a replacement for a protein carrier and our initial work appeared promising. In 1985, we began a long and tedious patent process to use branched poly-lysine and poly-ornithine as protein carriers for vaccine purposes (Figure 1) [7].

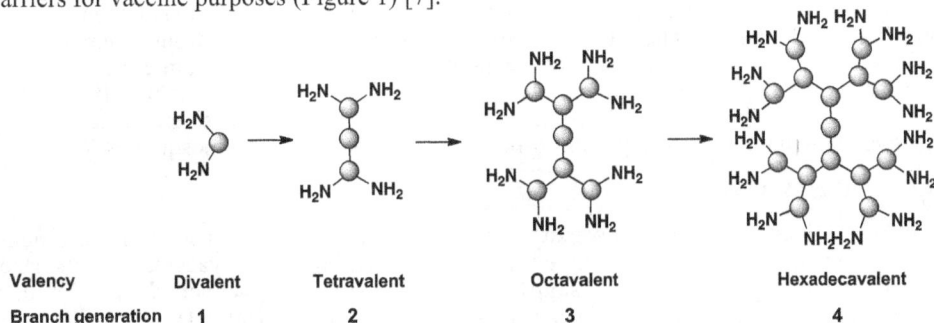

Valency	Divalent	Tetravalent	Octavalent	Hexadecavalent
Branch generation	1	2	3	4

Fig. 1. Multivalent lysine- or ornithine-based peptide dendrimers with increasing branch generations.

During mid-1980s, Tomalia described a new class of polymer called star-burst dendrimers [8]. Star-burst dendrimers are produced by controlled polymerization using a multifunctional group, such as a tri- or tetravalent amino core which is used to tether a defined generation of sequences. This results in producing a radically branched polymer of 10 or more levels with >100 reactive ends [9]. The difference between Tomalia and Denkewalter's approach is the nature of the building block.

In a peptide antigen-conjugated to protein-carrier, the peptide antigen represents only a small portion of the peptide conjugate. Thus the antibody response would be directed mostly against other antigenic sites of the complex. The carrier may suppress the B cell response to the peptide antigen of interest. And, conjugating to the carrier may radically alter the immunogenic determinants of the peptide. Peptide dendrimers are immunologically focused because of the clustered effects of peptide antigens whereas its lysine core is immunologically silent. This is in strong contrast to the peptide-protein carrier in which the desired peptide antigen is immunological scarce but the carrier is large and immunologically active with both T- and B-helper epitopes [10].

The question was whether a peptide immunogen can be produced using peptide alone without a protein carrier. Based on the chemical dendrimer approach, we introduced a cascade peptide dendrimer design in 1988 for preparing chemically-defined peptide vaccines [11]. We tested our concepts to elicit antibodies against specific proteins. We found the idea worked well and we call the design as multiple antigen peptides (MAP) [7,12].

The second challenge in producing a synthetic vaccine is immunogenicity. In our malaria work, we found that the quality and quantity of anti-peptide antibodies were highly dependent on the strains of the immunized inbred mice. In previous experiments, we showed that immunization with the *P. berghei* B epitope alone as a monomer or as an octameric MAP did not elicit antibody responses in A/J mice and several other inbred strains of mice. B-cell epitope alone did not elicit the desired sequence-dependent antibodies in certain strain of inbred mice and the inclusion of a T-helper epitope was necessary. Thus, to overcome this deficiency, we incorporated the malaria parasite-derived Th epitopes into the MAPs design, and which could be recognized by individuals of diverse genetic backgrounds. The question is the stoichiometry and the orientation of the B- and T-cell epitope [13].

To study these problems, we prepared various MAP models containing T and B epitopes of the CS protein of the rodent malaria parasite, *P. berghei*. For our diepitope MAP constructs, we chose the immunodominant B cell epitope of the CS protein which was known to be a 16-residue peptide (PPPPNPND)$_2$ and a T helper epitope KQIRDSITEEWS (aa 265-276 of CS protein). Our work is significant because it provides a validation for a chemically defined, peptide-based vaccine [14]. Again using one of our MAP design, we achieved 80% protection against malaria parasite challenged rodent models [13]. I left the malaria research in 1991.

Why no malaria vaccine is available today

There are several contributing factors; 1. The difficulty to express malarial recombinant proteins, especially the CS protein from *P. falciparum,* 2. Lack of harmonization - several groups, US Army and Navy, the French and EU research labs, each had their vaccine candidates and small trials, but the lack of harmonization and strong support from pharmaceutical companies as well as geopolitics have prevented a large field trial, 3. Disincentive for large pharmas to produce a vaccine for malaria because of the effectiveness of therapeutics. Artimesian which has its origin from traditional Chinese medicine is highly effective in disrupting regional person-to-person transmission through mosquitoes.

Lipidated HIV vaccine

In 1991, I joined Vanderbilt University and applied the lessons learned from developing malaria vaccine to HIV. The principle target for developing a HIV vaccine is the surface glycoprotein, consisting two subunits, gp120 and gp41. However, there are stark differences between the CS protein of malaria parasites and the S proteins of HIV. In malaria, the immunodominant epitopes can be found in the central repeating region. In HIV, the immunodominant epitopes are found in the variable loops, and thus it would be difficult to develop a vaccine that could be specific for different HIV strains. But, there were three pressing unresolved issues including (1) cytotoxic T lymphocytes (CTL) response, (2) adjuvant, (3) expressing mucosal immunity to elicit (Figure 2).

A peptide model based on the mimicry of surface coat protein of a pathogen was applied to developing HIV vaccines [15]. The key components of the model consisted of a tetravalent lysine scaffold to amplify peptide antigens covalently 4-fold. There was strong evidence that T-cytotoxic epitope is required to clear viral infection. Thus, we incorporated one or more peptide antigen derived from the third variable domain of glycoprotein gp120 in our multivalent antigen peptide (MAP) as HIV vaccine [16]. The antigen-conjugated peptide dendrimers were linked to tripalmitoyl-S-glycerylcysteine (P3C), a B-cell antigen that could induce CTL *in vivo* [17,18]. The complex MAP-P3C, in liposome or micelle, was used to immunize animals without any adjuvant, neutralize virus infectivity *in vitro*, elicit cytokine production, and prime CD8$^+$ cytotoxic T lymphocytes *in vivo*. Since HIV is a sexually transmitted disease, IgA production is important. The MAP-P3C HIV vaccine was found to induce mucosal immunization, systemic humoral and cellular responses irrespective of the route (oral or nasal administration) or method of delivery (liposome or microparticles) and without requiring the use of a carrier or an extraneous adjuvant [19,20].

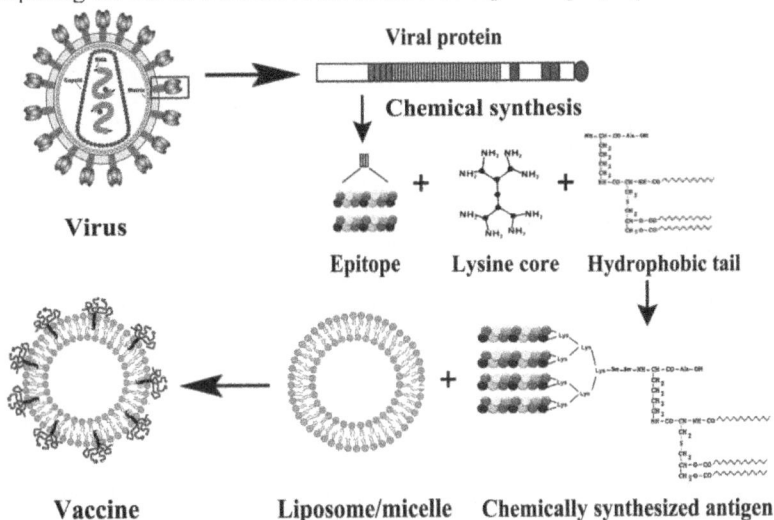

Fig. 2. Development of chemically defined Anti-HIV vaccine using lipidated peptide dendrimers.

A recurring problem in our development of peptide dendrimers as synthetic vaccines is the quality of the antibodies. Although we have used different lengths of peptide antigens, we found that it is difficult to obtain high-affinity antibodies. To a large extent, peptide antigens are largely disordered and lack the conformation of their proteins from which they are derived. To overcome this deficiency, we developed another set of HIV vaccine targeting the helical regions of gp41 using the dendrimeric approach to form the "quaternary protein mimetics" of the trimeric helical bundle of gp41 [21]. During our development of the quaternary mimetics of HIV vaccines, Peter Kim and other laboratories found that the HR1 and HR2 regions were effective anti-virals [22]. Wild's group reported the first anti-HIV peptide T20 isolated from gp41 helical region [23]. In the 18th and 20th American Peptide Symposiums, we reported the quaternary protein mimetics of T20 as a highly effective antiviral against HIV, which is >6,000-time more stable than T20 in the serum [24,25].

DNA vaccine

In 2000, I moved from Vanderbilt University to the Scripps Research Institute, Florida, where I continued my research work on HIV. During this time, I also took up the responsibility of founding the School of Biological Sciences and Bioscience research center at Nanyang Technological University. At NTU, I was interested in developing DNA-based vaccines and the challenge of its delivery, which requires condensing the extended structure of a DNA to a supercoiled DNA particle. It happened that one of the leading biochemists in DNA compactin, Lars Nordenskiold, joined NTU. Together with Lars, we exploited the dendrimer design for a synthetic vaccine using gene delivery. The spherical design of the conventional cascade type of peptide dendrimer was found not to be suitable to confer the compactin property but the pendant type of peptide dendrimer works well to compact multiple DNA segments as a synthetic vaccine (Figure 3) [26,27].

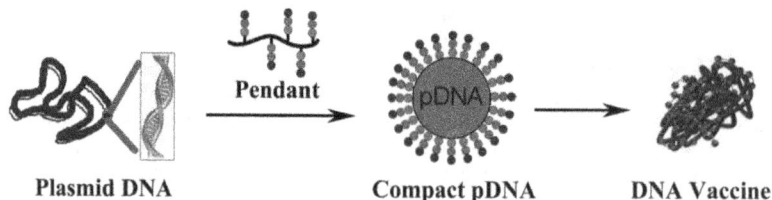

Plasmid DNA   **Compact pDNA**   **DNA Vaccine**

Fig. 3. Illustration of concept and synthetic scheme of DNA vaccine using pendant-type peptide dendrimers.

Antimicrobial dendrimers

In 2001-2002, we reported antimicrobial dendrimers. The R4 tetrapeptide (RLYR) contains a putative microbial surface recognition BHHB motif (B, basic, H, hydrophobic amino acid) found in protegrins and tachyplesins whereas the octapeptide R8 (RLYRKVYG) consists of an R4 and a degenerated R4 repeat. Antimicrobial assays against 10 organisms in high- and low-salt conditions showed that the R4 and R8 monomers as well as their divalent dendrimers contain no to low activity. In contrast, the tetra- and octavalent R4 and R8 dendrimers are broadly active under either conditions, exhibiting relatively similar potency with minimal inhibition concentrations <1 μM against both bacteria and fungi [28]. It is the first time to show a short tetrapeptide as dendrimer can be a potent antimicrobial.

Methodology development

During the 1970s and well into the 1980s, a focus of the Merrifield laboratory was to improve solid-phase methodology. One of the serious side reactions is the electrophilic alkylation of peptidyl side chains during acidic treatments in Boc chemistry during the repetitive steps by TFA to remove the Boc group and the final cleavage step by HF to remove all protecting groups. Alkylation side reactions and solutions to minimize them have been studied by many laboratories [29-32]. In the 1980s, I decided to examine these side reactions in a systematic manner based on acidity function and mechanistic viewpoints. The electrophilic alkylation reactions occur because of the acid-generated carbocations and can be classified as an S_N1 reaction. Thus, an S_N2 mechanism of deprotection method would eliminate carbocations, the source of the electrophilic alkylation side reactions.

A method to distinguish S_N1 from S_N2 reaction is by measuring the rate of the reaction. Kinetic studies of the deprotection rate-acid profiles of O-benzylserine in HF and TFMSA/TFA sets of experiments containing predetermined amounts of dimethyl sulfide (DMS) revealed sharp changeover points in the mechanism from S_N2 to S_N1 when the acidity was increased. Similar changeover points in mechanism were also found in the deprotection product-acid profiles of O-benzyltyrosine [33]. The S_N1 deprotection mechanism predominated at high acidities and low DMS activities. S_N2 reaction mechanisms were observed at moderate acidities and high DMS activities. On the basis of the changeover points, a mechanism-reagent composition diagram could be constructed in the form of an equilateral triangle in which the S_N1 and S_N2 regions could be defined as a function of reagent composition.

For deprotecting benzyl alcohol-derived protecting groups by an S_N2 mechanism, we developed low-high HF cleavage approach, in which HF is maintained at low concentration (25% by volume) and is complexed with DMS (75% by volume) in a 1:1 molar ratio. In the low-HF procedure, most of the precursors of harmful carbonium ions are removed by a S_N2 mechanism before the final strong-acid S_N1 step begins [34-38]. An additional advantage of low-HF is that it converts Met(O) to Met efficiently, and removes the "-formyl" group from Trp(For) concomitantly with other protecting groups [39,40].

Entropic orthogonal ligation

In the 1980s, Daniel Kemp and his coworkers at MIT were experimenting entropic ligation of peptide segments [41-43]. Their approach has three components: a template for a reversible attachment with one segment, a chemoselective capture of the other segment, and a proximity-driven acyl transfer reaction facilitated by the template to form a peptide bond. This concept was highly promising for making large peptide and proteins because unprotected peptide segments can be prepared by a stepwise solid-phase method or recombinant methods. It was particularly appealing to our goal to achieve a convergent approach to prepare chemically defined peptide dendrimers.

Our laboratory began a systematic approach to study ligation chemistry of peptide segments using chemoselective capture. Many of these reactions, thioalkylation, thiolene addition, imine formation, disulfide formation and thiol-thioester exchange reactions are well known in conjugation reactions to form non-amide bonds, and some of which we applied successfully for preparing peptide dendrimers. But I envisioned they could be exploited as capture reactions for peptide ligation. Dr. Chuan-Fa Liu, who joined my laboratory in 1989, was assigned to extend the chemoselective ligation to entropic orthogonal ligation, a reaction that would go an extra step involving an acyl transfer reaction to form a peptide bond, but without a template (Figure 4).

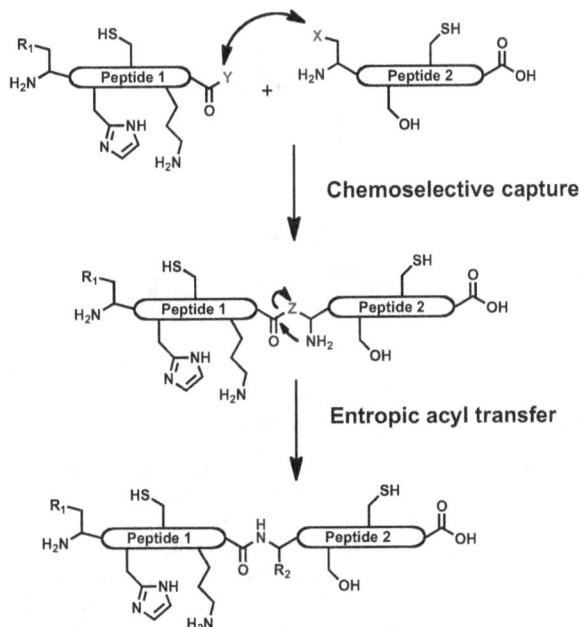

Fig. 4. Schematic representative of entropic orthogonal ligation.

In 1991, Chuan-Fa was successful in this endeavor and developed a pseudoproline ligation using an imine capture of an *N*-terminal Ser/Thr/Cys-segment (Figure 5), results which were presented in the 1993 American Peptide Symposium [44]. Thus, the general concept for an entropic orthogonal ligation would involve two unprotected peptides with two orthogonal pairs of NT-amines, whereas an electrophile at the CT of the first peptide as an O- or S-ester coupled to a nucleophile at the NT of the second peptide to give an intermediate Z followed by an acyl transfer reaction to α-amine of the first peptide. The key reaction, the acyl transfer reaction, is proximity-driven (entropic), invariable and a hallmark of most, if not all, peptide bond ligation reactions developed thus far.

The orthogonal ligation through thiaproline formation of two large, unprotected peptide segments was applied in the total synthesis of three active HIV-1 protease analogs. Three active analogs of HIV-1 protease were obtained in excellent yield by ligating two segments of 38 and 61 residues. Efficient ligation at pH 4 was attained at peptide segment concentrations as low as 50 μM, a concentration that is not feasible with conventional convergent methods using protected peptide segments [45].

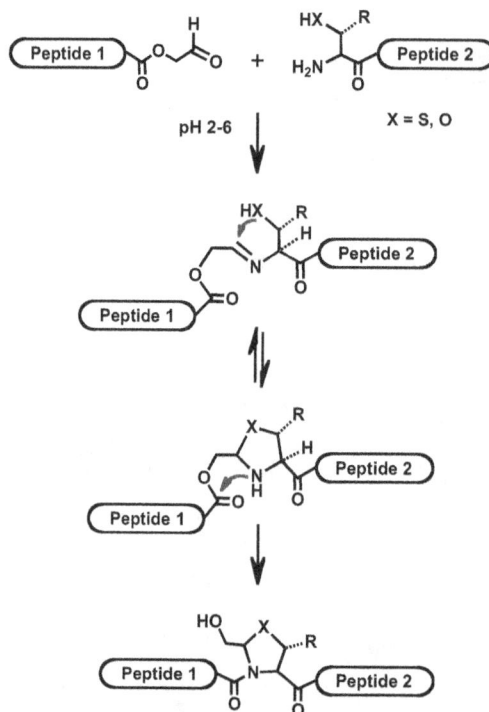

Fig. 5. Pseudo-proline ligation.

Chemical ligation and tandem ligation methods

In the same year we reported the pseudoproline ligation, Kent and co-workers reported a cysteine-based ligation with another segment bearing a *C*-terminal thioester and an S-N acyl shift through a five-member ring intermediate to form an amide bond. Because cysteine is regenerated at the ligation site, they called their ligation method native chemical ligation [46]. The supernucleophilic cysteinyl thiol of an unprotected peptide permits a rapid thiol-thioester exchange at a basic pH to form a covalent thioester intermediate, and the rate of S-N acyl shift is faster than the corresponding O-N acyl shift. Both factors contribute to the effectiveness of native chemical ligation, which our laboratory confirmed shortly after Kent's communication [47]. The elucidation of intein splicing mechanism [48], which involves N-S, S-S acyl transfer reactions, and an uncatalyzed S-N acyl transfer reactions added to the popularity of native chemical ligation. Indeed, the use of a *C*-terminal thioester in the chemoselective capture step, first reported by Wieland and Schneider in 1953 [49], is an excellent choice for chemical ligation. The combined work of Kemp's group and the early ligation studies published in the 1994 and 1995 period [46,47,50,51] provided the conceptual framework of the entropic ligation approach for nearly all subsequent chemical ligation methods developed afterwards. Throughout the 20 years, we have extended the ligation strategies on different ligation sites at His [52], Met [53], Ser/Thr [54], and X-Cys [55].

To prepare multipartite peptides with several functional cargoes including a cell-permeable sequence or transportant for intracellular delivery, tandem ligation of peptides is a convenient convergent approach with the fewest synthetic steps. It links three or four unprotected segments forming two or more regiospecific bonds consecutively without a deprotection step. A tandem ligation strategy to prepare multipartite peptides with normal and branched architectures carrying a novel transportant peptide that is rich in arginine and proline permits their cargoes to be translocated across membranes to affect their biological functions in cytoplasm. Our strategy consists of three ligation methods specific for amino terminal cysteine (Cys), serine/threonine (Ser/Thr), and NR-chloroacetylated amine to afford Xaa-Cys, Xaa-OPro (oxaproline) and Xaa-ψGly (pseudoglycine) at the ligation sites, respectively. Assembly of single-chain peptides from three different segments was achieved by the tandem Cys/OPro ligation to form two amide bonds, a Xaa-Cys and then a Xaa-OPro. Assembly of two- and three-chain peptides with branched architectures from four different segments was accomplished by tandem Cys/ψGly/OPro ligation. These NT-specific tandem ligation strategies were successful in generating cell-permeable multipartite peptides with one-, two-, and three-chain architectures, ranging in size from 52 to 75 residues and without the need of a protection or deprotection step [56,57].

Tandem thiol switch and thia zip cyclization

In the late 2000s, my laboratory at NTU has focused on the design and synthesis of peptide biologics, mostly in the MW range of 2 to 10 kDa. Our goal is to improve the metabolic stability of bioactive peptides to be potential drug candidates. In some cases we aim for their oral delivery after engineering them to be cyclic peptides.

We have developed simple and practical Fmoc-compatible Cys-thioester ligation strategies for preparing cyclic peptides using thioethylamido thioester surrogates including thiomethylthiazolidine, *N*-alkylated cysteine, thioethylalkylamide, [58-60]. The thioethylamido moiety mediated thioester formation via an N-S acyl shift reaction. We exploited this strategy to prepare cyclic peptides including sunflower trypsin inhibitors, cyclotides and cyclic conotoxins. The cysteine-rich peptides provided multiple internal thiols that facilitated cyclization by forming a series of thiolactone intermediates in the absence of external thiols (Figure 6).

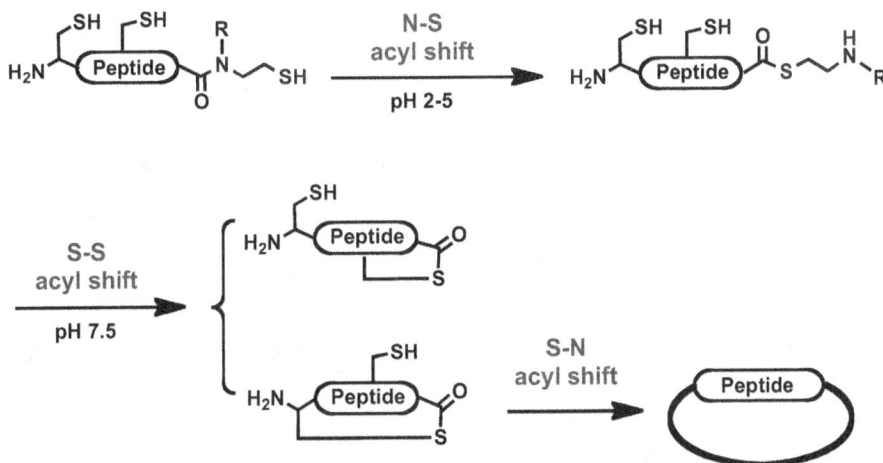

Fig. 6. Preparation of cyclic peptides by thioethylamido-mediated tandem thiol switch reactions.

This efficient cyclization that involves a series of intramolecular rearrangements in a cysteine-rich peptide for the synthesis of large end-to-end cyclic peptides containing multiple cysteine residues was designated as thia zip cyclization [61-63]. It involves two reactions: the reversible thiol-thioester exchanges through intramolecular transthioesterifications as thiolactones and the irreversible S-N acyl migration of the *N*-terminal thiolactone to the lactam (amide) (Figure 7). The facile and reversible intramolecular trans-thioesterifications lead to the ring expansion as discrete thiolactone intermediates, pulling the two ends successively into close proximity as the end-to-end *N*-terminal thiolactone to permit an S-N acyl shift via a five-member ring to form an end-to-end circular protein. As a result, the thia-zip-assisted cyclization, through small discrete intermediates, is entropy-driven and more efficient than the corresponding one-step end-to-end cyclization.

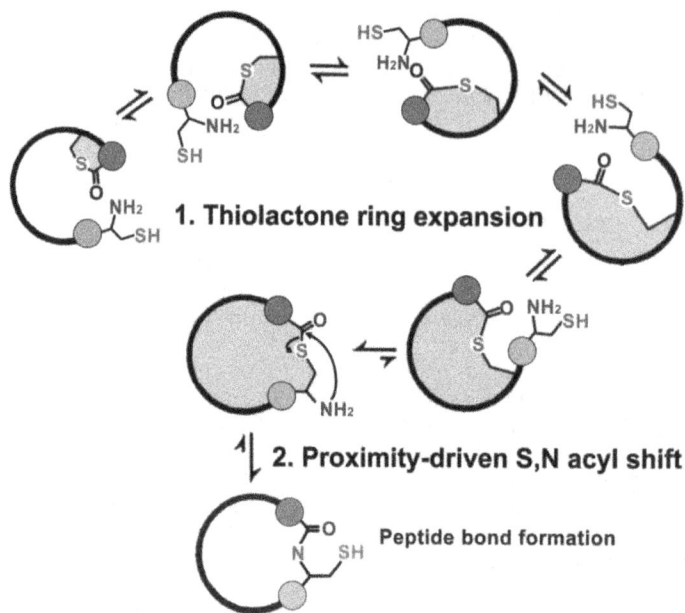

Fig. 7. Thia zip cyclization.

Conclusions

Our approaches in the design of peptide-based vaccines and biologics are loosely tied to themes related to protein deconstruction, a term that I borrowed from arts and humanity fields and used for teaching purpose. In design, protein deconstruction involves breaking a protein to parts of interest and then modifying them or in some cases, reassembling them to a simplified form to attain the desired bioactivity. This has been a guiding principle for many laboratories in the design of bioactive peptides and is broadly referred to as structure-activity-relationship study. In synthesis, the process of protein destruction has been exploited for developing new methods in peptide synthesis, and is generally known as a biomimetic approach. In particular, we have mimicked some of these proteins in making and breaking peptide bonds through the proximity-driven acyl transfer reactions, and modulating acidity function for deprotection reactions as well as preparing peptide thioesters. Our work on the design and synthesis, in part, is based on a protein deconstruction approach to study peptide biologics such as peptide hormones, synthetic vaccines, antibiotics, antivirals, and peptide dendrimers.

Acknowledgements

I have had the good fortune to have two great mentors in my career: my thesis advisor, Professor Daniel Rich who received the Merrifield Award in 2006, and Bruce Merrifield who was not only my postdoctoral adviser and a colleague at The Rockefeller University but also a friend for many years. I am very grateful to my colleagues and students for their contributions and hard work, and the National Institute of Health of USA, Agency for Science Technology and Research (A*STAR) and National Research Foundation (NRF) of Singapore for financial supports. My greatest thanks go to my wife, Sylvaine and my children, Greta and Jonathan. To them, I owe my special thanks for their support, encouragement and tolerance of my work and working schedule.

References

1. Arnon, R., et al. Immunological Cross-Reactivity of Antibodies to a Synthetic Undecapeptide Analogous to the Amino Terminal Segment of Carcinoembryonic Antigen, with the Intact Protein and with Human Sera *Proc. Nat. Acad. Sci. USA* **73**(6) 2123-2127 (1976).
2. Audibert, F., et al. Successful Immunization with a Totally Synthetic Diphtheria Vaccine *Proc. Nat. Acad. Sci. USA* **79**(16) 5042-5046 (1982).
3. Nussenzweig, R.S., et al. Protective Immunity Produced by the Injection of X-irradiated Sporozoites of Plasmodium berghei *Nature* **216**(5111) 160-162 (1967).
4. Eichinger, D.J., et al. Circumsporozoite Protein of Plasmodium berghei: Gene Cloning and Identification of the Immunodominant Epitopes *Mol. Cell. Biol.* **6**(11) 3965-3972 (1986).
5. Zavala, F., et al. Synthetic Peptide Vaccine Confers Protection Against Murine Malaria *J. Exp. Med.* **166**(5) 1591-1596 (1987).
6. Sadler, K., et al. Conformational and Conserved Epitope of gp120 Exposed During Viral Fusion in HIV-Vaccine Design *Biopolymers* **18** 974-975 (2004).
7. Posnett, D.N., McGrath, H., Tam, J.P. A Novel Method for Producing Anti-Peptide Antibodies. Production of Site-Specific Antibodies to the T cell Antigen Receptor Beta-Chain *J. Biol. Chem.* **263**(4) 1719-1725 (1988).
8. Tomalia, D.A., et al. A New Class of Polymers: Starburst-Dendritic Macromolecules *Polym. J.* **17**(1) 117-132 (1985).
9. Freemantle, M. Blossoming of Dendrimers *Chem. Eng. News Archive* **77**(44) 27-35 (1999).
10. Tam, J.P., Lu, Y.A. Vaccine Engineering: Enhancement of Immunogenicity of Synthetic Peptide Vaccines Related to Hepatitis in Chemically Defined Mmodels Consisting of T- and B-cell Epitopes *Proc. Nat. Acad. Sci. USA* **86**(23) 9084-9088 (1989).
11. Tam, J.P. Synthetic Peptide Vaccine Design: Synthesis and Properties of a High-Density Multiple Antigenic Peptide System *Proc. Nat. Acad. Sci. USA* **85**(15) 5409-5413 (1988).
12. Posnett, D.N., Tam, J.P. Multiple Antigenic Peptide Method for Producing Antipeptide Site-Specific Antibodies *Methods in Enzymology* **178** 739-746 (1989).
13. Tam, J.P., et al. Incorporation of T and B Epitopes of the Circumsporozoite Protein in a Chemically Defined Synthetic Vaccine Against Malaria *J. Exp. Med.* **171**(1) 299-306 (1990).
14. Romero, P.J., et al. Multiple T Helper Cell Epitopes of the Circumsporozoite Protein of Plasmodium berghei. *Eur. J. Immunol.* **18**(12) 1951-1957 (1988).
15. Defoort, J.-P., et al. A Rational Design of Synthetic Peptide Vaccine with a Built-in Adjuvant *Int.J. Pept. Prot. Res.* **40**(3-4) 214-221 (1992).
16. Defoort, J.P., et al. Macromolecular Assemblage in the Design of a Synthetic AIDS Vaccine *Proc. Nat. Acad. Sci. USA* **89**(9) 3879-3883 (1992).
17. Nardelli, B., et al. A Chemically Defined Synthetic Vaccine Model for HIV-1 *J. Immunol.* **148**(3) 914-920 (1992).
18. Nardelli, B., Tam, J.P. Cellular Immune Responses Induced by in vivo Priming with a Lipid-Conjugated Multimeric Antigen Peptide *Immunology* **79**(3) 355-361 (1993).
19. Mora, A.L., Tam, J.P. Controlled Lipidation and Encapsulation of Peptides as a Useful Approach to Mucosal Immunizations *J. Immunol.* **161**(7) 3616-3623 (1998).
20. Nardelli, B., Haser, P.B., Tam, J.P. Oral Administration of an Antigenic Synthetic Lipopeptide (MAP-P3C) Evokes Salivary Antibodies and Systemic Humoral and Cellular Responses *Vaccine* **12**(14) 1335-1339 (1994).
21. Sadler, K., et al. Quaternary Protein Mimetics of gp41 Elicit Neutralizing Antibodies Against HIV Fusion-Active Intermediate State *Biopolymers* **90**(3) 320-329 (2008).
22. Chan, D.C., et al. Core Structure of gp41 from the HIV Envelope Glycoprotein *Cell* **89**(2) 263-273 (1997).
23. Wild, C.T., Shugars, D.C., Matthews, T.J. Retroviruses *AIDS Res. Hum.* **9** 1051-1053.
24. Yang, J.-L., et al. Three-Helix Bundles of HIV Mimics Fusion State and Blocks M- and T-Tropic HIV-1 Entry into Host Cells *Proceedings of 18th American Peptide Symposium*, 2004, **18** 290-291.
25. Yu, Q., Li, L., Tam, J.P. Anti-HIV Dendrimeric Peptides *Proceedings of the 20th American Peptide Symposium* 2009, **611** 539-540.
26. Huang, D., et al. Design and Biophysical Characterization of Novel Polycationic ε-Peptides for DNA Compaction and Delivery *Biomacromolecules* **9**(1) 321-330 (2008).
27. Korolev, N., et al. A Universal Description for the Experimental Behavior of Salt-(in)Dependent Oligocation-Induced DNA Condensation *Nucleic Acids Research* **37**(21) 7137-7150 (2009).
28. Tam, J.P., Lu, Y.A., Yang, J.L. Antimicrobial Dendrimeric Peptides *Eur. J. Biochem. / FEBS* **269**(3) 923-932 (2002).
29. Erickson, B.W., Merrifield, R.B. Acid Stability of Several Benzylic Protecting Groups Used in Solid-Phase Peptide Synthesis. Rearrangement of O-benzyltyrosine to 3-benzyltyrosine *J. Amer. Chem.Soc.* **95**(11) 3750-3756 (1973).
30. Yamashiro, D., Li, C.H. Protection of Tyrosine in Solid-Phase Peptide Synthesis *J. Org. Chem.* **38**(3) 591-592 (1973).

31. Lundt, B.F., et al. Removal of t-butyl and t-butoxycarbonyl Protecting Groups with Trifluoroacetic Acid *Int. J. Pept. Prot. Res.* **12**(5) 258-268 (1978).
32. Bodanszky, M., Martinez, J. Side Reactions in Peptide Synthesis *Synthesis* **1981**(05) 333-356 (1981).
33. Tam, J.P., Heath, W.F., Merrifield, R.B. Mechanisms for the Removal of Benzyl Protecting Groups in Synthetic Peptides by Trifluoromethanesulfonic Acid-Trifluoroacetic Acid-Dimethyl Sulfide *J. Amer. Chem. Soc.* **108**(17) 5242-5251 (1986).
34. Heath, W.F., Tam, J.P., Merrifield, R.B. Improved Deprotection in Solid-Phase Peptide-Synthesis - Deprotection of Ni-Formyl-Tryptophan to Tryptophan in Low Concentrations of Hf in Me2s-P-Thiocresol Mixtures *J. Chem. Soc. Chem. Comm.*(16) 896-897 (1982).
35. Tam, J.P., Heath, W.F., Merrifield, R.B. Improved Deprotection in Solid-Phase Peptide-Synthesis - Quantitative Reduction of Methionine Sulfoxide to Methionine During HF Cleavage *Tetrahedron Letters* **23**(29) 2939-2942 (1982).
36. Tam, J.P., Heath, W.F., Merrifield, R.B. Improved Deprotection in Solid-Phase Peptide-Synthesis - Removal of Protecting Groups from Synthetic Peptides by an Sn2 Mechanism with Low Concentrations of Hf in Dimethylsulfide *Tetrahedron Letters* **23**(43) 4435-4438 (1982).
37. Tam, J.P., Tjoeng, F.S., Merrifield, R.B. Design and Synthesis of Multidetachable Resin Supports for Solid-Phase Peptide Synthesis *J. Amer. Chem. Soc.* **102**(19) 6117-6127 (1980).
38. Lu, G., et al. Improved Synthesis of 4-Alkoxybenzyl Alcohol Resin *J. Org. Chem.* **46**(17) 3433-3436 (1981).
39. Tam, J., Heath, W.F., Merrifield, R.B. SN2 Deprotection of Synthetic Peptides with a Low Concentration of HF in Dimethyl Sulfide: Evidence and Application in Peptide Synthesis *J. Amer. Chem. Soc.* 103, 6442-6455 (1983).
40. Heath, W.F., Tam, J.P., Merrifield, R.B. Improved Deprotection of Cysteine-Containing Peptides in HF *Int. J. Pept. Prot. Res.* **28**(5) 498-507 (1986).
41. Kemp, D.S. The Amine Capture Strategy for Peptide Bond Formation-An Outline of Progress *Biopolymers* **20**(9) 1793-1804 (1981).
42. Kemp, D.S., Galakatos, N.G. Peptide Synthesis by Prior Thiol Capture. 1. A Convenient Synthesis of 4-hydroxy-6-mercaptodibenzofuran and Novel Solid-Phase Synthesis of Peptide-Derived 4-(acyloxy)-6-mercaptodibenzofurans *J. Org. Chem.* **51**(10) 1821-1829 (1986).
43. Kemp, D.S., et al. Peptide Synthesis by Prior Thiol Capture. 4. Amide Bond Formation. The Effect of a Side-Chain Substituent on the Rates of Intramolecular O,N-acyl Transfer *J. Org. Chem.* **51**(17) 3320-3324 (1986).
44. Liu, C.F., et al. A Chemical Ligation Strategy to Form Peptide Bond Between Two Unprotected Peptides. in *Peptides: Chemistry, Structure and Biology, Proceedings of the 13th American Peptide Symposium.* 1994. Leiden, The Netherlands: ESCOM Science Publishers B.V.
45. Liu, C.F., Rao, C., Tam, J.P. Orthogonal Ligation of Unprotected Peptide Segments Through Pseudoproline Formation for the Synthesis of HIV-1 Protease *J. Amer. Chem. Soc.* **118**(2) 307-312 (1996).
46. Dawson, P.E., et al. Synthesis of Proteins by Native Chemical Ligation *Science* **266**(5186) 776-779 (1994).
47. Tam, J.P., et al. Peptide Synthesis Using Unprotected Peptides Through Orthogonal Coupling Methods *Proc. Nat. Acad. Sci. USA* **92**(26) 12485-12489 (1995).
48. Xu, M.Q., et al. Protein Splicing - An Analysis of the Branched Intermediate and Its Resolution by Succinimide Formation *EMBO J.* **13**(23) 5517-5522 (1994).
49. Wieland, T., Schneider, G. N-Acyl-Imidazole als energiereiche Acylverbindungen. *Justus Liebigs Annalen der Chemie* **580**(2) 159-168 (1953).
50. Liu, C.F., Tam, J.P. Chemical Ligation Approach to Form a Peptide-Bond between Unprotected Peptide Segments - Concept and Model Study *J. Amer. Chem. Soc.* **116**(10) 4149-4153 (1994).
51. Liu, C.F., Tam, J.P. Peptide Segment Ligation Strategy Without Use of Protecting Groups *Proc. Nat. Acad. Sci. USA* **91**(14) 6584-6588 (1994).
52. Zhang, L., J.P. Tam, J.P. Orthogonal Coupling of Unprotected Peptide Segments Through Histidyl Amino Terminus *Tetrahedron Letters* **38**(1) 3-6 (1997).
53. Tam, J.P., Yu, Q.T. Methionine Ligation Strategy in the Biomimetic Synthesis of Parathyroid Hormones *Biopolymers* **46**(5) 319-327 (1998).
54. Tam, J.P., Miao, Z.W. Stereospecific Pseudoproline Ligation of *N*-Terminal Serine, Threonine, or Cysteine-Containing Unprotected Peptides *J. Amer. Chem. Soc.* **121**(39) 9013-9022 (1999).
55. Lu, Y.-A., Tam, J.P. X-Cys Ligation. *Biopolymers: Proceedings of 18th American Peptide Symposium*, 2004. **18** 270-271.
56. Eom, K.D., et al. Tandem Ligation of Multipartite Peptides With Cell-Permeable Activity *J. Amer. Chem. Soc.* **125**(1) 73-82 (2003).
57. Tam, J.P., Eom, K.D. Mimicking Reverse Protein Splicing by Three-Segment Tandem Peptide Ligation *Protein and Peptide Letters* **12**(8) 743-749 (2005).
58. Sharma, R.K., Tam, J.P. Tandem Thiol Switch Synthesis of Peptide Thioesters via N-S Acyl Shift on Thiazolidine *Org. Lett.* **13**(19) 5176-5179 (2011).

59. Taichi, M., et al. A Thioethylalkylamido (TEA) Thioester Surrogate in the Synthesis of a Cyclic Peptide via a Tandem Acyl Shift *Org. Lett.* **15**(11) 2620-2623 (2013).
60. Hemu, X., et al. Biomimetic Synthesis of Cyclic Peptides Using Novel Thioester Surrogates. *Peptide Science* n/a-n/a (2013).
61. Tam, J.P., Lu, Y.A. A Biomimetic Strategy in the Synthesis and Fragmentation of Cyclic Protein *Protein Science* **7**(7) 1583-1592 (1998).
62. Tam, J.P., Lu, Y.A., Yu, Q.T. Thia Zip Reaction for Synthesis of Large Cyclic Peptides: Mechanisms and Applications *J. Amer. Chem. Soc.* **121**(18) 4316-4324 (1999).
63. Tam, J.P., Wong, C.T.T. Chemical Synthesis of Circular Proteins *J. Biol. Chem.* **287**(32) 27020-27025 (2012).

Proceedings of the 23rd American Peptide Symposium
Michal Lebl (Editor)
American Peptide Society, 2013

From Combinatorial Chemistry to Nanotechnology to Cancer Therapy

Kit S Lam[1-4] and Ruiwu Liu[1,4]

[1]Department of Biochemistry & Molecular Medicine, [2]Division of Hematology & Oncology, [3]Department of Internal Medicine, [4]UC Davis Cancer Center, University of California Davis, Sacramento, CA, 95817, U.S.A.

Introduction

In "one-bead-one-compound" (OBOC) combinatorial technology, diverse peptides, peptidomimetics or non-peptidic small molecules can be generated rapidly such that the library compound displays on the bead surface and coding tag resides inside each bead [1,2]. Such chemically encoded microbead libraries can then be rapidly screened via on-bead assays or solution-phase releasable assays. Many binding, biochemical and cell-based screening assays have been developed and compounds against a variety of biological targets have been discovered using this approach. For example, we have discovered LLP2A, a peptidomimetic ligand that targets activated $\alpha 4\beta 1$ integrin, which is overexpressed in lymphoma cells [3]. When conjugated to CB-TE2A metal chelator and loaded with ^{64}Cu, this ligand was found to target lymphoma xenografts with high specificity. LXY30, an optimized cyclic peptide ligand against $\alpha 3\beta 1$ integrin, which is overexpressed in several epithelial cancers, is able to target both orthotopically and subcutaneously implanted glioblastoma xenografts with high efficiency. We have also developed "one-bead-two-compound" (OB2C) technology to discover compounds that induce cell signaling or cell death [4]. Very recently, we have successfully applied OBOC technology to develop genetically encoded illuminants for functional imaging of living cells and animals. A few years ago, we reported the use of Fmoc-chemistry to construct a number of amphiphilic polymers, comprised of a cluster of cholic acids linked by a series of lysines and attached to one end of a linear polyethylene glycol chain. Under aqueous condition, such telodendrimers self-assemble to form highly stable monodisperse nanomicelles (15-150 nm diameter) [5]. Cancer targeting peptides decorated nanoparticles can deliver the drug not only to the tumor site but also into the tumor cells, making them more efficacious than nanoparticles without targeting peptides [6]. To minimize premature drug release, we have developed various strategies to reversibly stitch the telodendrimers together [7,8]. More recently, we have developed a novel nanoporphyrin platform that can be used as an effective carrier for targeted photodynamic therapy, hyperthermic therapy, systemic radiotherapy, systemic chemotherapy, optical imaging, radioimaging and magnetic resonance imaging.

Cell Surface Targeting Ligand Discovery with OBOC Combinatorial Technology

Cell surface targeting ligands can be discovered simply by mixing OBOC libraries with living cells. After incubation for a period of time (5 minutes to 3 hours), compound-beads coated with a single layer of cells can be readily identified and physically isolated for microsequencing or chemical decoding [9]. To facilitate the discovery of cell-type specific ligands, we may use dual color screening approach, in which one cell type is labeled with a fluorescent dye and the other cell type is unlabeled. Beads that bind to only one of the two cell types will be considered true positive beads against the specific cell line. Based on the chemical structure of the positive compounds, focused libraries can be designed, synthesized and screened under high stringency for stronger binding; this can be achieved by (i) down-substitution of the outer bead layer, (ii) using lower number of cells for screening, and (iii) incorporating soluble competing ligands into the screening medium. Using this approach we have discovered many different cancer cell surface targeting ligands [9]. Some of these ligands serve as excellent vehicles for the delivery of radionuclides, toxins, or nanoparticles to the tumor sites. LLP2A, described above, was discovered using this approach. When coupled to radiometals, it can be used as imaging agents for lymphoma [10], metastatic niche of breast cancer [11], and myeloma [12]. More recently, we were able to demonstrate that

LLP2A, when conjugated to alendronate (Ale, a bisphosphonate, Figure 1), was able to mobilize endogenous or exogenously added mesenchymal stem cells (MSC) to the bone matrix, preventing age-related trabecular bone loss *in vivo* [13]. In addition, exposure of hydroxyapatite bound LLP2A-Ale to MSC *in vitro* can increase MSC migration that was associated with an increase in phosphorylation of Akt kinase and osteoblastogenesis [14].

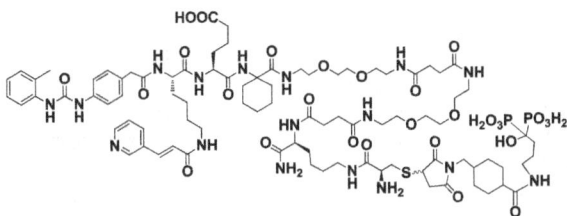

Fig. 1. Chemical structure of LLP2A-Alendronate.

In addition to screening for cell binding, one may also screen for cell function, such as apoptosis or specific cell signaling. On approach is to briefly fix the bead-bound cells with 4% paraformaldehyde, permeate the cells with non-ionic detergent, and incubate the cell-coated beads with antibody-enzyme conjugate (e.g. anti-phospho-Erk antibody-horse radish peroxidase). Upon addition of 3,3'-diaminobenzidine (DAB), cells with phosphorylated-Erk activation will turn brown. Those beads will be considered positive. Since hundreds of antibodies against various activation states of cell signaling proteins are commercially available, it is not difficult to envision that this simple method can afford discovery of functional ligands against various different signaling pathways.

OB2C Combinatorial Platform

We have recently reported the development of a OB2C combinatorial platform for the discovery of cell surface acting pro-apoptotic ligands [4]. In this method, TentaGel beads are first topographically segregated into outer layer and the inner core [1]. The library was constructed such that two compounds are displayed on the outer layer of each bead (the variable library compound and the fixed cell capturing ligand), and the chemical bar code resides in the bead interior. To screen for pro-apoptotic ligands, the target cells are added to the library beads. Because of the capturing ligand, every bead is coated with a layer of target cells. The free cells are then removed. After incubation for 24-48 hrs, apoptotic cells can be identified by (i) an immunocytochemistry method using anti-cleaved caspase 3 antibody-horse radish peroxidase conjugate and DAB as a substrate (cells turned brown), or (ii) propidium iodide (cells turn fluorescent red). Figure 2a shows the chemical structure of a benzimidazole-based peptidomimetic OB2C library with three diversity points and 74,088 permutations. Figure 2b shows the photomicrograph of positive beads detected by immunocytochemistry method. In this experiment, about 40,000 beads were screened for their pro-apoptotic functions against SKOV3 ovarian cancer cell line. Four positive beads were identified, isolated and decoded by Edman microsequencing. One of these compounds A2 (Figure 2c) was confirmed to be active in tetrameric form with an IC_{50} value of ~40 µM (Please refer to the chapter by Ruiwu Liu in this proceeding).

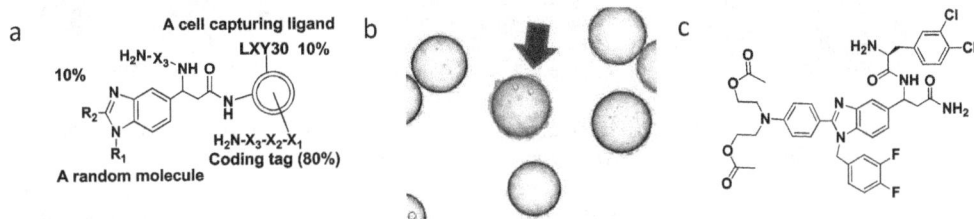

Fig. 2 a: OB2C-S3 library bead; b: Screening library for death ligands against SKOV3 cells. Red arrow points to a positive bead; c: Chemical structure of A2.

Cell penetrating peptides

Very recently, we have modified the OB2C combinatorial platform to discover cell-penetrating peptides. In this method, the library compound is tethered to the outer layer of each bead *via* a disulfide bond so that it can be released on demand by adding reducing agent such as dithiothreitol (DTT). The library is constructed such that a linker-Lys(biotin) is placed between the library compound and the disulfide bond. Like the OB2C platform outlined above, the cell capturing ligand is also displayed on the outer layer. To screen for cell penetrating peptides, we first incubate the bead library with streptavidin-quantum dot conjugate, followed by the addition of live target cells (e.g. brain endothelial cells). After removal of the free cells, the cell-coated library beads are then immobilized in methylcellulose. Upon addition of DTT, the biotinylated library compound-streptavidin-quantum dots are released from the bead surface. After 12 hours incubation, the beads are inspected under a fluorescent microscope. Beads coated with fluorescent-labeled cells were considered positive and isolated for chemical decoding. Using this approach, a linear octapeptide LBL7, comprised of both L- and D-amino acids was discovered. When prepared in biotinylated form, it was able to penetrate mouse brain endothelial cells in a time-dependent manner, and carry along with it streptavidin-Alexa 480 or streptavidin-quantum dot conjugate.

Solution Phase Releasable Assays

We previously reported the use of solution phase releasable assay to screen OBOC combinatorial peptide and small molecule libraries [15]. In this method, the library compounds are tethered to the outer layer of each bead *via* a cleavable linker (e.g. disulfide bond). The library beads are then loaded into a PDMS-based microbead cassette fabricated by soft lithography, with one bead per well and 10,000 wells per cassette. Target cells suspended in Matrigel® are then layered over entire cassette such that each and every bead is embedded in Matrigel surrounded by target cells. Upon addition of DTT, the library compounds are released and diffused into the surrounding 3D gel matrix containing the cells. A day later, MTT is then added to detect living cells (convert MTT to purple color formazan). Beads surrounded by a halo of dead cells (colorless) are considered positive (Figure 3). Other cleavable linkers such as photocleavable linker and reverse diketopiperazine linker can also be used. In addition to the cytotoxic assay outlined above, the OBOC solution phase releasable assays can be applied to most cell-based and biochemical assays commonly used in multi-titer plate format. Since the assay is highly compact with 10,000 compounds assayed concurrently in one single 3.5cm Petri dish, the amount of reagents needed is miniscule and therefore highly economical and efficient.

Fig 3. Screening of encoded OBOC N-acylated dipeptide combinatorial releasable library for the identification of cytotoxic compounds against Jurkat T-cell lymphoma. (a) Photomicrograph of a section of the mcirobead cassette inside a 3.5 mm Petri dish containing a bead in each well and Jurkat cells immobilized within the Matrigel matrix. (b) Photomicrograph of a larger section of the plate four hours after MTT addition, showing a halo surrounding a positive bead. (c) Higher power of the positive bead shown in (B).

Genetically Encoded Small Illuminant (GESI)

Genetically encoded fluorescent proteins such as green fluorescent proteins (GFP) and its relatives of color palette enable one to track in living cells the distribution, abundance, dynamics, interaction and conformational changes of essential signaling molecules in real time and space. This technology, while very powerful, is limited by the large size (27KDa) of GFP. Very recently, we have employed OBOC combinatorial peptide library approach to develop genetically encoded small illuminant (GESI) that can potentially replace GFP. Major advantage of GESI over GFP is that the former is much smaller (12 to16-mer peptide) and in principle can be inserted at the *N*-terminus, *C*-terminus or along the peptide sequence displayed on the surface of the target protein. To screen for GESI peptides that can activate the fluorescence of malachite green (MG) and report intracellular calcium, a linear 12-mer OBOC library with an EF hand motif can be designed and synthesized. Library beads are then immobilized on Petri dish with a thin film of PDMS or soften polystyrene, and scanned with a fluorescent microscope equipped with a motorized stage and a 650nm emission filter. MG (0.1 µM) is then added to the dish, incubated for an hour and then rescanned with the fluorescent microscope. EDTA is then added and scanned again. Beads that gain new fluorescence in the presence of calcium but lose fluorescence when EDTA is added are considered positive. These peptides are then genetically fused to the *C*-terminus of Cerulean (a model target protein with an added membrane localization peptide at the *N*-terminus) and expressed in target cells. Upon addition of MG, successful GESI should fluoresce red and co-localize with Cerulean (fluoresce blue/cyan). Of the 11 peptides we isolated from the OBOC library screen, 4 peptides clearly showed co-localization of infrared and blue/cyan fluorescence upon MG addition. One of these peptides was responsive to calcium changes inside living cells. (Please refer to the chapter by Sara Ahadi in this proceeding). Preliminary data in another set of experiments indicates that GESI peptide that report post-translational modification, such as protein phosphorylation, can also be discovered with this approach.

Targeted Delivery of Drugs

One major challenge of cancer therapy is that many chemotherapeutic agents are very toxic and have very narrow therapeutic index. To maximize the efficacy and minimize the adverse side effects of these drugs, one will need to develop methods to targeted delivery of these drugs to the tumor site while sparing normal tissues. One approach is to use cancer cell surface targeting ligands or monoclonal antibodies as vehicles for drug delivery. The cancer targeting ligands can be discovered via OBOC combinatorial chemistry as mentioned above. The therapeutic payloads can be peptide toxins such as MMAE or MMAF, highly potent chemotherapeutic agents, or radionuclides such as ^{131}I or ^{90}Y. One added advantage of radioconjugates is that even mutated tumor cells that no longer bind to the ligands will still be killed due to the "bystander effects." The second approach is to exploit the fact that tumor blood vessels are "leaky" and nanoparticles smaller than 100nm in diameter have the propensity to be preferentially concentrated at the tumor site after intravenous administration.

Nanotherapeutic and Nano Imaging Agents

In the last few years, we have developed a novel micellar nanoplatform for efficient drug delivery. The nanoplatform comprises of an amphiphilic polymer unit called telodendrimer, which can self-assemble with hydrophobic drugs in aqueous condition to form drug-encapsulated, size-tunable (10-100 nm), and monodispersed micelles [5,16]. To minimize premature release of drugs from the micelles during circulation due to lipoproteins such as HDL and LDL, the telodendrimers can be crosslinked by reversible covalent linkages such as disulfide bond [7] or boronate-catechol linkages [8]. Each telodendrimer consists of a cluster of cholic acids (e.g. eight) linked by a series of lysines and attached to one end of a linear polyethylene glycol chain (Figure 4). Unlike many amphiphilic polymers that are prepared by polymerization chemistry, telodendrimers are prepared step-wise by Fmoc-chemistry, resulting in a well-defined structure. The physicochemical properties of the final drug-loaded micelles are determined by the chemical structures of telodendrimer (size of PEG, number of cholic acids, and how they are linked together) and the nature of the drug to be encapsulated.

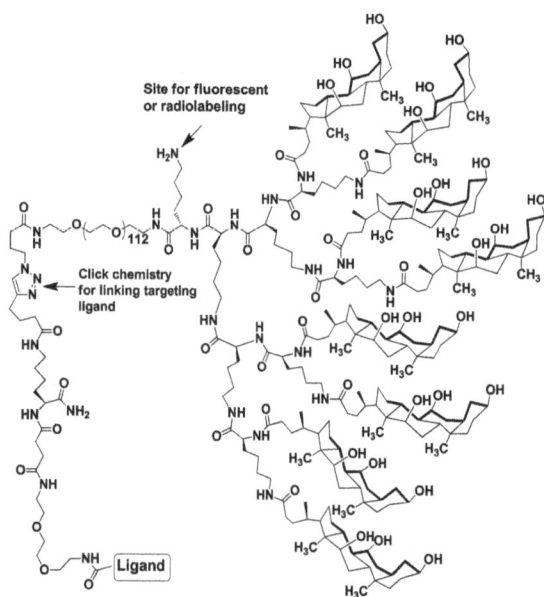

Fig. 4. Chemical structure of Ligand-telodendrimer.

We have successfully loaded many hydrophobic chemotherapeutic agents (paclitaxel, doxorubicin, vincristine, etoposide, BCNU, actinomycin D) as well as target-specific drugs (temsirolimus, bortezomib, sarafenib, lapatinib, and dexamethasone) into the above mentioned nanoplatform at high loading-capacity. *In vivo* optical and SPECT imaging studies demonstrated that the micellar-based nanoparticles can target the tumor with high efficiency in both a transgenic spontaneous mammary tumor model and in an ovarian cancer xenograft model [5]. *In vivo* therapeutic studies indicate that such nanoplatforms allows one to administer 3-4 fold more paclitaxel (compared to standard Cremophor formulated paclitaxel used in the clinic) to nude mice bearing ovarian cancer xenografts, with less side effects and much higher therapeutic efficacy. We have also demonstrated that addition of cancer cell surface targeting ligand [6] and crosslinking of the micelle [7] can further improve the therapeutic efficacy.

Very recently, we have developed a novel nanoporphyrin platform in which the telodendrimer is modified such that 4 of the 8 cholic acids are replaced with pyropheophorbide (a porphyrin). This nanocarrier is highly versatile and multifunctional. We have demonstrated that photodynamic therapy, photothermal therapy, and chemotherapy can all be delivered to transgenic mammary tumor model in one single nanocarrier. In addition, ^{64}Cu and Gd(III) can be readily loaded into the porphyrin moieties for PET and MRI imaging of the tumor.

Future Perspectives

The OBOC combinatorial technology is an enabling technology that exploits the powerful one-bead-one-compound concept in both library synthesis and screening. The technology has been applied successfully by many laboratories around the world to many fields: from the discovery of cell binding ligands for cancer imaging and therapy [9], to the discovery of pore-forming peptides in membranes [17], to the development of synthetic catalysts [18] and new materials. Because of the encoding methods that have been developed, the choices of chemistry and building blocks one may use in OBOC library synthesis have become much more plentiful and no longer limited to alpha amino acid containing peptides. With the advances of robotics and microfluidics, it is highly feasible that the entire process of library synthesis, screening and decoding can be automated. Only a limited number of commercially available solid supports are suitable for OBOC chemistry and screening. There is a need for the development of more solid supports, with a range of porosity, that are compatible with both chemical synthesis and biological screening. There is also a need for the development of novel inexpensive building blocks and efficient coupling chemistry that allow one to build complex natural product-like compounds on solid support. Microwave-assisted chemical reaction has become more popular and sophisticated. Undoubtedly, it will play a substantial role in the development of efficient coupling chemistry for OBOC technology. Although many cleavable linkers have been reported in the literature, few of them are compatible with cell-based assays. There is a need to develop unique cleavable linkers that are robust, compatible with many solid phase chemical synthesis, cleavage under mild conditions, and suitable for OBOC cell-based and biochemical screening.

The nanomedicine field has advanced rapidly over the past decade. Undoubtedly, it will play a major role in future cancer therapy and cancer imaging. Drugs that are too toxic to be clinically useful could now be reformulated with appropriate nanocarriers for clinical applications. Efficacies of nanoparticle drugs can be further enhanced with cancer targeting ligands discovered from OBOC method or other technologies. There is a need for the development of ligands that are highly specific against cancers or target tissues. Many laboratories in pharmaceutical industry and academia are working intensely on developing methods for efficient and non-toxic delivery of therapeutic siRNA to target cells *in vivo*. It is likely that both OBOC technology and nanotechnology will play a major role in achieving this goal. In addition, novel nanocarriers can potentially be developed with the enabling OBOC technology.

Acknowledgments

We appreciate expert editorial support from Mr. Joel Kugelmass. This work was supported by NIH R33CA160132, R01 CA115483, R01EB012569, R21 CA135345 and institutional funding from UC Davis.

References

1. Liu, R., Marik, J., Lam, K.S. *J. Am. Chem. Soc.* **124**, 7678-7680 (2002).
2. Lam, K.S., Salmon, S.E., Hersh, E.M., Hruby, V.J., Kazmierski, W.M., Knapp, R.J. *Nature* **354**, 82-84 (1991).
3. Peng, L., Liu, R., Marik, J., Wang, X., Takada, Y., Lam, K.S. *Nat. Chem. Biol.* **2**, 381-389 (2006).
4. Kumaresan, P.R., Wang, Y., Saunders, M., et al. *ACS Comb. Sci.* **13**, 259-264 (2011).
5. Xiao, K., Luo, J., Fowler, W.L., et al. *Biomaterials* **30**, 6006-6016 (2009).
6. Xiao, K., Li, Y., Lee, J.S., et al. *Cancer Res.* **72**, 2100-2110 (2012).
7. Li, Y., Xiao, K., Luo, J., et al. *Biomaterials* **32**, 6633-6645 (2011).
8. Li, Y., Xiao, W., Xiao, K., et al. *Angew. Chem. Int. Ed. Engl.* **51**, 2864-2869 (2012).
9. Aina, O.H., Liu, R., Sutcliffe, J.L., Marik, J., Pan, C.X., Lam, K.S. *Mol. Pharm.* **4**, 631-651 (2007).
10. Denardo, S.J., Liu, R., Albrecht, H., et al. *J. Nucl. Med.* **50**, 625-634 (2009).
11. Shokeen, M., Zheleznyak, A., Wilson, J.M., et al. *J. Nucl. Med.* **53**, 779-786 (2012).
12. Soodgupta, D., Hurchla, M.A., Jiang, M., et al. *PLoS One* **8**, e55841 (2013).
13. Guan, M., Yao, W., Liu, R., et al. *Nat. Med.* **18**, 456-462 (2012).
14. Yao, W., Guan, M., Jia, J., et al. *Stem Cells* Epub ahead of print, doi: 10.1002/stem.1461. (2013).
15. Townsend, J.B., Shaheen, F., Liu, R., Lam, K.S. *J. Comb. Chem.* **12**, 700-712 (2010).
16. Luo, J., Xiao, K., Li, Y., et al. *Bioconjug. Chem.* **21**, 1216-1224 (2010).
17. Krauson, A.J., He, J., Wimley, A.W., Hoffmann, A.R., Wimley, W.C. *ACS Chem. Biol.* **8**, 823-831 (2013).
18. Lichtor, P.A., Miller, S.J. *ACS Comb. Sci.* **13**, 321-326 (2011).

Proceedings of the 23rd American Peptide Symposium
Michal Lebl (Editor)
American Peptide Society, 2013

Peptides with Citrulline for Recognition of Autoantibodies in Rheumatoid Arthritis

Ferenc Hudecz[1,2], Fruzsina Babos[1], Eszter Szarka[3], Zsuzsa Majer[2], Anna Magyar[1], and Gabriella Sármay[3]

[1]Research Group of Peptide Chemistry, Hungarian Academy of Sciences, Eötvös L. University, Budapest, 1117, Hungary; [2]Dept. of Organic Chemistry, Institute of Chemistry, Eötvös L. University, Budapest, 1117, Hungary; [3]Dept. of Immunology, Eötvös L. University, Budapest, 1117, Hungary

Introduction

Posttranslational modifications (e.g. citrullination, deglycosylation) could have an essential role in the induction of protein core specific autoimmune responses. It is well known that citrullination of various proteins by peptidyl arginine deiminase (PAD) enzymes could be one of the main cause of Rheumatoid arthritis (RA). Thus the appearance of anti-citrullinated protein/peptide antibodies (ACPA) is sensitive and specific markers for diagnosis/prognosis of RA [1]. Therefore identification of RA specific peptide epitopes in the Arg-rich sequences of the relevant proteins (e.g. filaggrin, vimentin) could be an optimal strategy [2,3].

Results and Discussion

We have investigated the role of deamination of Arg to Cit in RA related proteins possessing multiple Arg residues. Oligopeptides from filaggrin, vimentin, collagen [4,5] as well as their *N*- and *C*-terminal biotin conjugates with single or multiple Arg/Cit changes and with different spacers were prepared using solid phase peptide synthesis and Fmoc/ᵗBu strategy [5] (Table 1).

The antibody binding properties of the epitope peptides were studied by indirect ELISA method using sera from RA patients and sera from healthy individuals as controls (Figure 1). Our data demonstrate that the autoantibodies of RA patients could not bind the Arg-containing epitopes but only their Cit-containing analogues. In order to obtain information on the secondary structure of the peptides ECD studies were performed in three different solvents: water, 50% trifluoroethanol (TFE) and TFE (Figure 2). Results demonstrated with the 19-mer filaggrin epitope region peptide show that in water only minor difference (near $\lambda \sim 220$ nm) could be observed in the ECD spectra of the peptides containing Arg or Cit at positions 312, 314 and 316 (Figure 2A).

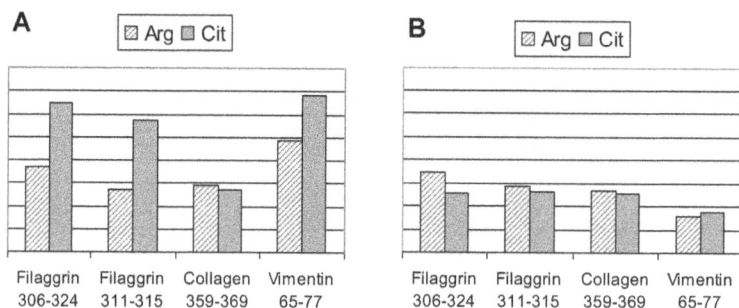

Fig. 1. A) Binding of antibodies from RA sera to C-biotinylated epitope peptides as detected by indirect ELISA. B) Sera from healthy subjects were used as control.

Table 1. Epitope peptides derived from RA specific proteins.

Protein	Peptide fragment	Original sequence	Citrulline containing sequence
filaggrin	306-324	SHQESTRGRSRGRSGRSGS	SHQESTXGXSXGRSGRSGS
	311-315	TRGRS	TXGRS
collagen	359-369	ARGLTGRPGDA	AXGLTGXPGDA
vimentin	65-77	SAVRARSSVPGVR	SAVRAXSSVPGVR

The spectra are indicative for the PPII and/or unstructured conformation. Comparison of the ECD spectra of the peptide pair in 50% TFE and also in TFE did not show major alterations in the pattern between the Arg- and Cit-containing analogues (data not shown).

In 50% TFE, the spectra have changed slightly, but in TFE we observed a totally different conformation (Figure 2B): conformer mixture with a high populations of ordered (mainly helical) structures (a strong positive band at $\lambda \sim 190$ nm, two negative bands $\lambda \sim 207$ nm and $\lambda \sim 220$ nm), namely α–helix, β-sheet and β-turn could be present. By Arg/Cit-peptides corresponding to the sequence of filaggrin 306-324 we have identified Cit residue(s) responsible for the antibody recognition of diseased individuals. We found that the position of Cit residue in the epitope core, the length of the *N*- and *C*-terminal flanking regions and the distance between the epitope core and biotin label could influence markedly antibody

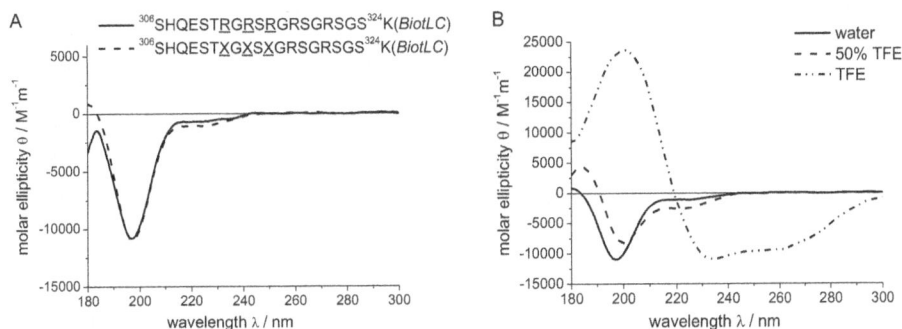

Fig. 2. ECD spectra of the C-terminal biotinylated A) filaggrin epitope region peptide with Arg or Cit in water and B) ${}^{306}SHQESTXGXSXGRSGRSGS^{324}$ epitope region peptide in water, 50% TFE and TFE.

binding. Based on these findings even a short 5-mer peptide with *C*-terminal biotin could be used as target antigen or as ligand of an antibody subset (biomarker) of the disease. Based on ECD experiments we demonstrate that the Arg/Cit change does not influence markedly the solution conformation of the 19-mer peptide analogues, but the 3D structure of these compounds was dependent on the solvent used.

Acknowledgments

This work was supported by grants from the OTKA, Hungary (A08-CK 80689, K81175), RAPEP_09 (OMFB-00135/2010) and from TÁMOP 4.2.1./B-09/KMR-2010-0003.

References

1. Vincent, C., et al. *Autoimmunity* **38**, 17-24 (2005).
2. Bang, H., et al. *Arth. Rheum.* **56**, 2503-2511 (2007).
3. Schellekens, G.A., et al. *J. Clin. Invest.* **101**, 273-281 (1998).
4. Vossenaar, E.R., et al. *Arthritis Res. Ther.* **6**, 86-89 (2004).
5. Babos, F., et al. *Bioconjugate Chem.* **24**, 817-827 (2013).

Proceedings of the 23rd American Peptide Symposium
Michal Lebl (Editor)
American Peptide Society, 2013

Designed Peptide Libraries for Cell Analyzing Microarrays

Hisakazu Mihara[1], Kenji Usui[2], Hiroshi Tsutsumi[1], and Kiyoshi Nokihara[3]

[1]Department of Bioengineering, Tokyo Institute of Technology, Yokohama, 226-8501, Japan;
[2]FIRST (Faculty of Frontiers of Innovative Research in Science and Technology) and FIBER
(Frontier Institute for Biomolecular Engineering Research), Konan University, Kobe, 650-0047,
Japan; [3]HiPep Laboratories, Kyoto, 602-8158, Japan

Introduction

As advances in genome-wide sciences, the protein-detection microarrays have been promising technologies providing high-throughput (HT) detection for proteins of interest [1]. To realize such a practical protein-detection system, the development of capturing agents is one of the key steps. Peptides are useful candidates for capturing agents of protein-detection chips [2]. The strategy using *de novo* designed peptides involving α-helix, β-sheet and β-loop enables peptide microarrays to mimic protein-protein interactions and discriminate protein structures. Thus, we have attempted to develop designed peptide microarrays for protein detection [3]. Fluorescently labelled peptide libraries with designed sequences for α-helix, β-sheet, and β-loop were constructed [4-7], and the fluorescence data of protein binding were demonstrated to show color-coded patterns regarded as protein fingerprints (PFP). The PFP analyses using the designed peptide arrays well discriminated various proteins using the clustering method and the principal component analysis. Moreover, peptides can interact with various molecules displayed on cell surfaces and cause various types of cell responses. Cell-based analyses using compounds will open the novel approaches for new biomedical techniques applicable to cell diagnosis and therapy. Our designed peptide array would be valuable to develop such a cell-based assay system. To this goal, we utilized peptide libraries expressing a various cell-penetrating (CP) activities. The cell fingerprint (CFP) color-coded patterns generated together with the CP activities of libraries discriminated and profiled cultured cancer cell types.

Results and Discussion

α-Helical model peptide libraries with 202 sequences were constructed with TAMRA fluorophore at the *N* terminus, in which cationic LK series and anionic LE series were systematically designed [8]. Among the model libraries, 54 representative peptides with varied net charges (-2 to +5 at side chains), hydrophobicity, and α-helicity were selected to examine the CP activity to HeLa cell using confocal fluorescent microscopy. As results, most of the cationic peptides in LK series with +3 and +5 net charges at side chains showed higher scored (++ or +++) CP activities, and some of the anionic peptides in LE series showed also high activities scored as ++, in which the latter anionic peptides had one Arg and one Trp/Phe residue in the sequence. In summary of the CP activity for HeLa cell, more helical, hydrophobic and cationic peptides showed a higher activity in general, and some aromatic residues such as Trp and Phe had an important role for the activity.

Next, a further 24 selected peptides from LK series with +3 and +5 charges were examined for the CP activity to 4 different cell lines, HeLa, A549, 3T3-L1, and PC12, to investigate the cell selectivity of the peptides. As a result, for example, peptide 018 (named as LKARFH) showed high CP activity for both HeLa (+++) and A549 (+++) cells, but low activity for both 3T3-L1 (+) and PC12 (-), indicating that the same peptide shows a varied activity for each cell and thus the CP activity is much different by cell types. Summarizing the CP activities of 24 peptides, the CFP patterns for each cell line were generated, and the color patterns well discriminated and characterized the 4 different cell lines (Figure 1). Furthermore, some peptides showing cell selectivity were elucidated using the clustering analyses of peptide sequences in the CP activities. LK peptides having RFH amino acids showed the selective activity for HeLa and A549, and peptides having WEF and WRF amino acids showed the high activity for all 4 cell lines. Peptide 018(LKARFH) showed the high activity for HeLa cell, but no activity for PC12 cell, and peptide 008(LKARAH), in which only one amino acid was replaced by Ala for Phe in peptide 018, showed the high activity for both HeLa and PC12 cells. Using the clustering analysis of cell types in the CP activity, the human cancer cells, HeLa and A549, were well discriminated from the mouse and rat cell lines, 3T3-L1 and PC12. These results demonstrated that the CFP in an array format using the CP activity of helical designed peptides could be applied to a cancer cell typing technology.

In addition to the CP activity, other CFPs were generated using cell toxicity profiles of the LK peptides at a higher concentration to characterize the different 4 cell lines, HeLa, 10T1/2, COS7, CHO [9]. Moreover, the selected CP peptides with cell selectivity were used for conjugation with gold nanosphere (GNS) [10]. These CP peptide-GNS conjugates maintained the cell selectivity of the peptides, and were utilized for the cell-selective labeling and the cancer drug delivery. Throughout these studies, it has been implied that the cell responses profiling using the designed peptide library will develop a microarray system useful for novel cell analyses and diagnostics.

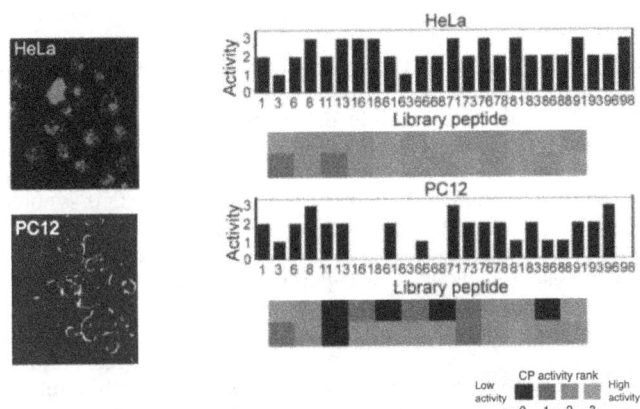

Fig. 1. *Confocal fluorescent microscopic analyses of peptide 018(LKARFH) for HeLa and PC12 cells, and cell fingerprint (CFP) analyses of the selected 24 peptides using the cell penetration (CP) activities.*

Acknowledgments

These studies were, in part, supported by grants of JSPS KAKENHI and JST A-STEP.

References

1. Tomizaki, K.-Y., Usui, K., Mihara, H. *ChemBioChem* **6**, 782-799 (2005).
2. Tomizaki, K.-Y., Usui, K., Mihara, H. *FEBS J.* **277**, 1996-2005 (2010).
3. Usui, K., Tomizaki, K.-Y., Mihara, H. *Methods in Molecular Biology* **570**, 273-284 (2009).
4. Takahashi, M., Nokihara, K., Mihara, H. *Chem. Biol.* **10**, 53-60 (2003).
5. Usui, K., Takahashi, M., Nokihara, K., Mihara, H. *Mol. Divers.* **8**, 209-218 (2004).
6. Usui, K., Ojima, T., Takahashi, M., Nokihara, K., Mihara, H. *Biopolymers* **76**, 129-139 (2004).
7. Usui, K., Ojima, T., Tomizaki, K., Mihara, H. *NanoBiotechnology* **1**, 191-200 (2005).
8. Usui, K., Kikuchi, T., Mie, M., Kobatake, E., Mihara, H. *Bioorg. Med. Chem.* **21**, 2560-2567 (2013).
9. Usui, K., Kakiyama, T., Tomizaki, K.-Y., Mie, M., Kobatake, E., Mihara, H. *Bioorg. Med. Chem. Lett.* **21**, 6281-6284 (2011).
10. Park, H.J., Tsutsumi, H., Mihara, H. *Biomaterials* **34**, 4872-4879 (2013).

Proceedings of the 23rd American Peptide Symposium
Michal Lebl (Editor)
American Peptide Society, 2013

Self-Assembled Cyclic D,L-α-Peptide Architectures with Potent Anti-Amyloidogenic Activity

Marina Chemerovski, Michal Richamn, Sarah Wilk, and Shai Rahimipour

Department of Chemistry, Bar-Ilan University, Ramat-Gan, 52900, Israel

Introduction

Misfolding of proteins and their subsequent transformation to amyloids is believed to be the major cause of more than 20 devastating diseases, including Alzheimer's disease (AD) and Parkinson's disease (PD). Recent studies have shown that pathogenic amyloids share common structural and functional features despite being composed of different proteins and amino acids. The immense structural similarity between different amyloids is highlighted by their equal reaction with polyclonal antibodies raised against prefibrillar assemblies of amyloid beta (Aβ) peptide [1]. Moreover, prefibrillar assemblies of different proteins interact similarly with synthetic phospholipid bilayers and with cell membranes, causing their destabilization by a mechanism resembling that of membrane-active antimicrobial peptides [2]. The strong structural and functional similarities between the amyloidogenic protein aggregates may also be related to their cross-reactivity. For example, Aβ interacts specifically with various amyloidogenic proteins, including tau, α-synuclein (α-syn), and islet amyloid polypeptide, and modulates their aggregation and toxicity [3].

We have recently shown that the structural and biochemical characteristics of amyloids strikingly resemble those of cyclic D,L-α-peptide nanotubes [4]. These supramolecular structures are generated from the self-assembly of simple cyclic peptides comprising an even number of alternating D- and L-α-amino acids that can form flat and ring-shaped conformations (Figure 1A). Under conditions that favor hydrogen-bond formation, cyclic D,L-α-peptides stack on top of each other to form hollow β-sheet-like tubular structures that can act as effective antibacterial and antiviral agents by interacting with different membranes [5,6]. The strong similarities between amyloids and self-assembled cyclic D,L-α-peptides led us to hypothesize that appropriately designed cyclic D,L-α-peptides may cross-react with different amyloids through a complementary sequence of hydrogen-bond donors and acceptors to modulate their aggregation and toxicity.

Results and Discussion

We employed the "one-bead-one-compound" combinatorial approach to synthesize an unbiased library of cyclic D,L-α-peptides hexamer. Screening of 500 members of the library against $A\beta_{40}$, as a model amyloidogenic protein, led to discovery of two potent cyclic peptides, which after sequence optimization generated the cyclic D,L-α-peptide [lJwHsK] (**CP2**; upper and lower case letters represent L- and D-α-amino acids, respectively. Square brackets indicate a cyclic structure, and J denotes norleucine). **CP2** caused a dose-dependent reduction in $A\beta_{40}$ aggregation and its activity was superior to that of antiamyloidogenic Ac-KLVFF-NH$_2$ peptide. Transmission electron microscopy (TEM) analysis also suggested that coincubation of **CP2** with soluble $A\beta_{40}$ drastically reduced fibril formation. **CP2** also dose-dependently decreased the aggregation kinetics of $A\beta_{40}$, while it completely arrested the aggregation of $A\beta_{42}$ even at a 1:5 Aβ:**CP2** concentration ratio, suggesting that **CP2** has a somewhat stronger inhibitory effect on the more neurotoxic $A\beta_{42}$ peptide.

To shed light on the mechanism of action of **CP2**, we evaluated its effect on the kinetics of seed-induced Aβ aggregation. We found that coincubation of **CP2** with 5% $A\beta_{42}$ seeds and monomeric $A\beta_{42}$ completely inhibited the formation of Aβ amyloids, suggesting that **CP2** interacts either with the monomers, and/or binds and remodels the seeds to incompetent structures, most likely *via* an "off-pathway" mechanism [7]. By using dot blotting assay and ELISA, employing the amyloid oligomer-specific polyclonal antibody A11 and the Aβ oligomer-specific monoclonal antibody (OMAB), we showed that **CP2** dramatically reduced the amount of toxic oligomers. These findings support our seeding results that interaction of **CP2** with Aβ may alter its aggregation to an "off-pathway" mechanism.

Our ThT results together with TEM analysis also demonstrate that **CP2** can effectively disassemble preformed fibrils in a near-stoichiometric ratio, suggesting that **CP2** generates relatively stable 1:1 coaggregates with monomeric or small $A\beta$ oligomers (1-3 mers according to our PICUP results) and shifts the aggregation equilibrium toward these species. To confirm that the disaggregation of $A\beta$ fibrils does not involve the formation of a new pool of toxic oligomers, **CP2**-treated $A\beta_{40}$ fibrils were tested for their reactivity toward the A11 and OMAB antibodies. The results suggest that disaggregation of the $A\beta$ fibrils by **CP2** leads to the generation of coaggregates that are not recognized by either antibodies further confirming the "off-pathway" mechanism.

Because self-assembly and aggregation of $A\beta$ are associated with AD, we next probed whether **CP2** could reduce $A\beta$-induced toxicity in PC12 cells. Our cell-based viability assays showed that **CP2** significantly decreased $A\beta$-induced toxicity. A maximal increase in cell

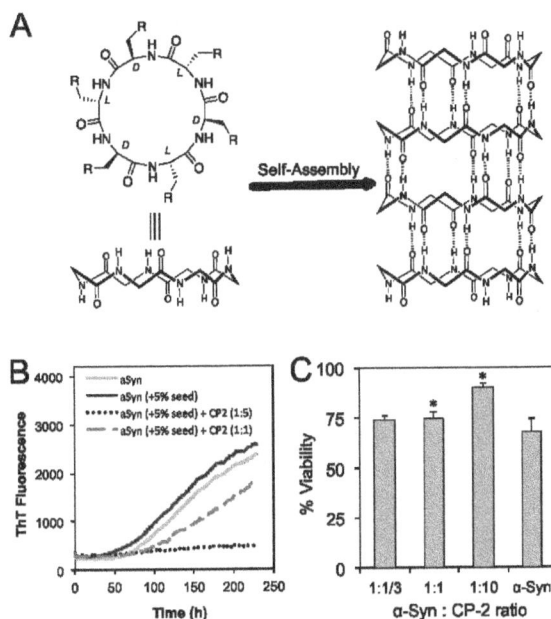

Fig. 1. A. Schematic representation of nanotube assembly from cyclic D,L-α-peptides. B. Effect of CP-2 on the kinetics of α-syn aggregation under seeding conditions. C. Dose-dependent effect of CP-2 on α-syn-mediated toxicity to PC12 cells.

viability was observed when $A\beta$ was incubated with a 5-10-fold excess of the cyclic peptide.

In order to show that antiamyloidogenic activity of **CP2** is stemmed from its structural and functional similarities to that of pathogenic amyloids, we have tested its effect on aggregation and toxicity of PD-associated α-syn. The ThT-based assay (Figure 1B) together with TEM analysis demonstrated that **CP2** effectively inhibits the aggregation of α-syn and disassembles preformed fibrils also by an "off-pathway" mechanism. Moreover, **CP2** significantly reduced the toxicity of α-syn to PC12 cells (Figure 1C).

In conclusion, we have shown that a simple self assembled system of cyclic D,L-α-peptides may be used to modulate the aggregation and toxicity of pathogenic amyloids due to their immense structural and functional similarities. Our results support previous studies, proposing that amyloids may have common structural and functional properties, and therefore they may cross-react and modulate the aggregation and toxicity of each other.

Acknowledgments

This work was supported in part by a grant from the Chief Scientist of the Israel Ministry of Health and Israel Ministry for Senior Citizens.

References

1. Kayed, R., Head, E., Thompson, J.L., McIntire, T.M., Milton, S.C., Cotman, C.W., Glabe, C.G. *Science* **300**, 486-489 (2003).
2. Quist, A., et al. *Proc. Nat. Acad. Sci. USA* **102**, 10427-10432 (2005).
3. Yan, L.M., et al. *Angew. Chem., Int. Ed.* **46**, 1246-1252 (2007).
4. Richman, M., et al. *J. Am. Chem. Soc.* **135**, 3474-3484 (2013).
5. Fernandez-Lopez, S., et al. *Nature* **412**, 452-455 (2001).
6. Horne, W.S., et al. *Bioorg. Med. Chem.* **13**, 5145-5153 (2005).
7. Ehrnhoefer, D.E., et al. *Nat. Struct. Mol. Biol.* **15**, 558-566 (2008).

Proceedings of the 23rd American Peptide Symposium
Michal Lebl (Editor)
American Peptide Society, 2013

Multivalent Peptide-Mimetics of Apolipoprotein A-I That Modulate HDL Particles to Combat Atherosclerosis

Yannan Zhao[1], Tomohiro Imura[1], Luke J. Leman[1], Audrey S. Black[2], David J. Bonnet[2], Bruce E. Maryanoff[1], Linda K. Curtiss[2], and M. Reza Ghadiri[1,3,*]

[1]Department of Chemistry; [2]Department of Immunology and Microbial Science; [3]The Skaggs Institute for Chemical Biology, The Scripps Research Institute, La Jolla, CA, 92037, U.S.A.

Introduction

High-density lipoproteins (HDLs) protect against atherosclerosis [1]. While the anti-atherogenic properties of HDL and its major protein, apolipoprotein A-I (apoA-I) have been documented in numerous studies [2], high production costs and lack of oral bioavailability have made apoA-I impractical for chronic use in the management of atherosclerosis. Here we describe the design, synthesis, and functional characterization of novel molecules that mimic apoA-I based on multivalent presentation of short, synthetic, amphiphilic, α-helical peptides.

Results and Discussion

We utilized native chemical ligation [3] to ligate an amphiphilic peptide to small-molecule scaffolds that differed in the number of reactive arms. Thus, we prepared two distinct families of monomer, dimer, trimer, and tetramer peptide constructs with different lengths of peptide building blocks, 23- and 16-mers, to systematically examine the functional effects of helix multimerization. Lipid nanoparticles were prepared by incubating each peptide construct with an aqueous suspension of DMPC multilamellar vesicles at 22°C with shaking. Despite having unnatural, branched topologies, all of the multivalent constructs generated discoidal nanoparticles of similar size and morphology as discoidal HDLs (Figure 1 a,b).

HDL particles promote the efflux of cholesterol from macrophage cells in the early, rate-limiting step of reverse cholesterol transport (RCT) [4]. We measured cholesterol efflux mediated by each peptide, alone and in the context of their nanolipid particles. In both series, the multivalent constructs promoted cholesterol efflux more efficiently than the monomeric peptide. In the 16-mer family, the observed efflux EC_{50} values followed the trend of tetramer16 > trimer16 > dimer16, while monomer16 failed to cause any notable efflux.

Small, dense HDLs are key initial players in RCT as the initial acceptors of cellular cholesterol [5]. As determined by immunoblotting for apoA-I, all the peptide nanoparticles induced an increase in small, dense HDL levels, with the exception of monomer16. We independently tracked the peptide and lipid components of synthetic nanoparticles as they interacted with lipoproteins in the plasma sample by using fluorescently labeled analogs. We found that the multivalent peptides were readily transferred from the synthetic nanoparticles to native HDL particles, caused HDL remodeling, and at that point coexisted in the particles with apoA-I. Whereas the fluorescent monomer23 was largely bound to human albumin, the dimer23 and trimer23 constructs associated almost exclusively with HDL particles, suggesting that multimerization of amphiphilic α-helical peptides results in more selective, higher affinity binding to lipoproteins.

By following the disappearance of our intact peptides in mouse serum or in the presence of isolated proteases (chymotrypsin, thermolysin, pronase, pepsin), we found that the monomer was degraded quickly, whereas the multivalent constructs were remarkably more stable. When administered intraperitoneally (i.p.) to mice (BALB/cByJ), the multivalent nanoparticles exhibited superior blood concentrations and much longer plasma residence times (e.g., ~8 h and ~6 h for trimer23 or trimer16 nanoparticles, respectively) compared with the nanolipid particles from monomers (~2 h and <1 h for monomer23 or monomer16 nanoparticles, respectively) for both series of constructs (Figure 1c). There was a marked enhancement of small, dense HDL bands from 2-8 h post-injection for dimer23 and trimer23, indicating successful remodeling of mouse HDLs in vivo. We hypothesize that inter- and intramolecular self-assembly of the multivalent constructs might be partly responsible for the high proteolytic stabilities and prolonged circulatory half-lives observed.

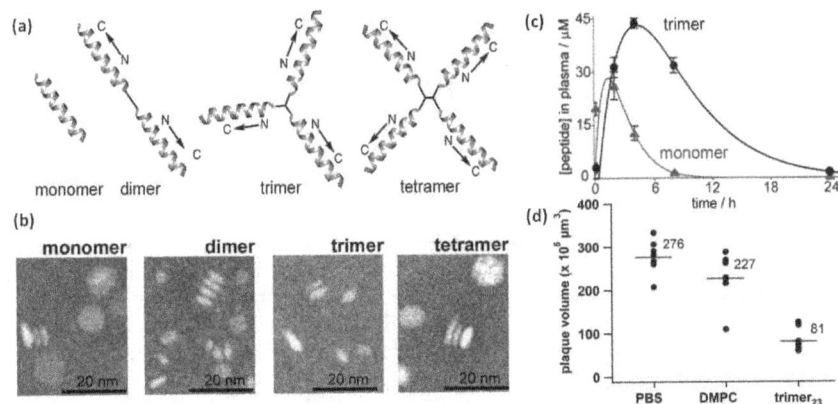

Fig. 1. (a) Multivalent 23-mer peptides; (b) peptide DMPC nanoparticles; (c) Pharmacokinetics (i.p.) of monomer23 and trimer23 nanoparticles in mice; (d) Aortic sinus lesion volumes of LDLr-/- mice after 10-week treatment with trimer23 DMPC nanoparticles in drinking water compared to PBS and DMPC control.

To determine the efficacy of the peptides in an animal model of atherosclerosis, we carried out a series of studies involving low-density lipoprotein receptor knockout (LDLr-/-) mice [7]. At 10 weeks of age, the mice were switched to a high-fat diet, and daily 40-mg/kg i.p. injections of the trimer23/DMPC nanoparticles were commenced. After two weeks, treatment with the trimer23 nanoparticles reduced plasma total cholesterol levels (mainly from VLDL and LDL) by 40%, compared to controls involving i.p. injection of phosphate-buffered saline (PBS). The plasma HDLs were also remodeled in favor of smaller HDL particle sizes. Encouraged by these results, we carried out further cholesterol reduction studies in which the trimer23/DMPC nanoparticles were administered orally in the drinking water (~50 mg/kg/day). After two weeks of treatment, the trimer nanoparticles again reduced plasma total cholesterol levels by 40% compared to PBS control.

To establish the effect of the trimer23/DMPC nanoparticles on the development of atherosclerotic plaques in the LDLr-/- mice, groups of mice were treated with daily i.p. injections (40 mg/kg) or oral administration in the drinking water (~50 mg/kg/day) for 10 weeks. The daily i.p. treatment with the trimer23 nanoparticles reduced whole aorta lesion areas and heart aortic valve plaque volumes by 55% and 61%, respectively. Likewise, oral administration of the trimer23/DMPC nanoparticles in the drinking water (~50 mg/kg) for 10 weeks significantly reduced the development of whole aorta lesion areas and heart aortic valve plaque volumes by 50% and 70%, respectively (Figure 1d). Our apoA-I mimetics provide a promising path toward the development of new therapeutics to combat heart disease.

Acknowledgments

We thank the NIH for financial support (NHLBI HL104462), the American Heart Association Western States Affiliate for a postdoctoral fellowship (12POST12040298) to Dr. Y. Zhao, and the National Institute of Advanced Science and Technology (AIST) for a visiting research fellowship to Dr. T. Imura. We also thank the 23rd American Peptide Symposium for the travel grant and the Young Investigator Presentation award to Dr. Y. Zhao. Reprinted (adapted) with permission from Zhao, Y., et al. *J. Am. Chem. Soc.* **2013**, http://dx.doi.org/10.1021/ja404714a.

References

1. Kontush, A., Chapman, M.J. *High-Density Lipoproteins. Structure, Metabolism, Function, and Therapeutics*, John Wiley & Sons, Inc.: Hoboken, NJ, 2012.
2. Rubin, E.M., Krauss, R.M., Spangler, E.A., Verstuyft, J.G., Clift, S.M. *Nature* **353**, 265-267 (1991).
3. Dawson, P.E., Muir, T.W., Clark-Lewis, I., Kent, S.B. *Science* **266**, 776-779 (1994).
4. Kane, J.P., Malloy, M.J. *Curr. Opin. Lipidol.* **23**, 367-371 (2012).
5. Oram, J.F., Heinecke, J.W. *Physiol. Rev.* **85**, 1343-1372 (2005).
6. Werle, M., Bernkop-Schnuerch, A. *Amino Acids* **30**, 351-367 (2006).
7. Van Craeyveld, E., Gordts, S.C., Singh, N., Jacobs, F., De Geest, B. *Acta Cardiol.* **67**, 11-21 (2012).

Proceedings of the 23rd American Peptide Symposium
Michal Lebl (Editor)
American Peptide Society, 2013

$n{\rightarrow}\pi^*$ Interactions in Helices

Ronald T. Raines

Departments of Biochemistry and Chemistry, University of Wisconsin–Madison,
Madison, WI, 53706, U.S.A.

Introduction

Helices are prevalent molecular structures in natural and synthetic polymers, including peptides and proteins. The topology of helices brings atoms that are nearby in a sequence into close proximity in a three-dimensional structure. This proximity can be reinforced by hydrogen bonds, like those in an α or 3_{10} helix. Other helices, however, lack hydrogen bonds (*e.g.*, polyproline type-I and type-II, and poly(L-lactic acid)), and do not have a clear basis for their conformational stability.

We have proposed that a quantum mechanical interaction stabilizes common helices - an $n{\rightarrow}\pi^*$ interaction between backbone carbonyl groups [1,2]. This interaction is between a lone pair (n) of the oxygen (O_{i-1}) of one carbonyl group and the antibonding orbital (π^*) of the subsequent carbonyl group ($C'_i{=}O'_i$) [3-5]. Here, the structures of natural and synthetic helices are analyzed from this perspective.

Results and Discussion

Models of AcAlaNH$_2$ in an α, 3_{10}, and polyproline type-I and type-II helix, and a model of an Ac(L-lactic acid)OH helix were created based on the ω, ϕ, and ψ backbone dihedral angles listed in Table 1. These models are depicted in Figure 1. A key aspect of these helices is that $r_{O{\cdots}C{=}O}$ is near or within the sum of the van der Waals radii of oxygen and carbon (3.22 Å) and $\angle_{O{\cdots}C{=}O}$ is close to the Bürgi–Dunitz trajectory for the approach of a nucleophile to a carbonyl group (~109°). Accordingly, the parameters indicate that $n{\rightarrow}\pi^*$ interactions make a significant contribution to the conformational stability to each of these helices. The energy of the interaction is 0.3 kcal/mol in AcProNH$_2$, and is enhanced with thiocarbonyl groups [6,7].

This analysis has numerous implications. For example, the data provide a new view of the α helix, which is the most prevalent structural element in folded proteins. This view is depicted in Figure 2 [1]. In addition, the data establish a physicochemical basis for the conformational stability of polyproline type-I and type-II helices, and the helical structure of poly(L-lactic acid) [2,5].

Table 1. Structural parameters of common helices with backbone carbonyl groups.[a]

Helix	ω	ϕ	ψ	$r_{O{\cdots}C{=}O}$	$\angle_{O{\cdots}C{=}O}$
A	180°	−60°	−45°	2.91 Å	97.9°
3_{10}	180°	−49°	−26°	2.73 Å	116.1°
polyproline type-I[b]	0°	−75°	150°	3.35 Å	127.5°
polyproline type-II	180°	−75°	160°	3.18 Å	89.4°
poly(L-lactic acid)[c]	180°	−64°	154°	2.86 Å	94.4°

[a]ω: $C^a_i{-}C'_i{-}N_{i+1}{-}C^a_{i+1}$, ϕ: $C'_{i-1}{-}N_i{-}C^a_i{-}C'_i$, ψ: $N_i{-}C^a_i{-}C'_i{-}N_{i+1}$; [b]*In the polyproline type-I helix, $O{\cdots}C{=}O$ refers to $O'_i{\cdots}C'_{i-1}{=}O'_{i-1}$; [c]In poly(L-lactic acid), N_i and N_{i+1} are replaced with oxygens.*

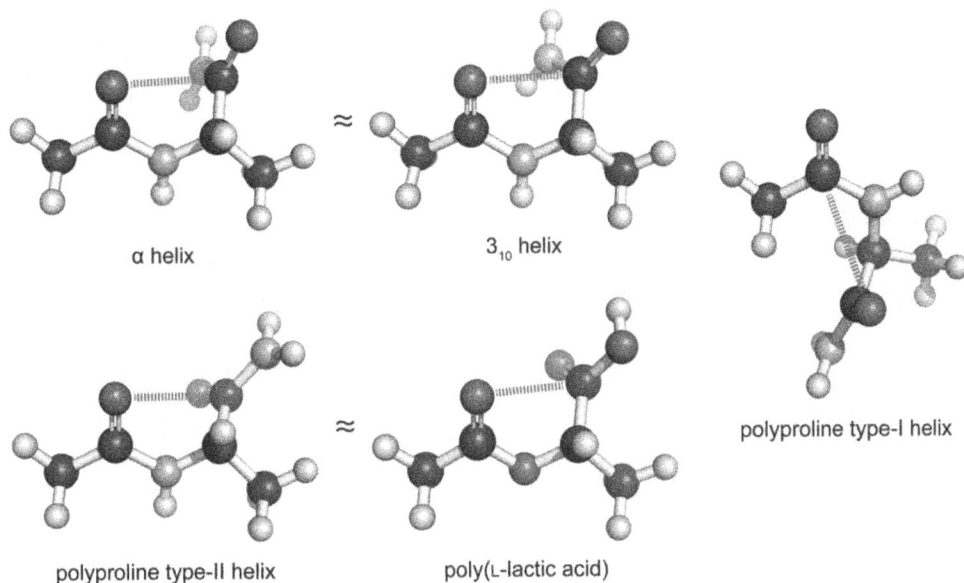

α helix   ≈   3₁₀ helix

polyproline type-I helix

polyproline type-II helix   ≈   poly(L-lactic acid)

Fig. 1. $n{\rightarrow}\pi^*$ Interactions (hatched lines) between backbone carbonyl groups in common helices. The depicted models are AcAlaNH₂ or Ac(L-lactic acid)OH with structural parameters listed in Table 1.

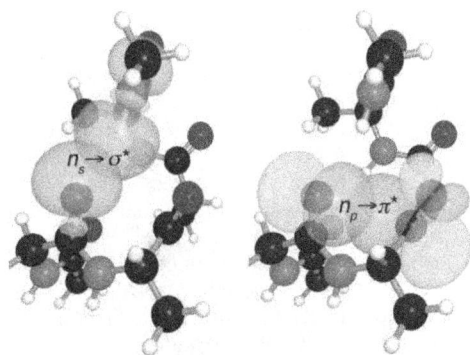

Fig. 2. A new view of the α helix. The s-rich lone pair (n_s) forms the canonical $i{\rightarrow}i{+}4$ hydrogen bond; the p-rich lone pair (n_p) forms an $i{\rightarrow}i{+}1$ $n{\rightarrow}\pi^*$ interaction. The depicted model is AcAla₄NH₂ in an α helix.

We conclude that the $n{\rightarrow}\pi^*$ interaction must be considered along with other noncovalent interactions when examining and considering the structure, stability, engineering, and design of common helices, and that $n{\rightarrow}\pi^*$ interactions be included in relevant computational force fields.

Acknowledgments

This work was supported by grants R01 AR044276 (NIH) and CHE-1124944 (NSF).

References

1. Bartlett, G.J., Choudhary, A., Raines, R.T., Woolfson, D.N. *Nat. Chem. Biol.* **6**, 615-620 (2010).
2. Newberry, R.W., Raines, R.T. *Chem. Commun.* **49**, 7699-7701 (2013).
3. Bretscher, L.E., Jenkins, C.L., Taylor, K.M., DeRider, M.L., Raines, R.T. *J. Am. Chem. Soc.* **123**, 777-778 (2001).
4. DeRider, M.L., Wilkens, S.J., Waddell, M.J., Bretscher, L.E., Weinhold, F., Raines, R.T., Markley, J. L. *J. Am. Chem. Soc.* **124**, 2497-2505 (2002).
5. Hinderaker, M.P., Raines, R.T. *Protein Sci.* **12**, 1188-1194 (2003).
6. Choudhary, A., Gandla, D., Krow, G. R., Raines, R. T. *J. Am. Chem. Soc.* **131**, 7244-7246 (2009).
7. Newberry, R.W., VanVeller, B., Guzei, I.A., Raines, R.T. *J. Am. Chem. Soc.* **135**, 7843-7846 (2013).

Proceedings of the 23rd American Peptide Symposium
Michal Lebl (Editor)
American Peptide Society, 2013

Novel Formulations for Non-Invasive Delivery and Stabilization of Peptides

Edward T. Maggio

Aegis Therapeutics LLC, San Diego, CA, 92127, U.S.A.

Introduction

Certain proprietary GRAS alkylsaccharides designated as Intravail® or ProTek® excipients can dramatically increase transmucosal absorption of certain peptides and proteins up to about 30 kDa in size allowing non-invasive delivery via the intranasal, oral and buccal administration routes [1]. These same excipients have been shown to effectively prevent aggregation during manufacturing and in final formulations and may serve as non-oxidizing and non-damaging replacements for polysorbates [2].

Results and Discussion

Alkylsaccharides, comprised of disaccharides and alkyl chain substituents with lengths between 10 and 16 carbons have been shown to be among the most effective excipients in increasing transmucosal absorption of peptides, proteins, as well as small molecule drugs.

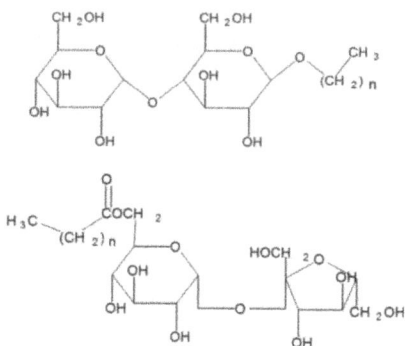

Fig. 1. Examples of alkyl glycosides and alkyl esters exhibiting enhancement of transmucosal absorption and prevention of aggregation (n=10-16).

Two classes of alkylsaccharides in particular, namely, the alkylglycosides and alkyl esters, as shown in Figure 1, have been found to be particularly useful for delivery of peptides and proteins up to approximately 30,000 Da molecular weight. Unlike many other surfactants which have been previously explored as transmucosal absorption enhancers, alkyl saccharide surfactants combine unmatched absorption enhancement with a high degree of safety and lack of toxicity and metabolize rapidly to the corresponding free sugars and fatty acid or corresponding long chain fatty alcohol. The general characteristics of alkylsaccharides suitable for pharmaceutical formulations are summarized in Table 1. Published examples of absorption enhancement for proteins ranging from 4000 Da to 30,000 Da in preclinical studies are summarized in Figure 2. Surfactants are incorporated into many peptide therapeutic formulations to prevent aggregation, improve reproducibility upon reconstitution of lyophilizates, and prevent loss due to stickiness on filters, columns, and container surfaces thus minimizing loss of efficacy, induction of unwanted immunogenicity, altered pharmacokinetics, and reduced shelf life. Unlike polysorbates and other polyoxyethylene-based surfactants, alkylsaccharides do not contain ether linkages (polyoxyethylene moieties) and unsaturated alkyl chains that spontaneously and rapidly auto-oxidize in aqueous solution to protein-damaging peroxides, epoxy acids, and reactive aldehydes inducing unwanted immunogenicity and in some instances promoting re-aggregation. These contaminating reactive chemical species modify methionine, histidine, tryptophan, as well as any primary amines or accessible nucleophiles such as those found in cysteine and tyrosine, creating neoantigens that cause unwanted immunogenicity. Immunogenicity of peptide therapeutics is a significant and growing concern of the FDA and EMEA and will have significant impact on the clinical trial and regulatory approval processes.

Table 1. General characteristics of Intravail®/ProTek® alkylsaccharides.

➢ *Safe, odorless, tasteless, non-toxic, non-mutagenic, and non-irritating*

➢ *Synthetic pure chemical entities prepared under GMP*

➢ *Provides unmatched bioavailability - comparable to subcutaneous injection, via the intranasal and other mucosal membrane administration routes (up to ~30KDa MW)*

➢ *Allows controlled transient mucosal permeation by both paracellular (tight-junction) and transcellular routes*

➢ *Soluble in water or oils – compatible with routine liquid formulation and dispensing processes for ease of scale-up and production*

➢ *Shown to be highly effective (orally) for BCS Class III/IV small peptides and small molecules*

➢ *Shown to greatly increase oral bioavailability in tablets, oils (i.e., soft-gel compatible), and other formats.*

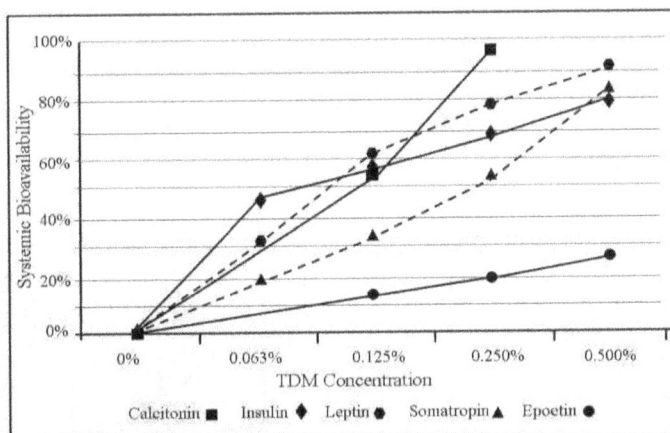

Fig. 2. Selected literature examples of bio absorption enhancement of peptides with tetradecyl maltoside (TDM) [3,4].

References

1. Maggio, E.T., Pillion, D.J. *Drug Delivery and Translational Research* **3**, 16-25 (2013).
2. Maggio, E.T. *J. Excipients and Food Chemicals* **3**(2), 45-53 (2012).
3. Ahsan, F., Arnold, J.J., Meezan, E., Pillion, D.J. *Pharm. Res.* **18**(12), 1742-1746 (2001).
4. Arnold, J.J., Fakhrul, A., Meezan, E., Pillion, D.J. *J. Pharm. Sci.* **93**(9), 2205-2213 (2004).

Proceedings of the 23rd American Peptide Symposium
Michal Lebl (Editor)
American Peptide Society, 2013

Two Putative Holin-Like Peptide with Anti-MRSA Activity and Their Potential Application

Ratchaneewan Aunpad[1], Siriporn Kaewklom[1], and Watanalai Panbangred[2]

[1]Graduate Program in Biomedical Sciences, Faculty of Allied Health Sciences, Thammasat University, Pathumthani, 12121, Thailand; [2]Department of Biotechnology, Faculty of Science, Mahidol University, Bangkok, 10400, Thailand

Introduction

Novel antibacterial agents are urgently needed to combat the drug resistance problem especially methicillin resistant *Staphylococcus aureus* (MRSA). Members of the genus *Bacillus* are known to produce a wide variety of antimicrobial agents with potential applications in human and animal health as alternatives to conventional antibiotics [1]. *Bacillus pumilus* strain WAPB4 produced bacteriocin with remarkable activity against MRSA [2]. Analysis of the whole genome sequence of *B. pumilus* strain SAFR032 revealed the presence of a possible bacteriocin precursor gene, *bhl*A. The gene product is holin BhlA. In bacteriophage, holins are small membrane proteins which accumulate and oligomerize to form non-specific lesions in the cytoplasmic membrane allowing the release of the second proteins, endolysins, to access the peptidoglycan [3]. In this study, we report the characterization of two putative holin-like peptides from *B. pumilus* strain WAPB4.

Results and Discussion

By using primers designed from a possible bacteriocin precursor gene, *bhl*A, from the genome of *B. pumilus* SAFR-032, there were multiple PCR products with size ranging from 50-1,500 bp. The BLASTN analysis of cloned 400 bp PCR product showed 89% identity with *bhl*A gene of *B. pumilus* SAFR-032. The BLASTP analysis of deduced amino acid sequence showed 98% identity to holin BhlA of *B. pumilus* SAFR-032. It also showed 90-97% similarity to bacteriocin UviB of *Clostridium botulinum* and *C. perfringens*. The BLASTP analysis of second PCR product (260 bp) showed 100% identity with holin SPP1 from *B. pumilus* ATCC 7061, holin XhlB from *B. pumilus* SAFR-032, XpaF2 protein from *B. licheniformis* and holin from *Bacillus* bacteriophage PBSX.

BhlA of *B. pumilus* strain WAPB4 consists of 71 amino acid residues with a calculated molecular mass of 8.4 kDa and a predicted pI of 4.2. Analysis of BhlA using TMHMM server suggested one putative transmembrane domain (TMD) at the *N*-terminal part and a number of highly charged amino acid residues at the *C*-terminal part. An ORF located 10 bp downstream of the stop codon (TAG) of *bhl*A was found to be a holin-like peptide encoding gene, *xhl*B. Its gene product consisting of 86 amino acids has a calculated molecular mass of 9.6 kDa and a predicted

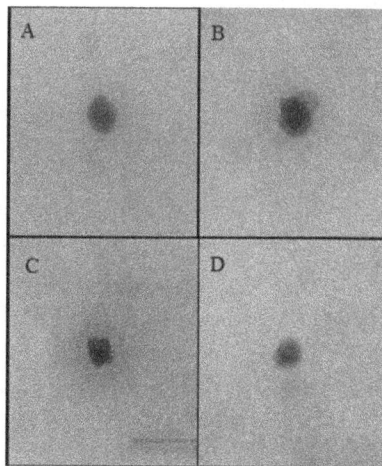

Fig. 1. Leakage assay of E. coli transformants harboring pTTQ-H (A) or pTTQ-X (B) or pTTQ-AB (C) growing on the agar plate containing the chromogenic sub-strate X-Gal (printed in grayscale). (D) is control (no plasmid).

pI of 9.63. XhlB of *B. pumilus* strain WAPB4 composed of two putative transmembrane domains separated by a β-turn, and numerous charged residues in the *C*-terminus. The dual start motifs were found in both BhlA and XhlB. Analysis of the region upstream of *bhl*A gene revealed the presence of a possible promoter and a possible Shine-Dalgarno sequence. These results showed the existence of two putative holin-like peptide encoding genes, *bhl*A and *xhl*B, in the same operon in *B. pumilus* strain WAPB4. There are few holin-like peptides found in bacteria such as XpaG2 of *B. licheniformis* [4], BhlA of *B. licheniformis* AnBa9 [5], HolNu3-1 of methicillin-resistant *S. aureus* [6], TcdE of *Clostridium difficile* [7] and STY1365 of *Salmonella enteric* serovar Typhi [8]. It is to our knowledge, BhlA and XhlB are first two bacterial putative holin-like peptides found in the same operon.

Holin-like activity was evaluated by determination of enzyme β-galactosidase leakage into the surrounding agar resulted from holin-like damage to the cytoplasmic membrane [9]. The holin structural genes, *bhl*A (H) and *xhl*B (X), were cloned separately and together (AB) into plasmid vector pTTQ18 and expressed in *E. coli* C600. Our hypothesis that *bhl*A and *xhl*B genes function as an operon was supported by the observation that the production of both holins in *E. coli* C600 (Figure 1C) caused more release of β-galactosidase into the periplasm and eventually into the surround medium as visualized by the formation of wide blue zone surrounding colonies than the colonies containing holin BhlA (Figure 1A) or XhlB (Figure 1B) alone. These results also confirmed a membrane-damaging effect of both BhlA and XhlB holin-like peptide.

The expression of holin in phage causes bacterial cell death due to damage and disruption of the cell membrane [10]. The mode of action of BhlA on bacterial cell was clearly demonstrated by TEM. The obvious effects of holin BhlA on *E. coli* cells became visible starting with the formation of membrane bleb, the condensation and release of cytoplasmic material, detachment of the outer membrane form the plasma membrane and finally, the completely disappearance of cytoplasm (Figure 2).

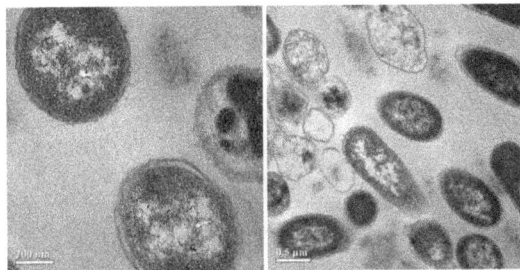

Fig. 2. Morphological changes in E. coli by BhlA.

In conclusion, two putative holin-like peptides, BhlA and XhlB, were found in *B. pumilus* strain WAPB4. The site of action of BhlA is on the cell membrane and caused bacterial death by cell membrane disruption. Therapeutic applications of BhlA could be considered as a combination therapy to promote the drug efficacy.

Acknowledgments

This work was supported by a research grant for new scholars from the Commission on Higher Education jointed with Thailand Research Fund.

References

1. Abriouel, H., Franz, C.M., Omar, N.B., Gálvez, A. *FEMS. Microbiol. Rev.* **35**, 201-232 (2011).
2. Aunpad, R., Na-Bangchang, K. *Curr. Micobiol.* **55**, 308-313 (2007).
3. Young, R.Y. *Microbiol. Rev.* **56**, 430-481 (1992).
4. Kyogoku, K., Sekiguchi, J. *Gene* **168**, 61-65 (1999).
5. Anthony, T., Stalin, C.G., Rajesh, T., Gunasekaran, P. *Arch. Microbiol.* **192**, 51-56 (2010).
6. Horii, T., Suzuki, Y., Kobayashi, M. *FEMS. Immunol. Med. Microbiol.* **34**, 307-310 (2002).
7. Tan, K.S., Wee, B.Y., Song, K.P. *J. Med. Microbiol.* **50**, 613-619 (2001).
8. Rodas, P., Trombert, A.N., Mora, G.C. *FEMS. Microbiol. Lett.* **321**, 58-66 (2011).
9. Delisle, A.L., Gerard, J.B., Guo, M. *Appl. Environ. Microbiol.* **72**, 1110-1117 (2006).
10. Oki, M., Kakikawa, M., Nakamura, S., Yamamura, E.T., Watanabe, K., Sasamoto, M., Taketo, A., Kodaira, K.I. *Gene* **197**, 137-145 (1997).

Proceedings of the 23rd American Peptide Symposium
Michal Lebl (Editor)
American Peptide Society, 2013

Effects of Selective NADPH Oxidase Inhibitors on Real-Time Blood Nitric Oxide and Hydrogen Peroxide Release in Acute Hyperglycemia

Matthew L. Bertolet, Michael Minni, Tyler Galbreath, Robert Barsotti, Lindon H. Young, and Qian Chen

Department of Bio-Medical Sciences, Philadelphia College of Osteopathic Medicine (PCOM), Philadelphia, PA, 19131, U.S.A.

Introduction

Hyperglycemia (blood glucose ≥ 5.5 mM) has been definitively linked to the development of vascular and neurologic complications in diabetic patients. Moreover, even in non-diabetic subjects, acute hyperglycemia during oral glucose tolerance tests or postprandially can temporarily induce vascular endothelial dysfunction, which is characterized by decreased endothelium-derived nitric oxide (NO) release and increased reactive oxygen species (ROS): superoxide (SO) and hydrogen peroxide (H_2O_2). Vascular NO is produced by endothelial NO synthase (eNOS) and plays critical roles in maintaining blood flow and suppressing inflammatory and coagulant vascular signals. It has been proposed that hyperglycemia induces eNOS enzymatic uncoupling resulting in SO production instead of NO leading to oxidative stress. Our lab has established an acute hyperglycemia animal model to monitor blood NO and H_2O_2 levels via free radical microsensors in real-time. We found that acute hyperglycemia had significantly higher blood H_2O_2 and lower blood NO levels compared to euglycemia [1]. However, the initial or trigger source of oxidative stress under acute hyperglycemic conditions is still unclear. One strong candidate is non-phagocytic nicotinamide adenine dinucleotide phosphate (NADPH) oxidase a multi-subunit enzyme that catalyzes SO production (see Figure 1), is widely distributed in various cells types, normally participates in various signaling pathways and has been shown to be activated following protein kinase C activation in chronic hyperglycemia [2,3]. To better understand the role of NADPH oxidase in acute hyperglycemia induced vascular dysfunction, two NADPH oxidase inhibitors, gp91 ds-tat (RKKRRQRRR-CSTRIRRQL-amide, MW=2452 g/mol, 1.2 mg/kg, Genemed Synthesis Inc., San Antonio, TX) and apocynin (MW=166 g/mol, Sigma Chemicals), were tested to determine whether they will increase NO and reduce H_2O_2 blood levels under acute hyperglycemic conditions.

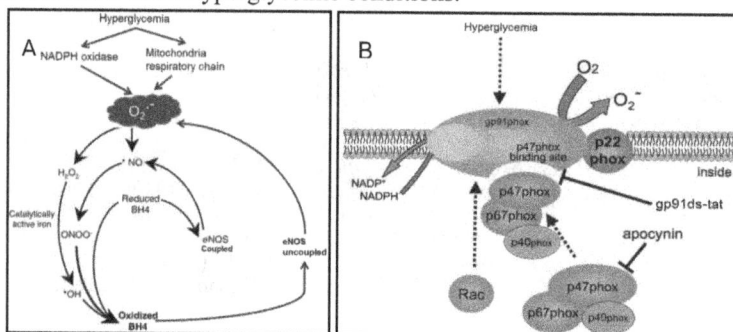

Fig. 1. Proposed vicious cycle induced by NADPH oxidase activation (A) and NADPH oxidase structure and inhibitors (B).

Results and Discussion

Male Sprague-Dawley rats (275-325g) were anesthetized and the jugular vein was catheterized and infused with saline, 20% D-glucose, 20% D-glucose with 1.2 mg/kg gp91ds-tat, or 20% D-glucose with 14 mg/kg apocynin (see Table 1). The continuous infusion of 20% D-glucose solution maintained hyperglycemia at 200 mg/dL for 180 min. To measure blood NO and H_2O_2 levels in real-time, NO and H_2O_2 microsensors (100 μm, WPI Inc., Sarasota, FL) connected to a free radical analyzer (Apollo 4000, WPI Inc., Sarasota,

FL) were inserted into individual catheters placed randomly in both femoral veins. NO, H_2O_2, and glucose levels were then recorded at baseline and at every 20 min interval throughout a 180 min infusion period. The time course of changes in blood NO and H_2O_2 levels were expressed as the relative change to the saline group and are shown in Figure 2. All the data were represented as means ± SEM, with $p<0.05$ were considered statistically significant.

Table 1. Experimental groups and i.v. solutions.

saline	20% glucose	20% glucose+gp 91 ds-tat (1.2 mg/kg, approx. 20 µM in blood)	20% glucose+apocynin (14 mg/kg, approx. 1mM in blood)
n=6 or 7	n=6	n=5 or 6	n=5 or 6

The blood glucose in the saline group remained at baseline values of 80-100 mg/dl throughout the experiment. By contrast, after 20 min infusion of 20% D-glucose with or without the drug, blood glucose levels rose to hyperglycemic levels at ~ 200 mg/dl. All hyperglycemic rats urinated between 20-40 min after glucose infusion. The changes of blood NO levels in different experimental groups relative to saline are shown in Figure 2, left panel. We found that acute hyperglycemia (200 mg/dl) significantly reduced blood NO compared to saline control. By contrast, the infusion of gp91ds-tat or apocynin with glucose inhibited the significant decrease in blood NO levels observed with glucose infusion alone ($p<0.05$). The right panel in Figure 2 illustrates the changes in blood H_2O_2 levels in different experimental groups relative to saline controls. Acute hyperglycemia resulted in a higher level of H_2O_2 in blood compared to saline control. By contrast, gp91ds-tat or apocynin significantly attenuated the hyperglycemia-induced increase in blood H_2O_2 levels ($p<0.05$).

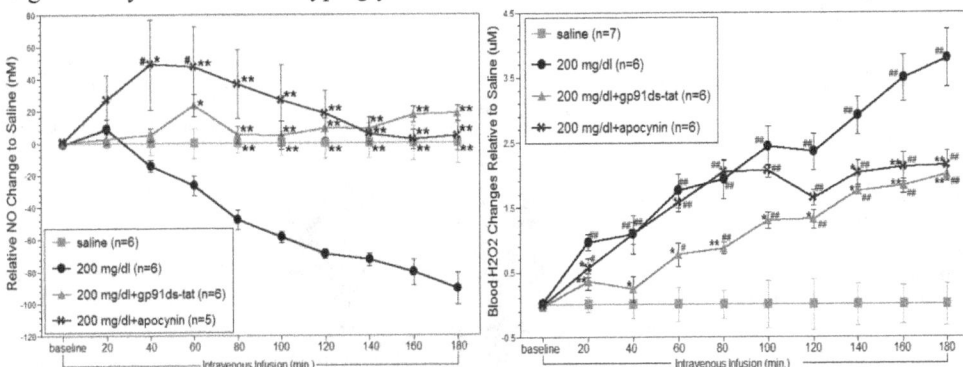

Fig. 2. The changes of blood NO levels (left panel) and H_2O_2 levels (right panel) relative to saline group in different experimental groups. *$p<0.05$, **$p<0.01$ vs glucose, # $p<0.05$, ## $p<0.01$ vs saline. All data were analyzed using ANOVA with Student Newman Keuls test.

These results suggest that under acute hyperglycemic conditions NADPH oxidase is a significant source of ROS that leads to vascular endothelial dysfunction. Treatment with gp91ds-tat or apocynin may be beneficial to attenuate hyperglycemia induced vascular endothelial dysfunction. Moreover, blood H_2O_2 levels in hyperglycemic gp91ds-tat or apocynin treated groups were still significantly higher compared to that in saline groups. This suggests that other sources of ROS exist under hyperglycemia such as mitochondria, which needs to be verified with further investigation.

Acknowledgments

This study was supported by the Center for the Chronic Disorders of Aging and the Department of Bio-Medical Sciences at PCOM.

References

1. Minni, M., et al. *Proceedings of the 22nd American Peptide Symposium* 280-281 (2011).
2. Fatehi-Hassanabad, Z., et al. *Eur. J. Pharmacol.* **636**, 8-17 (2010).
3. Inoguchi, T., et al. *Diabetes* **49**, 1939-1945 (2000).

Proceedings of the 23rd American Peptide Symposium
Michal Lebl (Editor)
American Peptide Society, 2013

SAR Study on CN17β a Potent and Selective CaMKII Peptide Inhibitor

**Alfonso Carotenuto[1], Maria R. Rusciano[2], Marina Sala[3],
Ermelinda Vernieri[3], Isabel Gomez-Monterrey[1], Antonio Limatola[1],
Diego Brancaccio[1], Alessia Bertamino[3], Paolo Grieco[1],
Ettore Novellino[1], Maddalena Illario[2], and Pietro Campiglia[3]**

[1]Department of Pharmacy, University of Naples, Naples, 80131, Italy; [2]Department of Cellular and Molecular Biology, University of Naples, Naples, 80131, Italy; [3]Department of Pharmacy, University of Salerno, Fisciano, 84084, Italy

Introduction

Ca^{2+}/Calmodulin-dependent protein kinase II (CaMKII) constitutes a family of kinases that transduces elevated Ca^{2+} signals in cells to a number of targets [1]. CaMKII regulates diverse cellular functions, including Ca^{2+} homeostasis [2]. Misregulation of this enzyme is involved in cardiovascular diseases including hypertrophy and other types of ischemia/reperfusion injury [3], obesity and diabetes [4]. A natural inhibitor of CaMKII, protein CaM-KIIN and a 27-residue peptide derived from it, (CaM-KNtide, Table 1), are highly selective for inhibition of CaMKII [5]. We have recently demonstrated that a 17-mer peptide (named CN17β) derived from CaM-KNtide still retained useful inhibitory potency (Table 1) [6].

Results and Discussion

The X-ray structure of CaM-KNtide bound to the kinase domain of CaMKII [7] indicates that *C*-terminal region (residues 11-17) of CN17β is very important for the binding to CaMKII since it occupies a very large area on CaMKII. This area was defined by the authors [7] as docking site A (Figure 1). *C*-terminal region of CaM-KNtide crystal structure is in extended conformation. To demonstrate that an extended conformation is needed at the *C*-terminal of an inhibitory peptide, we compared the inhibitory activity of compounds **1** and **2** (Table 1). Compound **1** misses the *C*-terminal Ile residue of CN17β and its inhibitory activity drops to 13% of the parent peptide.

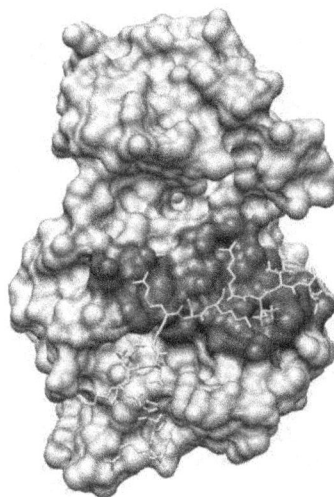

Fig. 1. X-ray Structure of CaM-KIINtide bound to CaMKII (pdb code 3KL8). Docking site A is highlighted.

Docking Site A

Replacing *C*-terminal Val of **1** with a Pro residue (peptide **2**) the activity raised again to 75% of CN17β. Since Pro residue is missing amide hydrogen, it should not stabilize helical conformation. In fact, CD spectra (50% HFA) demonstrate that helical content of peptide **2** is lower (30% vs 40%, SELCON analysis) compared to peptide **1** (Figure 2a).

Table 1. Structure and activity of compounds CaM-KNtide, CN17β and peptides **1-7**.

Frag	Peptide	Sequence	Inhib (%±SD)*
1-27	**CaM-KNtide**	KRPPKLGQIGRAKRVVIEDDRIDDVLK	94.8± 5.0
1-17	**CN17β**	KRPPKLGQIGRAKRVVI	76.0±4.1
1-16	**1**	KRPPKLGQIGRAKRVVV	10.2±2.0
1-16	**2**	KRPPKLGQIGRAKRVVP	57.0±1.0
11-17	**3**	RAKRVVI	6.0±2.0
11-17	**4**	RaKRVVI	5.0±1.0
11-17	**5**	RAKrVVI	9.0±1.0
11-16	**6**	RaKRVVI	5.0±1.0
11-16	**7**	RAKrVVI	8.0±2.0

*All peptides are amidated at the C-terminus. Peptides 3-7 are acetylated at the N-terminus. Peptide concentration 5 µM. Lower case letters indicate D-residues.

Starting from these results, we tried to address docking site A of CaMKII using short peptides whose sequences encompass the C-terminal region of CN17β (peptides 3-7, Table 1). First, we tested the C-terminal peptide fragment of CN17β (peptide **3**) to see if it kept some of the inhibitory activity of the parent peptide. Unfortunately, it lost most of the inhibitory potency (Table 1). Then, we replaced residues Ala2 and Arg4 (one by one) with the corresponding D-residue (peptides **4** and **5**). These substitutions were driven by the observation that the φ angle values of the corresponding residues in the crystal structure are both positive. Finally, we considered the derivatives of peptides **4** and **5** in which the terminal Val-Ile dipeptide was replaced by a Pro residue as in peptide **2** (peptides **6** and **7**, respectively). All these short peptides show the expected disordered conformation (Figure 2b), but they were all virtually inactive (Table 1). Actually, peptides **5** and **7**, bearing a DArg4 residue, recover part of the inhibitory activity against CaMKII. Therefore these peptides will be the starting points for the development of novel short peptides aimed at inhibiting CaMKII activity.

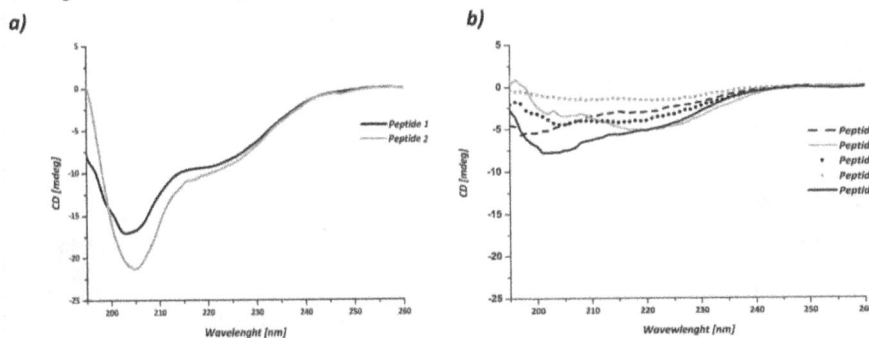

Fig. 2. a) CD spectra of peptides **1** (black line) and **2** (gray line); b) CD spectra of peptides **3-7** (**3** black line, **4** gray line, **5** black circle, **6** gray circle, **7** dotted black line).

Acknowledgments

Supported by grant from the Italian Ministry of Education (MIUR) (PRIN n° 2009EL5WBP).

References

1. Soderling, T.R., et al. *J. Biol. Chem.* **76**, 3719-3722 (2002).
2. Braun, A.P., et al. *Ann. Rev. Physiol.* **57**, 417-445 (1995).
3. Zhang, T., et al. *Cardiovascular Res.* **63**, 476-486 (2004).
4. Højlund, K., et al. *J. Clin. Endocrinol. Metab.* **94**, 4547-4556 (2009).
5. Chang, B.H., et al. *Proc. Nat. Acad. Sci. USA* **95**, 10890-10895 (1998).
6. Gomez-Monterrey, I., et al. *Eur. J. Med. Chem.* **22**, 425-434 (2013).
7. Chao, L.H., et al. *Nat. Struct. Mol. Biol.* **17**, 264-272 (2010).

Proceedings of the 23rd American Peptide Symposium
Michal Lebl (Editor)
American Peptide Society, 2013

Effects of Modulating eNOS Activity on Leukocyte-Endothelial Interactions in Rat Mesenteric Postcapillary Venules

Amber N. Koon, Maria A. Kern, Lindon H. Young, Edward Iames, Robert Barsotti, and Qian Chen

Department of Bio-Medical Sciences, Philadelphia College of Osteopathic Medicine (PCOM),
4170 City Avenue, Philadelphia, PA, 19131, U.S.A.

Introduction

Inflammation following vascular endothelial dysfunction is a common feature for the pathogenesis of various vascular diseases, such as ischemia/reperfusion injury, hypertension and atherosclerosis. Normally, vascular endothelium maintains an anti-thrombotic and anti-inflammatory surface to facilitate blood flow principally by producing endothelial-derived nitric oxide (NO). Endothelial NO synthase (eNOS) produces NO from L-arginine in the presence of the essential cofactor tetrahydrobiopterin (BH_4). Under oxidative stress, BH_4 is oxidized to dihydrobiopterin (BH_2), and the ratio of BH_2/BH_4 is increased, eNOS is enzymatically uncoupled, producing superoxide (SO) instead of NO because molecular oxygen, rather than L-arginine, accepts the electron [1]. Our previous studies have shown that administration of BH_2 promotes leukocyte-endothelial interactions in the mesenteric circulation *in vivo* [2]. To better understand the roles of coupled/uncoupled eNOS activity in inflammation, eNOS modulators, protein kinase C epsilon (PKC ε) peptide activator (+) or inhibitor (-) was administered with BH_4 or BH_2 to evaluate leukocyte-endothelial interactions *in vivo*. PKC ε increases eNOS activity *via* phosphorylation of eNOS serine 1177 [3]. We hypothesized that PKC ε+ (N-Myr-HDAPIGYD, MW=1097 g/mol, Genemed Synthesis Inc., San Antonio, TX) combined with BH_2 would result in increased uncoupled eNOS activity and thus augment or sustain the BH_2-induced leukocyte-endothelial interactions, while PKC ε+ combined with BH_4 would attenuate BH_2-induced inflammation. In contrast, PKC ε- (N-Myr-EAVSLKPT, MW=1054 g/mol), Genemed Synthesis Inc.) will attenuate BH_2 induced inflammation in the presence or absence of BH_4 (Figure 1).

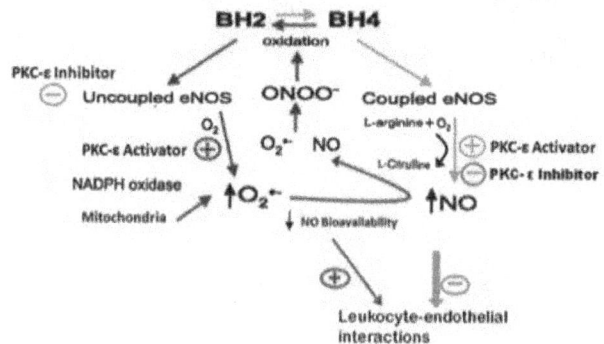

Fig. 1. The hypothesis diagram.

Results and Discussion

Intravital microscopy was performed on anesthetized male SD rats (275-325 g, Ace Animals, Boyertown, PA) and one loop of mesentery was placed on a viewing pedestal to observe microcirculation in postcapillary venules in real-time as previously described [4]. During the experiment, test solutions (listed in the following experimental groups) were superfused over the mesentery and leukocyte-endothelial interactions (i.e., leukocyte rolling, adherence, and transmigration) were recorded for 2 min. at 30 min intervals after baseline. Superfused mesenteric tissue was harvested at the end of experiment for later hematoxylin & eosin (H&E) staining to confirm intravital microscopy observations. All data were represented as means ± SEM and analysis by ANOVA with the Fisher's PLSD test. $p<0.05$ is considered to be statistically significant.

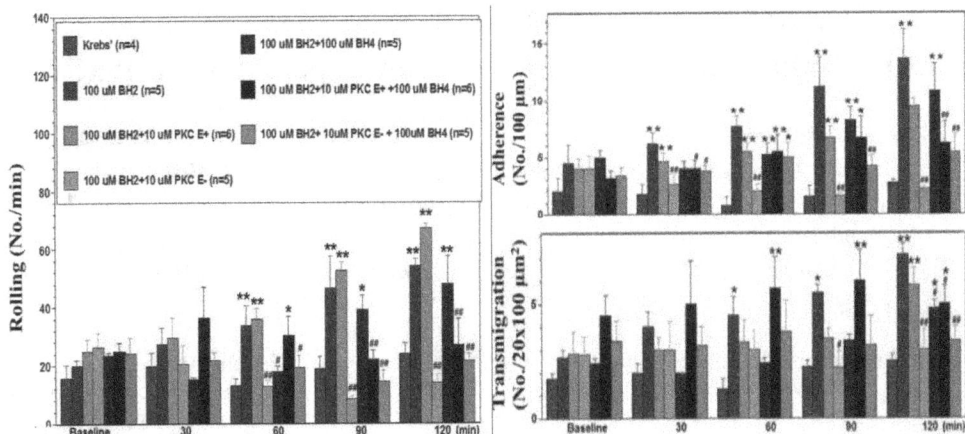

Fig. 2. *Leukocyte-endothelial interactions in different experimental groups. *p<0.05,**p<0.01 vs Krebs'; #p<0.05,##p<0.01 vs 100 μM BH₂.*

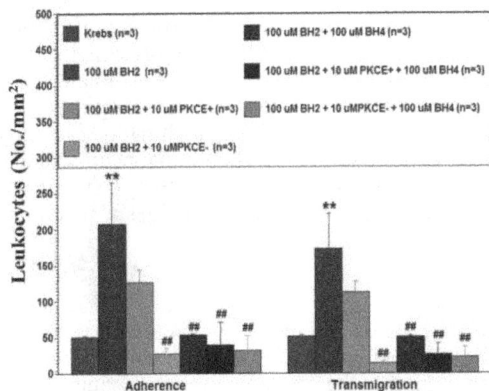

Fig. 3. *Leukocyte vascular adherence/transmigration in different experimental groups by H&E staining. *p<0.05, **p<0.01 vs Krebs'; #p<0.05, ##p<0.01 vs 100 μM BH₂.*

We found that BH_2, an eNOS uncoupling cofactor, significantly increased leukocyte-endothelial interactions compared to Krebs' control, and this effect was similar to that with BH_2+PKC ε+. In contrast, PKC ε+ with BH_4, PKC ε- alone, or PKC ε- with BH_4 significantly attenuated BH_2-induced inflammatory responses (Figure 2). Moreover, leukocyte adherence/ transmigration in superfused mesenteric tissue by H&E staining (Figure 3) were consistent with results obtained from intravital microscopy.

The data suggest that facilitation of eNOS uncoupling *via* BH_2 addition or an increased uncoupled eNOS activity by PKC ε+ induces inflammatory responses when endothelial-derived NO bioavailability is reduced. However, inhibiting uncoupled eNOS activity by PKC ε- or increasing coupled eNOS activity by addition PKC ε+ with BH_4 significantly attenuates the inflammatory responses. This study suggests the importance of maintaining normal BH_4/BH_2 ratio for preserving anti-inflammatory properties of normal vascular endothelial function. Moreover, PKC ε+ combined with BH_4 or PKC ε- may serve as potential strategies to mitigate the pathogenesis of inflammation-related vascular diseases.

Acknowledgments

This study was supported by the Center for the Chronic Disorders of Aging and the Department of Bio-Medical Sciences at PCOM.

References

1. Schmidt, T.S., Alp, N.J. *Clinical Science* **113**, 47-63 (2007).
2. Kern, M.A., et al., In Lebl, M. (Ed.) *Proceedings of the 22nd American Peptide Symposium,* American Peptide Society, San Diego, 2011, p. 286 (2011).
3. Perkins, K.A., et al. *Naunym-Schmiedeberg's Arch. Pharmacol.* **385**, 27-38 (2012).
4. Chen, Q., et al. *Curr. Topics Pharmacol.* **14**, 11-24 (2010).

Proceedings of the 23rd American Peptide Symposium
Michal Lebl (Editor)
American Peptide Society, 2013

Design and Synthesis of Novel Bivalent Ligands (μ and δ) of Enkephalin Analogues with 4-Anilidopiperidine Derivatives

Srinivas Deekonda[1], Jacob Cole[1], Vinod Kulkarni[1], Lauren Wugalter[1], David Rankin[2], Peg Davis[2], Josephine Lai[2], Frank Porreca[2], and Victor J. Hruby[1]

[1]Department of Chemistry and Biochemistry, University of Arizona, Tucson, AZ, 85721, U.S.A.;
[2]Department of Pharmacology, University of Arizona, Tucson, AZ, 85721, U.S.A.

Introduction

Opioid receptors are an important class of GPCRs which deal with the analgesic effects in humans, and these are classified into three different types, μ, δ and κ receptors. Opioid analgesics are widely used in the treatment of moderate to severe acute and chronic pain. The clinical use of the opioids is limited by serious side effects such as respiratory depression, constipation, development of tolerance, and physical dependence and addiction liabilities. μ receptors are the most important receptor target for almost all commercially available potent opioid agonists.

Depending on their nature all opiates can be broadly divided into two categories: non-peptide and peptide based. Mostly Morphine and Fentanyl derivatives represent the non-peptide based opiates. Endogenous opioid peptides such as endomorphins, enkephalins, and dynorphins occurring naturally in the brain represent the second class of opiates. The 4-anilidopiperidines represent the most powerful synthetic analgesics, which include fentanyl and related compounds. Fentanyl is a well-known μ-selective synthetic analgesic which is 50-100 times more potent than morphine, has a short duration of action. Fentanyl, sufentanyl, and alfentanyl currently the three most popular compounds used for analgesia in clinical practice.

Results and Discussion

Here we designed and synthesized novel bivalent ligands (μ and δ) of enkephalin analogues with 4-anilido piperidine derivatives. Basically our designs are hybrid molecules which are derived from two distinct classes of opioid ligands: the peptide moiety containing an enkephalin analogue and the non-peptide moiety from a 4-anilidopiperidine series. In the non-peptide 4-anilidopiperidine series we replaced the phenethyl group with a tetrahydro-naphthalen-2yl) methyl moiety with an amino substitution at the 5th position, and then we

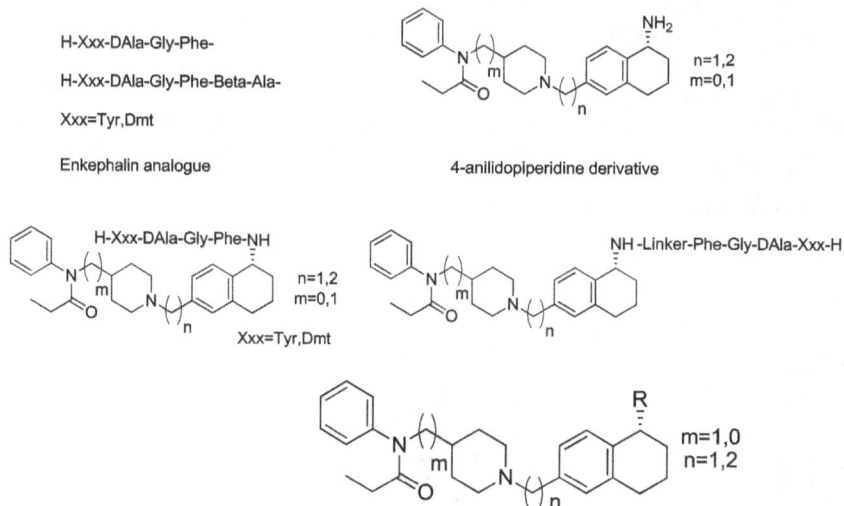

Fig. 1. Design principle.

Table 1. In vitro data of bivalent ligands.

Compound	R	m	n	K_i^{μ} (nM)	K_i^{δ} (nM)	μ/δ
DS-125	NH$_2$	0	1	40	10000	1/250
DS-118	H-Tyr-DAla-Gly-Phe-NH	1	1	10	320	1/32
DS-123	H-Tyr-DAla-Gly-Phe-NH	0	2	10	723	1/72
DS-802-08	H-Tyr-DAla-Gly-Phe-NH	0	1	3	368	1/122
DS-802-07	H-Tyr-DAla-Gly-Phe-βAla-NH	0	1	1	34	1/34
DS-802-32	H-Dmt-DAla-Gly-Phe-βAla-NH	0	1	0.1	0.5	1/5

attached the enkephalin opioid peptide to the amino group. Here we introduced a linker, beta alanine, in between the peptide and small molecule. The enkephalin peptide analogue was prepared by solution phase peptide synthesis by using HBTU as a coupling reagent, DIPEA and acetonitrile as a solvent. The final compounds were purified by prep. RP-HPLC (10-40% acetonitrile gradient in water containing 0.1%TFA in 18 min, 3mL/min). We then coupled the peptides to the non-peptide moiety.

A series of novel bivalent ligands have been synthesized and tested for biological activities at the µ and δ opioid receptors. The designed bivalent ligands showed very good affinity towards both (µ and δ) opioid receptor. The small molecule (DS-125, Ki 40nm) highly selective (250 fold) towards the µ receptor. The enkephalin analogues attached to the amino group of small molecule (DS-802-08) 13 fold increases in binding affinity of µ receptor and 27 fold increase towards the δ comparatively to the small molecule (DS-125). The introduction of linker in between enkephalin analogue and small molecule, and the replacement of Tyr with Dmt (DS-802-07 and DS-802-32) showed very good binding affinity towards both µ and δ opioid receptors. These novel bivalent ligands showed good affinity towards both µ and δ opioid receptors, among these molecules the compound DS-802-032 exhibits the highest potency in the series. Functional assays and *in vivo* assays are in progress.

Acknowledgments

Support by grants from USPHS, NIDA.

References

1. Porecca, F., et al. *J. Pharmacol. Exp. Ther.* **263**,147-152 (1992).
2. Vuckovic, S., et al. *Curr. Med. Chem.* **16**, 2468-2474 (2009).
3. WO 2006041888 A2 **2006**.
4. Ananthan, S., et al. *AAPS J.* **8**, E118-25 (2006).

Proceedings of the 23rd American Peptide Symposium
Michal Lebl (Editor)
American Peptide Society, 2013

Cardioprotective Effects of Cell Permeable NADPH Oxidase Inhibitors in Myocardial Ischemia/Reperfusion (I/R) Injury

**Issachar Devine, Qian Chen, Regina Ondrasik, William Chau,
Katelyn Navitsky, On Say Lau, Christopher W. Parker, Kyle D. Bartol,
Brendan Casey, Robert Barsotti, and Lindon H. Young**

*Department of Bio-Medical Sciences, Philadelphia College of Osteopathic Medicine,
Philadelphia, PA, 19131, U.S.A.*

Introduction

In myocardial infarction, reperfusion can compound ischemic damage and cause additional injury. Reperfusion injury is closely related to oxidative stress [1]. Clinical trials suggest nonselective antioxidants do not effectively attenuate reperfusion injury, which may be due to lack of specifically targeting the source of oxidative stress. It is proposed that overproduction of superoxide (SO) by activated NADPH oxidase can serve as a principle source of reactive oxygen species (ROS) under I/R conditions. Overproduction of SO *via* NADPH oxidase can result in mitochondrial dysfunction, cell/tissue damage, endothelial nitric oxide synthase (eNOS) uncoupling, and vessel constriction [2]. Therefore, we hypothesize that administration of selective NADPH oxidase inhibitors, gp91 ds-tat (RKKRRQRRR-CSTRIRRQL-amide, MW=2452 g/mol, Genemed Synthesis Inc. San Antonio, TX) or apocynin (MW=166 g/mol, Sigma Chemicals), will improve postreperfused cardiac function and reduce infarct size [3,4].

Results and Discussion

Hearts were excised from anesthesized male Sprague Dawley rats (275-325g) and perfused with modified Krebs' buffer by Langendorff preparation [1]. Isolated hearts were subjected to 15 minutes (min) of baseline perfusion, 30 min of global ischemia, and a 45 min reperfusion period with infusion of 5 ml of plasma (control), or plasma containing apocynin (40, 400 and 1000 μM) or gp91 ds-tat (10, 40 and 80 μM) during the first 5 min of reperfusion (1 ml/min) (Figure 1 and 2). At the end of reperfusion, two left ventricle (LV) cross sections were subjected to 1% triphenyltetrazolium chloride (TTC) to detect infarct size (viable= red, infarct =white) (Figure 3 and Table 1). All data were analyzed by ANOVA with the Student-Newman-Keuls test, $p < 0.05$ were considered to be statistically significant.

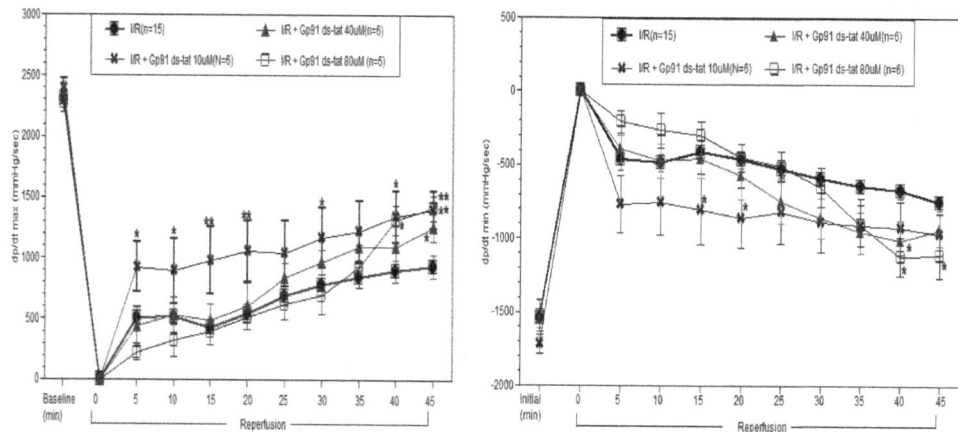

Fig. 1. Time course of maximal rate of left ventricular developed pressure (LVDP) +dP/dt$_{max}$ (left), and minimal rate of LVDP -dP/dt$_{min}$ (right) for gp91-ds-tat treated I/R hearts. *p<0.05, **p<0.01 vs. I/R control.

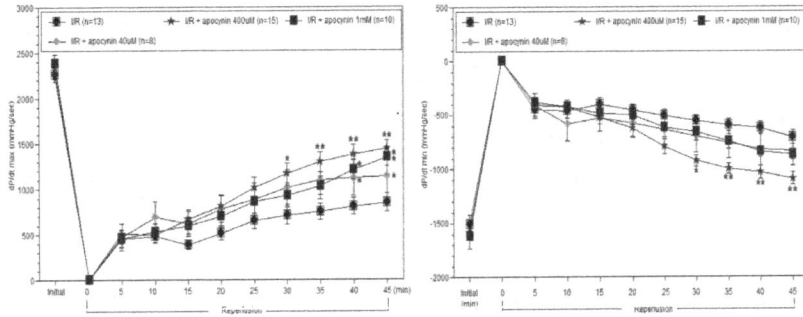

Fig. 2. Time course of $+dP/dt_{max}$ (left), and $-dP/dt_{min}$ (right) for apocynin treated I/R hearts. $*p<0.05$, $**p<0.01$ vs. I/R control.

Table 1. Infarct Size and Percent Recovery.

		Infarct Size[a]	Percent Recovery[b]	
			dP/dt_{max}	dP/dt_{min}
Control I/R		45.7±2.14	40.4±1.46	48.9±0.78
Gp91 ds-tat	10 μM	14.8±1.35*	58.6±1.89**	56.2±2.15
	40 μM	22.6±2.03*	53.6±1.56*	60.6±0.87
	80 μM	18.6±1.62*	62.1±1.27**	73.2±1.38*
Apocynin	40 μM	40.2±8.42	48.5± 4.61*	52.7± 3.28
	400 μM	28.1±3.69*	63.3±1.30**	68.8±0.86**
	1000 μM	23.9±5.07*	56.5±0.95**	53.8±0.41

[a]Percentage of infarcted tissue to the LV area at risk as determined by TTC staining; [b]Final value at reperfusion as compared to each initial baseline value ($*p<0.05$, $**p<0.01$ compared to control I/R).

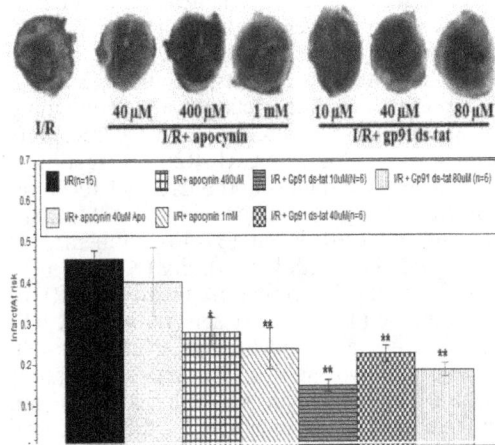

Fig. 3. Pictures (left top) and summary (left bottom) graph of TTC staining $*p<0.05$, $**p<0.01$ vs. I/R control.

We found that both NADPH oxidase inhibitors significantly improved post-reperfused cardiac contractile and diastolic functions and reduced infarct size compared to control I/R hearts. This study showed that NADPH oxidase mainly in vascular endothelium and myocytes, may be a principal ROS source to induce myocardial I/R injury. Therefore, both NADPH oxidase inhibitors may be potential agents to reduce ROS production and mitigate reperfusion injury in myocardial infarction, angioplasty and organ transplantation patients.

Acknowledgments

This study was supported by the Center for the Chronic Disorders of Aging and the Department of Bio-Medical Sciences at PCOM.

References

1. Chen, Q., et al. *Advances In Pharmacological Sciences* 2010 (2010).
2. Schramm, A., et al. *Vascular Pharmacology* **56**, 216-231 (2012).
3. Rey, F.E., et al. *Circulation Research* **89**, 408-414 (2001).
4. Mora-Pale, M., et al. *Free Radical Biology and Medicine* **52**, 962-969 (2012).

Proceedings of the 23rd American Peptide Symposium
Michal Lebl (Editor)
American Peptide Society, 2013

Corticotropin Releasing Factor (CRF) and Urocortins (Ucns) Chimeras

Judit Erchegyi, Jozsef Gulyas, Marilyn Perrin, Charleen Miller, Kathy Lewis, Wolfgang Fischer, Cindy Donaldson, and Jean Rivier

Clayton Foundation Laboratories for Peptide Biology, The Salk Institute, La Jolla, CA, 92037, U.S.A.

Introduction

Replacing residues at positions 30 and 33 (hCRF numbering) of [D-Phe^{12}, Nle^{21,38}]-hCRF(12-41) with a glutamic acid and lysine, respectively and linking these two residues through a lactam bridge produced astressin {cyclo(30-33)[D-Phe^{12}, Nle^{21,38}, Glu^{30}, Lys^{33}]-h/rCRF(12-41)} a more potent antagonist than its linear analog [1]. The cyclic peptide showed enhanced α-helical conformation [2]. Beyermann, et al. [3] showed that the α-helicity rather than amino acid composition of a central linker (residues 22 – 33) in CRF-like peptides played an important role in their biological activity. Mazur, et al. [4] demonstrated that amino acids in the *N*-terminus and *C*-terminus of chimeras formed by linking partial sequences of CRF, Ucn1, Ucn2 and sauvagine together influenced CRF receptor selectivity. To further explore the region(s) of the CRF receptors, which is (are) responsible for the high affinity binding and receptor selectivity, we have constructed chimeras covalently coupling the *N*- and *C*-terminal regions of hCRF, hUcn1, and hUcn2 through the helix inducing cyclic (i, i+3) tetrapeptide connector cyclo(Glu-Ala-Aib-Lys). The parent chimeras without the connector were also synthesized and their binding affinity and selectivity were compared. Compounds **12, 3,** and **16** (Table 1) were tested for their ability to stimulate the accumulation of the cAMP from A7r5 cells endogenously expressing CRF-R2β. The enzymatic stability of the analogs was also examined and compared to that of hCRF, hUcn1 and hUcn2.

Results and Discussion

Connecting *N*- and *C*-terminal fragments of hCRF, hUcn1, and hUcn2 through a cyclopeptide linker, cyclo(Glu-Ala-Aib-Lys), resulted in chimeras with enhanced binding affinity for both CRF-R1 and CRF-R2β (**1** vs. **2, 7** vs. **8,** and **13** vs. **14;** see Table 1). The selectivity of the new ligands for the two receptors was also affected. Analogs **2** and **6** showed enhanced selectivity for CRF-R1. The presence of the lactam-bridge containing tetrapeptide in **8** and **10** doubled the binding affinity of the peptides for both receptors compared to that of **7** and **9** and they became less selective towards CRF-R2β. Interestingly, chimera **12** showed increased binding affinity and selectivity favoring the CRF-R1 receptor. A positive influence of the bridged connector in the binding of analogs with the *N*-terminus of hUcn2 (**14, 16** and **18**) was also observed. hUcn2(1-30)-hUcn1(31-40) chimera (**17**) showed no displacement at CRF-R1 and bound to CRF-R2β with low affinity (K$_i$ = 34 nM). When the *N*-terminal fragment of hUcn2 was coupled to the *C*-terminal fragment of hCRF or hUcn1 through cyclo(Glu-Ala-Aib-Lys), these chimeras (**16, 18**) lost selectivity for CRF-R2β; they bound with high affinity to CRF-R1 (K$_i$s = 8 and 1 nM, respectively) as well. These results suggest a change in the conformation favoring binding and support our previous findings that the absence or presence of a helix-inducing kink in the ligand influences the binding affinity and might affect the signaling pathway [2]. Further structural studies are needed. Chimeras **12, 3,** and **16** showed weaker agonistic properties than hUcn2 in relation to their ability to stimulate the accumulation of intracellular cAMP from A7r5 cells, which are endogenously expressing CRF-R2β. hUcn2 stimulated the release of cAMP with EC$_{50}$ = 0.09 nM. The EC$_{50}$ value of the analogs **12, 3** and **16** was 0.65 nM, 10.9 nM and 4.1 nM, respectively. The order of potency was similar to that of their binding affinity for CRF-R2β. (Table 1) We found that chimeras with the cyclic-peptide linker were more resistant to trypsin digestion than the native peptides or the linear chimeras followed by HPLC, despite the presence of several arginines in the sequences. It can be assumed that the constrained bioactive structure is less accessible to this enzyme and most importantly to serum enzymes, suggesting that cyclic analogs may be longer acting than their linear forms.

Table 1. hCRF/hUcns Chimeras.

ID	Analogs	Receptor Binding Affinity	
		CRF-R1 (K_i nM)	CRF-R2β (K_i nM)
1	hCRF(1-41)	1.0 [2]	6.2 [2]
2	hCRF(1-29)-X-hCRF(34-41)	0.04	0.8
3	hCRF(1-31)-hUcn1(31-40) chimera	0.3	6.7
4	hCRF(1-29)-X-hUcn1(33-40) chimera	0.1	0.6
5	hCRF(1-31)-hUcn2(29-38) chimera	1.4	6.5
6	hCRF(1-29)-X-hUcn2(31-38) chimera	0.2	2.0
7	hUcn1(1-40)	0.4 [2]	0.5 [2]
8	hUcn1(1-28)-X-hUcn1(33-40)	0.06	0.35
9	hUcn1(1-30)-hCRF(32-41) chimera	0.8	1.0
10	hUcn1(1-28)-X-hCRF(34-41) chimera	0.06	0.22
11	hUcn1(1-30)-hUcn2(29-38) chimera	0.9	0.6
12	hUcn1(1-28)-X-hUcn2(31-38) chimera	0.1	0.3
13	hUcn2(1-38)	ND [2]	0.5 [2]
14	hUcn2(1-26)-X-hUcn2(31-38)	ND	0.3
15	hUcn2(1-30)-hCRF(32-41) chimera	ND	31.9
16	hUcn2(1-29)-X-hCRF(34-41) chimera	7.8	2.4
17	hUcn2(1-30)-hUcn1(31-40) chimera	ND	34.3
18	hUcn2(1-28)-X-hUcn1(33-40) chimera	1.0	2.6

X = -cyclo[Glu-Ala-Aib-Lys]-; Aib = 2-Aminoisobutyric acid; ND = No displacement of bound [^{125}I-Tyr^0,Glu^1,Nle^{17}]-sauvagine; the competitive displacement data were analyzed by GraphPad Prism program from which the Ki values were obtained, each assay included at least triplicate wells for each concentration, and the assays were repeated 3 or 2 times.

Acknowledgments

We thank W. Low for mass spectra analysis, J. Vaughan for radioligand preparations, and D. Doan for assistance in manuscript preparation. Supported by NIH/NIDDK Grant DK026741.

References

1. Gulyas, J., et al. *Proc. Nat. Acad. Sci. USA* **92**, 10575-10579 (1995).
2. Grace, C.R.R., et al. *J. Am. Chem. Soc.* **129**, 16102-16114 (2007).
3. Beyermann, M., et al. *J. Biol. Chem.* **275**, 5702-5709 (2000).
4. Mazur, A.W., et al. *J. Med. Chem.* **47**, 3450-3454 (2004).

Proceedings of the 23rd American Peptide Symposium
Michal Lebl (Editor)
American Peptide Society, 2013

Effects of Mitochondrial-Targeted Antioxidants on Real-Time Nitric Oxide and Hydrogen Peroxide Release in Hind Limb Ischemia and Reperfusion (I/R)

T. Galbreath, Q. Chen, R. Ondrasik, M. Bertolet, R. Barsotti, and L. Young

Department of Bio-Medical Sciences, Philadelphia College of Osteopathic Medicine, 4170 City Avenue, Philadelphia, PA, 19131, U.S.A.

Introduction

In the body, reperfusion of ischemic tissue with blood causes the release of reactive oxygen species (ROS), in part, from damaged mitochondria leading to endothelial and organ dysfunction. Endothelial dysfunction occurs within 5 min of reperfusion, is common to all vascular beds, and is characterized by increased hydrogen peroxide (H_2O_2) and decreased nitric oxide (NO) levels in the blood that further exacerbate I/R injury. Previous studies have shown that promoting endothelial NO synthase coupling during reperfusion increases blood NO and decreases blood H_2O_2 levels in hind limb I/R and attenuates myocardial I/R injury [1]. This study examines the effectiveness of mitochondria-targeted antioxidants, mitoquinone (mitoQ), a cell permeable coenzyme Q analogue or SS-31 ((D-Arg)-Dmt-Lys-Phe-Amide; Genemed Synthesis, San Antonio, TX), a cell permeable peptide, in inhibiting H_2O_2 release and increasing NO bioavailability in hind limb I/R. MitoQ [2] and SS-31 [3] are able to concentrate into the inner mitochondrial membrane via an electrical potential gradient or selective diffusion respectively (Figure 1). We hypothesized that these mitochondria-targeted agents will attenuate superoxide and subsequent H_2O_2 production thus allowing an increase in NO bioavailability, reducing I/R injury.

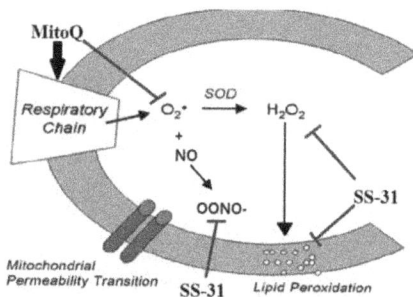

Fig. 1. Schematic showing the mitochondrial mode of action of both mitoQ and SS-31 peptide. Red-lines denote areas of inhibition. Adapted from Szeto [3].

Results and Discussion

We measured blood H_2O_2 or NO release from femoral veins in real-time: one vein was subjected to I/R while the other was used as a non-ischemic sham control. The H_2O_2 or NO microsensors (100 μm, WPI Inc., Sarasota, FL) were connected to a free radical analyzer (Apollo 4000, WPI Inc.) and were inserted into a catheter placed in each femoral vein. Ischemia was induced by clamping the femoral artery/vein of one limb for 30 min of ischemia followed by 45 min of reperfusion. MitoQ (2 mg/kg), SS-31 (2.5 mg/kg), or saline (for non-drug control group) was administered as a bolus injection via the jugular vein at the beginning of reperfusion. We continuously recorded the H_2O_2 or NO release and collected the data at 5 min intervals during a 15 min baseline period, 30 min of ischemia and 45 min of reperfusion. The changes in H_2O_2 or NO release during reperfusion are expressed as relative change to baseline. Results show that blood H_2O_2 increased significantly by ~1-3μM in I/R compared to sham limb in saline controls (n=6, p≤0.05). NO bioavailability decreased significantly by ~100-155nM in I/R compared to sham limb in saline controls (n=6, p<0.05). SS-31 or mitoQ, given at reperfusion, significantly decreased H_2O_2 release by ~2-2.5μM and ~1-1.5μM respectively (n=6, p<0.05; Figure 2) compared to the saline controls. Furthermore, SS-31 significantly increased NO blood levels by ~200-250nM and mitoQ increased NO blood levels by ~60nM (n=6, p<0.05; Figure 3) compared to the saline controls. In summary, when either mitoQ or SS-31 was given at the beginning of reperfusion, there was a

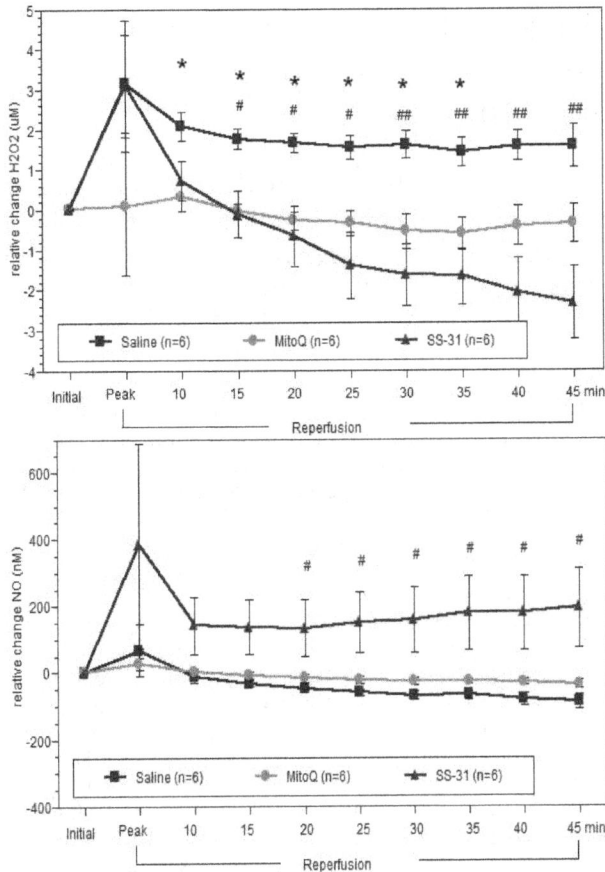

*Fig. 2. Comparison of relative difference in H_2O_2 release between I/R and sham limbs during reperfusion in saline, mitoQ, or SS-31 groups. There was a significant decrease in H_2O_2 release in the mitoQ-treated and SS-31 treated groups from 10-35 min and 15-45 min of reperfusion respectively (MitoQ * p<0.05, SS-31 # p<0.05, ## p<0.01 from saline controls by ANOVA analysis using Student Newman Keuls test).*

Fig. 3. Comparison of relative difference in NO release between I/R and sham limbs during reperfusion in saline, mitoQ, or SS-31 groups. There was a significant increase in NO release in the SS-31 treated groups from 20-45 min of reperfusion (SS-31 # p<0.05 from saline controls by ANOVA analysis using Student Newman Keuls test).

significant reduction of blood H_2O_2. Moreover, SS-31 showed a significant increase in endothelial-derived NO bioavailability compared to saline controls. The results of this study indicate that the mitochondria-targeted antioxidant agents mitoQ and SS-31 are able to attenuate the changes in blood H_2O_2 and NO levels during I/R and support our hypothesis that mitochondrial derived ROS are major contributors to oxidative stress in I/R injury. Collectively, the data suggests that mitoQ or SS-31 can be effective tools in the clinical setting for attenuating reperfusion injury and endothelial dysfunction.

Acknowledgments

This study was supported by the Center for the Chronic Disorders of Aging and the Department of Bio-Medical Sciences at the Philadelphia College of Osteopathic Medicine.

References

1. Perkins, K., et al. *Naunyn Schmiedebergs Arch Parmacol.* **385**, 27-38 (2012).
2. Adlam, V.J., Harrison, J.C., Porteous, C.M., et al. *The FASEB Journal* **19**, 1088-1095 (2005).
3. Szeto, H.H. *The American Association of Pharmaceutical Scientists* **8**, E277-E283 (2006).

Proceedings of the 23rd American Peptide Symposium
Michal Lebl (Editor)
American Peptide Society, 2013

Design, Synthesis and Biological Evaluation of Multivalent Ligands with μ/δ Opioid Agonist (μ-preferring) /NK-1 Antagonist Activities

Aswini Kumar Giri[1], Qiong Xie[1], Christopher R. Apostol[1],
David Rankin[2], Peg Davis[2], Eva Varga[2], Frank Porreca[2],
Josephine Lai[2], and Victor J. Hruby[1]

[1]Department of Chemistry and Biochemistry, University of Arizona, Tucson, AZ, 85721, U.S.A;
[2]Department of Pharmacology, University of Arizona, Tucson, AZ, 85721, U.S.A.

Introduction

The management of pain is a major challenge and millions of people all over the world suffer from various kinds of pain every day. Opioids continue to be the backbone for the treatment of these pain states. However, constant opioid treatment is accompanied with serious side effects. Persistent use of opioid therapy also develops analgesic tolerance in many patients. These unwanted effects significantly diminish the patients' quality of life. Sustained pain states lead to overexpression of substance P and the corresponding receptors that escalate pain [1]. Current drugs used for the treatment of pain cannot treat the prolonged pain state.

It was observed that co-administrations of a δ/μ opioid agonist and a neurokinin-1 (NK1) antagonist gave beneficiary results in the substance P-NK1 system in opioid signal transmission [2]. This drug cocktail showed enhanced potency in acute pain models and inhibition of opioid-induced tolerance in chronic tests using rats [3]. A study revealed that NK1 knockout mice did not show the rewarding properties of morphine. Drug combinations have restrictions as therapeutics because of poor patient compliance and difficulties in drug metabolism, distribution, and possible drug-drug interactions. Here, a new approach has been taken to combine these two different activities in one ligand which should have good metabolic and pharmacological properties [4,5]. The ligand would have potent analgesic affects in both acute pain and in neuropathic pain states without the development of unwanted side effects [6]. The present approach of our drug-design is based on the use of adjacent and/or overlapping pharmacophores, in which an opioid agonist pharmacophore is placed at the *N*-terminus and the NK1 antagonist pharmacophore at the *C*-terminus of a single peptide derived ligand. The opioid pharmacophore of these multivalent ligands were designed based on opioid ligands, enkephalins and morphiceptin, while the NK1 pharmacophore was adopted from our early lead compound TY-032 [4]. The two pharmacophores are joined directly or by a linker/address moiety. It should be emphasized that the designed multivalent ligands have additional advantages over a cocktail of individual drugs for easy administration, a simple ADME property, and no drug-drug interactions. Earlier studies have shown that agonist activities at MOR and DOR, and antagonist activity at NK1 is beneficiary over targeting a single receptor [6]. However, it is still largely unclear what binding ratio(s) for these receptors would be ideal to achieve the desired biological profile. To address these highly challenging issues an approach has been taken to design, synthesize and evaluate in detail the biological profile of the ligands showing selectivity for MOR over DOR. As ligands with *C*-terminal amide showed more stability in rat plasma, we took that as reference. To enhance ligands' BBB permeability we used *N*-methylated benzyl amine derivatives.

Results and Discussion

The ligands were anticipated to interact with each receptor separately, to show an improved biological profile, and to provide better synergy compared to the coadministration of two or more drugs. Removal of 5th residue (Nle) from **1** (δ-selective) resulted in a μ-selective ligand **3**. When Trp in ligand **1** was substituted by Gly at 4th position (ligand **2**), it maintained its selectivity and became more potent. Removal of Pro from NK1 pharmacophore not only affected the NK1 binding affinity, but also altered the selectivity (ligands **4** and **5**). Truncation of Trp from ligand **3** formed ligand **6**, which reversed the selectivity. All three morphiceptin and NK1 derived ligands (**7**, **8**, and **9**) showed similar selectivity between

MOR and DOR (Table 1). Functional assays are in progress to determine the agonist and antagonist activities of the ligands at the appropriate receptors.

Table 1. Binding affinity results.

Ligand No.	Ligand structures	K_i^μ $(nM)^a$	K_i^δ $(nM)^a$	$K_i^\mu:K_i^\delta$	$hNK1^b$
[Leu⁵]Enkephalin	Tyr-Gly-Gly-Phe-Leu-NH$_2$	9.4	2.5		
1 (QXP01)	Tyr-D-Ala-Trp-Phe-Nle-Pro-Leu-Trp-NMeBn(CF$_3$)$_2$	25	5	5:1	2.6
2 (AKG-29)	Tyr-D-Ala-Gly-Phe-Pro-Leu-Trp-NMeBn(CF$_3$)$_2$	40	200	1:5	2.5
3 (QXP14)	Tyr-D-Ala-Trp-Phe-Pro-Leu-Trp-NMeBn(CF$_3$)$_2$	300	1500	1:5	1.8
4 (QXP15)	Tyr-D-Ala-Trp-Phe-Leu-Trp-NMeBn(CF$_3$)$_2$	400	200	2:1	11
5 (QXP16)	Tyr-D-Ala-Phe-Phe-Leu-Trp-NMeBn(CF$_3$)$_2$	38	28	1.4:1	3.8
6 (AKG-30)	Tyr-D-Ala-Phe-Pro-Leu-Trp-NMeBn(CF$_3$)$_2$	1300	250	5:1	3.5
Morphiceptin	Tyr-Pro-Phe-Pro-NH$_2$	19.8	>10000		
7 (QXP06)	Tyr-Pro-Phe-Gly-Nle-Pro-Leu-Trp-NMeBn(CF$_3$)$_2$	360	550	1:1.5	5.6
8 (QXP08)	Tyr-Pro-Phe-Pro-Leu-Trp-NMeBn(CF$_3$)$_2$	1300	2100	1:1.6	0.8
9 (AKG-33)	Tyr-Pro-Phe-Trp-NMeBn(CF$_3$)$_2$	3000	5600	1:2	1.3

[a]Competition analyses were carried out using membrane preparations from transfected HN9.10 cells that constitutively expressed the DOR and MOR, respectively. [b]Competition analyses were carried out using membrane preparations from transfected CHO cells that constitutively expressed human NK1 receptor. [³H]DAMGO, [³H]Deltorphin II, and [³H]-substance P were used as standard ligands for MOR, DOR and NK1 receptors, respectively.

Conclusions and future directions

Gly at the 3rd position has been found to be promising for enhanced binding affinity. The presence and absence of a 5th residue plays an important role in opioid receptor subtype selectivity. The SAR studies have shown that the *C*-terminus modification has appreciable impact on opioid receptors binding. Design, synthesis and SAR of more ligands are required to achieve desired potency and selectivity.

Acknowledgements

Authors thank the US Public Health Service, National Institute of Health (NIH) for supporting this project. Special thanks from A. K. Giri to the American Peptide Society (http://www.aps.com) for the "APS Travel Award".

References

1. Michael, T.K., Ossipov, H., Vanderah, T.W., Porreca, F., Lai, J. *Neurosignals* **14**, 194-205 (2005).
2. Misterek, K, Maszczynska, I., Dorociak, A., Gumulka, S.W., Carr, D.B., Szyfelbein, S.K., Lipkowski, A.W. *Life Sci.* **54**, 939-944 (1994).
3. Ripley, T.L., Gadd, C.A., Felipe, C.D., Hunt, S.P., Stephens, D.N. *Neuropharmacology* **43**, 1258-1268 (2002).
4. Hruby, V.J., Nair, P., Yamamoto, T. *U.S. Patent*, US 8026218 B2, (2011).
5. Yamamoto, T., Nair, P., Largent-Milnes, T.M., Jacobsen, N.E., Davis, P., Ma, S.-W., Yamamura, H.I., Vanderah, T.W., Porreca, F., Lai, J., Hruby, V.J. *J. Med. Chem.* **54**, 2029-2038 (2011) and references there in.
6. Largent-Milnes, T.M., et al. *Br. J. Pharmacol.* **161**, 986-1001 (2010) and references there in.

Proceedings of the 23rd American Peptide Symposium
Michal Lebl (Editor)
American Peptide Society, 2013

Fighting Bacterial Resistance: Modifying the Antimicrobial Peptide Tachyplesin

Hareesh Mukkisa, Lauren Crisman, Sarah Davis, Stacie Wood, and Deborah Heyl

Department of Chemistry, Eastern Michigan University, Ypsilanti, MI, 48197, U.S.A.

Introduction

Microorganisms develop resistance to many antibiotics over time. This has become a serious health issue as "superbugs" have become more prevalent with increasing antibiotic use. Antimicrobial peptides have bacteriocidal activity through various modes, including pore formation and membrane permeabilization in a detergent-like mechanism, essentially by pulling the membrane apart. The positively charged arginines and lysines of these peptides are attracted to the negatively charged bacterial membrane. Once associated, the hydrophobic amino acids of the peptide can insert into the nonpolar region of the lipid bilayer, forming pores or micelles and destroying the membrane. Human cell membranes are more zwitterionic so the peptides do not affect them to the same extent. Due to the mechanism by which these peptides act, bacteria should not be able to develop resistance to them.

Results and Discussion

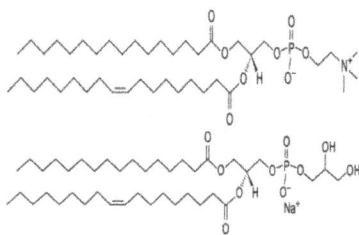

Fig. 1. POPC, palmitoyl-oleoyl-phosphocholine (top) and POPG palmitoyl-oleoylphosphoglycerol (sodium salt, bottom).

Various changes were made to the native sequence of Tachyplesin, an antimicrobial peptide from horseshoe crab, KWCFRVCYRGICYRRCR (amide) [1]. First, the cysteines were removed in a previous study to give the linear and activity-retaining cysteine-deleted Tachyplesin, CDT, KWFRVYRGIYRRR (amide) [2]. In this study, we replaced all arginines with lysine (CDT-K5), followed by removing one arginine (CDT-R), two arginines (CDT-2R) and three arginines (CDT-3R) from the C-terminus of CDT, and then replaced hydrophobic residues of CDT with aromatic phenylalanine (KWFRFYRFFYRRR, CDT-F3). The peptides were synthesized by standard fluorenylmethyl-oxycarbonyl (Fmoc)-based solid phase techniques, cleaved with trifluoroacetic acid, and purified by reverse-phase high performance liquid chromatography (RP-HPLC). Electrospray mass spectrometry confirmed the targeted molecular weights.

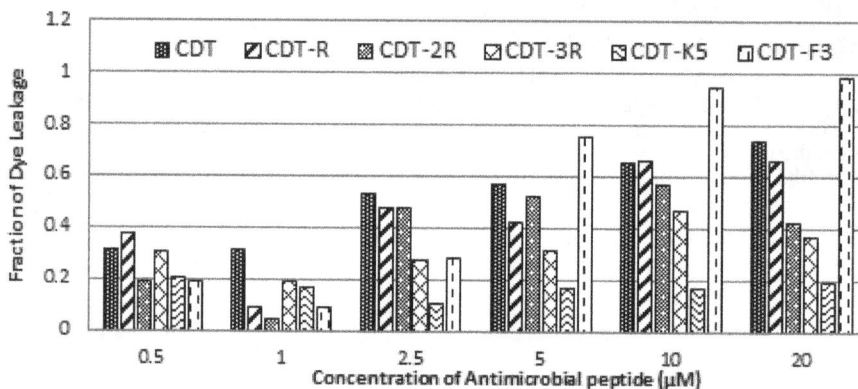

Fig. 2. Fluorescence readings shown as fraction of dye leaked relative to detergent control at various concentrations of peptide.

Table 1. Effective minimum concentrations of the varying antimicrobial peptides to inhibit bacterial growth as well as effective membrane damage in the model system.

Peptide	S. aureus MIC (mg/mL)	B. subtillis MIC (mg/mL)	E. coli MIC (mg/mL)	Percent Dye Leakage(10 μM)
Tachyplesin	Not tested	>200	11.5	Not tested
CDT	26.2	26.2	13.1	66
CDT-K$_5$	56.2	450	14.1	17
CDT-R	26.2	26.2	13.1	67
CDT-2R	26.2	26.2	13.1	57
CDT-3R	250	125	≥ 1000	47
CDT-F$_3$	15.6	15.6	62.5	94

Carboxyfluorescein-encapsulating vesicles which mimic the bacterial membrane were created using a 3:1 ratio of the lipids POPC and POPG, respectively (Figure 1), and the percent leakage of fluorescent dye from the vesicles in the presence of varying concentrations of the peptides was calculated as compared to a 100% Triton-X detergent-treated control. Fluorescence values were recorded by an FLx fluorescence microplate reader. Assays were run in triplicate in a 96-well plate, and average values are reported (Figure 2 and Table 1). To determine minimum inhibitory concentrations, serial dilutions of the peptides were added to wells containing specific concentrations of three strains of bacteria (Table 1). The cultures were allowed to grow overnight and then turbidity readings were taken to measure bacterial growth in the presence of varying peptide concentrations.

Lysine replaced CDT (CDT-K$_5$) showed a slight decrease in membrane damage in the dye leakage assay relative to CDT but did retain much of its antibacterial activity against *S. aureus* and *E. coli*. However, replacing positively charged Arg with similarly positively charged Lys was detrimental to activity (~20-fold reduction) against the gram positive spore-forming *B. subtillus*.

As the Arg residues were removed from the *C*-terminus, it was clear that deleting the first and second positively charged residues was well tolerated; CDT-R and CDT-2R retained activity relative to CDT against all strains tested. In addition, dye leakage from the model membranes remained relatively constant. However, upon removal of the 3rd positively charged R residue from the *C*-terminus, a dramatic loss of activity (≥10-fold) was observed in all strains, especially gram negative *E. coli*. Membrane damage in the model system also was lower, but the effect was not as significant. An Arg at this position, or perhaps a *C*-terminal positive charge, appears necessary for optimal activity.

Replacing hydrophobic residues with aromatic Phe (to pick up possible base stacking interactions between peptides) resulted in a slightly more effective analogue against gram positive *B. subtillis* and *S. aureus*, and caused more damage in the model membrane system. However, this analogue was less effective against *E. coli*.

Acknowledgments

Sincere thanks to Dr. Jamie Scaglione for training and supervision on the bacterial assays and Dr. Ruth Ann Armitage for mass spectral data. Supported by the EMU Graduate School and Chemistry Department and an EMU Faculty Research Fellowship.

References

1. Nakamura, T., Furunaka, H., Miyata, T., Tokunaga, F., Muta, T., Iwanaga, S., Niwa, M., Takao, T., Shimonishi, T. *J. Biol. Chem.* **263**, 16709-16713 (1988).
2. Ramamoorthy, A., Thennarasu, S., Tan, A., Gottipati, K., Sreekumar, S., Heyl, D., An, F., Shelburne, C. *Biochemistry* **45**, 6529-6540 (2006).

Proceedings of the 23rd American Peptide Symposium
Michal Lebl (Editor)
American Peptide Society, 2013

Surface Plasmon Resonance Analysis of Beetle Defensin-Derived Antimicrobial Peptide-Membrane Interactions

Jun Ishibashi[1], Ai Asaoka[1], Takashi Iwasaki[1], Hideaki Suzuki[2], Tomio Nagano[2], Makoto Nakamura[3], Mitsuhiro Miyazawa[1], and Minoru Yamakawa[1]

[1]Insect Mimetics Research Unit, National Institute of Agrobiological Sciences, Tsukuba, Ibaraki 305-8634, Japan, [2]JITSUBO Co., Ltd., Koganei, Tokyo 184-0012, Japan, [3]Industrial Technology Center of Wakayama Prefecture, Wakayama, Wakayama 649-6261, Japan

Introduction

We developed 9-mer antimicrobial peptides (AMPs) derived from a defensin of the rhinoceros beetle, *Allomyrina dichotoma* (peptide A: RLYLRIGRR-NH$_2$, peptide B: RLRLRIGRR-NH$_2$, peptide C: ALYLAIRRR-NH$_2$, peptide D: RLLLRLIGRR-NH$_2$) [1]. The defensin-derived AMPs exhibited negatively charged membrane selective disruptive activity [2]. The AMP-lipid bilayer interactions were analyzed by surface plasmon resonance (SPR) spectroscopy and the mode of action was speculated.

Results and Discussion

Liposomes were captured on sensor chip L1 to form lipid bilayer on the surface of the chip and the peptide solutions were injected onto the lipid surface using Biacore 1000 analytical system (GE Healthcare). In phosphate buffered saline (x1 PBS: 10 mM sodium phosphate, 130 mM NaCl, pH 7.0), responses of the AMPs with lipid bilayer of acidic phospholipid, egg yolk phosphatidylglycerol (PG) were much stronger than those of zwitterionic phospholipid, egg yolk phosphatidylcholine (PC). Increase of ionic strength (x2 PBS: 20 mM sodium phosphate, 260 mM NaCl, pH 7.0) or change of pH (10 mM sodium acetate, 130 mM NaCl, pH5.0; 10 mM sodium phosphate, 130 mM NaCl, pH 6.0; 10 mM Tris-HCl, 130 mM NaCl, pH 8.0) also reduced responses. These results supported the idea that the negatively charged membrane selective disruptive activity of the defensin-derived AMPs attributes to electrostatic interaction.

Sensorgrams for pore-forming AMPs (*Oryctes rhinoceros* defensin, *Bombyx mori* cecropin B and *Bombyx mori* moricin) after end of injection did not decrease rapidly as defensin-derived AMPs (Figure 1). During the dissociation of the peptides from the lipid

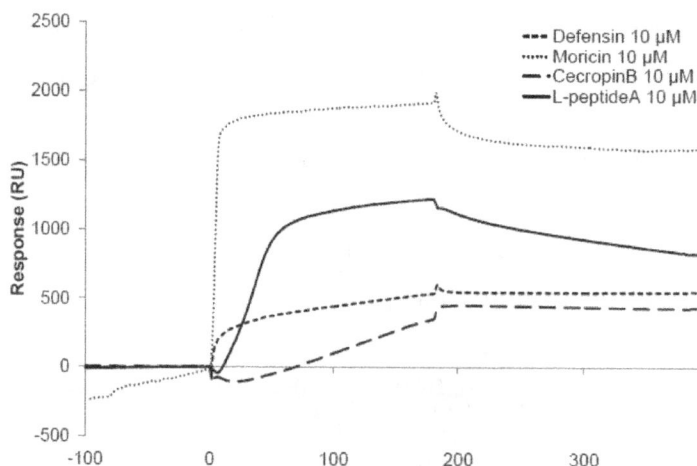

Fig. 1. Comparative peptide sensorgrams for the binding of L-peptideA and pore-forming AMPs to PG in x1 PBS.

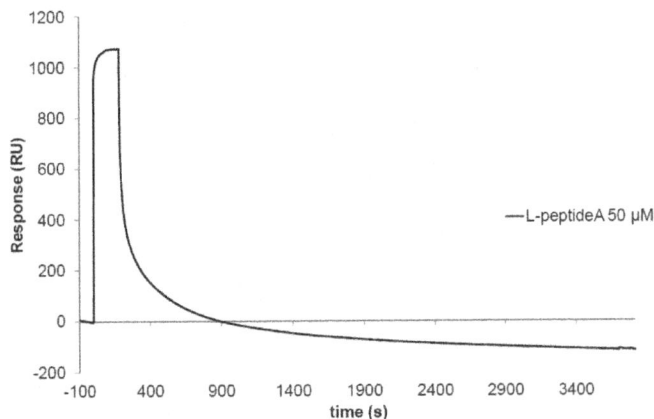

Fig. 2. Peptide sensorgram for the dissociation of L-peptideA from PG at higher concentration in long period in x1 PBS.

bilayer at a higher peptide concentration (50 μM) in a longer period, the responses decreased to the level lower than at the point of peptides injection (Figure 2). These results suggested that the defensin-derived peptides stripped lipid from the surface of sensor chip. This result indicated that the defensin-derived AMPs permeabilize membrane with "carpet model" [3].

Acknowledgments

This work was supported by the Programme for Promotion of Basic and Applied Researches for Innovations in Bio-oriented Industry.

References

1. Saido-Sakanaka, H., Ishibashi, J., Momotani, E., Amano, F., and Yamakawa, M. *Peptides* **25**, 19-27 (2004).
2. Iwasaki, T., Ishibashi, J., Tanaka, H., Sato, M., Asaoka, A., Taylor, D., Yamakawa, M. *Peptides* **30**, 660-668 (2009).
3. Chen, H.M., Clayton, A.H., Wang, W., Sawyer, W.H. *Eur. J. Biochem.* **268**, 1659-1669 (2001).

Proceedings of the 23rd American Peptide Symposium
Michal Lebl (Editor)
American Peptide Society, 2013

Fragmentation Pathways of Sweet Dipeptides by High Resolution (+) ESI Mass Spectrometry

Maroula G. Kokotou[1], Eleni Siapi[2], Nikolaos S. Thomaidis[1], and George Kokotos[2,3]

[1]Laboratory of Analytical Chemistry, Department of Chemistry, University of Athens, Panepistimiopolis, Athens, 15771, Greece; [2]Institute of Biology, Medicinal Chemistry and Biotechnology, National Hellenic Research Foundation, Athens, Greece; [3]Laboratory of Organic Chemistry, Department of Chemistry, University of Athens, Panepistimiopolis, Athens, 15771, Greece

Introduction

Sweeteners are divided in two main groups: caloric or nutritive, and non-caloric or non-nutritive compounds. Nutritive sweeteners are carbohydrates such as glucose, fructose and maltose. Non-nutritive sweeteners belong to various chemical classes and they are usually known as artificial sweeteners [1]. The food industry is heavily promoting its artificially-sweetened products (frequently called "diet" or "light"), highlighting their benefits. Low-calorie or reduced-calorie food products and beverages can help in treatment of obesity, maintaining body weight and management of diabetes. The wide application of aspartame (each year 16,000 t for worldwide consumption) as an artificial sweetener in low caloric products led to the discovery of new sweet dipeptides, namely alitame, neotame and most recently advantame. Alitame, neotame and advantame are 2000, 6000 and 37000 sweeter than sucrose, respectively [1,2]. Due to increased consumption, sweeteners are widely distributed in the aquatic environment and are characterized as emerging contaminants [3]. We have recently reported a method for the simultaneous determination of the sweet dipeptides by HILIC-ESI-MS/MS [4]. In this work, the fragmentation pathways of neotame (NEO) and advantame (ADV), in comparison to those of aspartame (ASP) and aspartame-d3 (ASP-d3), were studied by positive ion electrospray ionization (ESI) high resolution mass spectrometry (Thermo Orbitrap mass analyzer). Elucidation of the fragmentation is very useful for the trace-level determination of the artificial dipeptide sweeteners in complex matrices [3,5].

Results and Discussion

Accurate mass spectra of the dipeptides allowed proposing specific fragment ions. Neotame and advantame are the N-(3,3-dimethylbutyl) and N-[3-(3-hydroxy-4-methoxyphenyl)-propyl] derivatives of aspartame (Figure 1).

Measurements were performed with Thermo Orbitrap Velos and a Thermo TSQ Quantum Access system with triple quadrupole, for comparison purposes. The data acquisition was carried out with XCalibur Data System software.

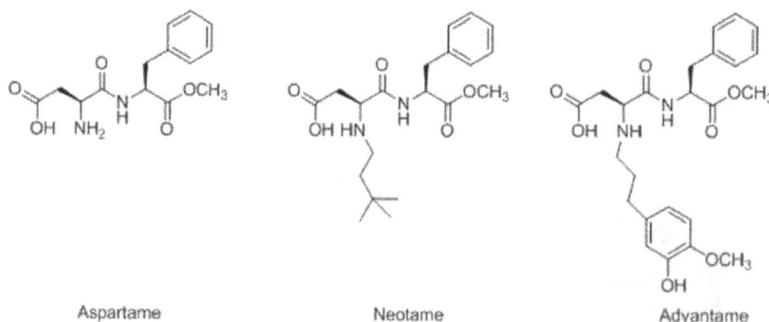

Fig. 1. Structures of aspartame, neotame, and advantame.

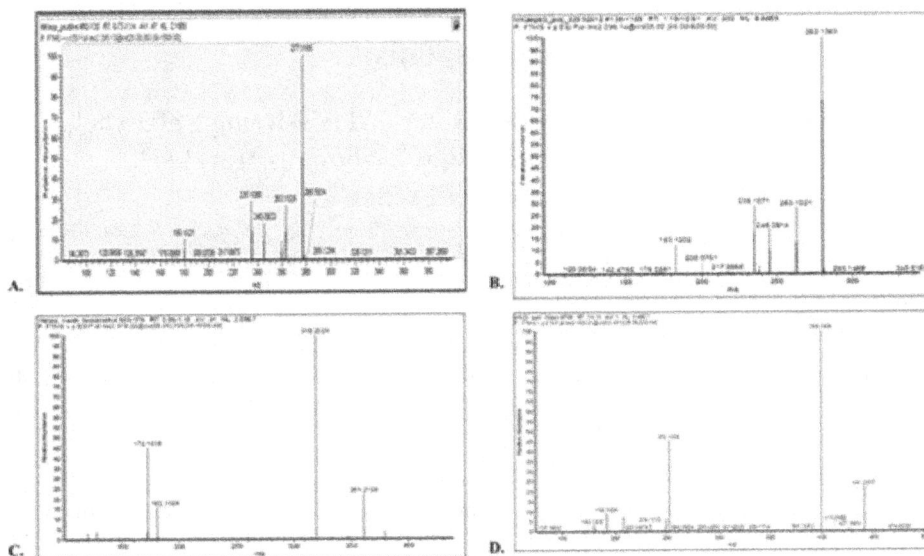

Fig. 2. MS² Spectra of A. aspartame, B. aspartame-d3, C. neotame, and D. advantame recorded in Orbitrap Velos.

The MS² spectra of all the dipeptides (aspartame, aspartame-d3, neotame and advantame) recorded in Orbitrap Velos are shown in Figure 2. The first fragmentation step is the loss of either both CH_3OH (CD_3OH) and CO_2 (ions with m/z 235.1080, 235.1071, 319.2029, 399.1929, respectively), or H_2O (ions with m/z 277.1185, 280.1363, 361.2138, 441.2037, respectively). Then, the loss of CH_3COOH (ions with m/z 175.0868, 175.0861, 259.1813, 339.1714, respectively) or both CH_3OH (CD_3OH) and CO_2 (ions with m/z 217.0975, 217.0966, 301.1921, 381.1823, respectively) was observed in both MS² and MS³ spectra.

Acknowledgments

Maroula G. Kokotou as a recipient of Peptisyntha S.A.Travel Award.

References

1. Kokotou, M.G., Asimakopoulos, A.G., Thomaidis, N.S., In Nollet, L.M.L. and Toldra, F. (Eds.) "Sweeteners" in "Food Analysis by HPLC", CRC Press: Boca Raton, Florida, 2012, p. 493.
2. Otabe, A., Fujieba, T., Masuyama, T., Ubukata, K., Lee, C. *Food Chem. Toxicol.* **49**, S2-7 (2011).
3. Kokotou, M.G., Asimakopoulos, A.G., Thomaidis, N.S. *Analytical Methods* **4**, 3057-3070 (2012).
4. Kokotou, M.G., Kokotos, C.G., Thomaidis, N.S., In Kokotos, G., Constantinou-Kokotou, V., Matsoukas, J. (Eds.) *Peptides 2012 (Proceedings of the 32nd European Peptide Symposium)*, Athens, 2012, p. 226.
5. Kokotou, M.G., Thomaidis, N.S. *Anal. Methods* **5**, 3825- 3833 (2013).

Proceedings of the 23rd American Peptide Symposium
Michal Lebl (Editor)
American Peptide Society, 2013

Structural Determinants of Gi Activation by Peptidic CXCR4 Agonists

Marilou Lefrancois[1], Christine Mona[1], Marie-Reine Lefebvre[1], Richard Leduc[1], Pierre Lavigne[1], Nikolaus Heveker[2], and Emanuel Escher[1]

[1]Département de Pharmacologie, Université de Sherbrooke, Sherbrooke, Québec, J1H 5N4, Canada;
[2]Département de Biochimie, Université de Montréal, Montréal, Québec, H3T 1C5, Canada

Introduction

Therapeutic intervention targeting the CXCR4/CXCL12 axis is a tricky endeavor, considering their many intricate physiological roles. This receptor is vastly expressed in the majority of tissues, and is of great importance in multiple essential processes, such as hematopoiesis and angiogenesis. The use of CXCR4 antagonists is limited by the numerous adverse effects, such as cardiotoxicity [1]. The elaboration of CXCR4 agonists is therefore highly interesting for potential treatment options. Our team has elaborated synthetic CXCR4 agonists [2] by combining the N-terminus of the endogenous ligand, CXCL12, with a high affinity inverse agonist of CXCR4, T140 [3], to produce high affinity agonists. These compounds have subsequently proven efficacious for both in vitro and in vivo CXCR4-mediated chemotaxis. Signaling assays have shown that none of these agonists induce beta-arrestin-2 recruitment. Previous assessment of the affinities of the compounds showed improvement by lengthening of the chain to up to 10 amino acids, and that correct positioning of the chain on the T140 scaffold was essential. The present study aims to evaluate if the main CXCR4 signaling pathway, $G\alpha_i$, is activated by the compounds, and establish which of their structural elements are associated with higher $G\alpha_i$ response.

Results and Discussion

$G\alpha_i$ activation was assessed with a cAMP Bioluminescence Resonance Energy Transfer (BRET) sensor, Epac. $G\alpha_i$ i inhibition of cAMP production after forskolin treatment (Figure 1). Resulting BRET signal was measured after 10 minutes incubation at room temperature with coelenterazine 400A. EC_{50} as well as Emax values were obtained for each compound. The compounds that we have tested vary in the CXCL12 amino-terminus chain length (from 6aa to 10aa), as well as in the positioning of the chain at either Arg-2, Arg-12 or Arg-14 of T140 (see Table 1 for structures). **1281, 1251Y, 1251R, 1280, 1288** and **1282** have the chain attached to Arg-14 of T140, **1294** has the chain located at Arg-12 and **1292, 1290, 1289, 1291, 1293** have the chain at Arg-2.

The concentration needed for half maximal $G\alpha_i$ activation, EC_{50}, is shown in Table 1. All compounds have EC_{50}s in the nanomolar range, varying between 10 and 56 nM. The compounds marked with an asterisk (*) did not show sufficient $G\alpha_i$ response to allow the determination of EC50 values (Table 1). $G\alpha_i$ responses are greatly influenced by the length of the side chain: an increase of (EC50) correlates with increase in chain length. A chain with 7 amino acids and less appears to be too short to induce significant $G\alpha_i$ activation. Figure 1 shows the results of maximal $G\alpha_i$ induction (E_{max}) as the percentage of the maximal response induced by CXCL12. The induction of $G\alpha_i$ response is proportional to the chemotactic activity observed in vitro, a longer chain length appears to improve $G\alpha_i$ activation. The maximum $G\alpha_i$ response correlates with the capacity of the compounds of inducing chemo-

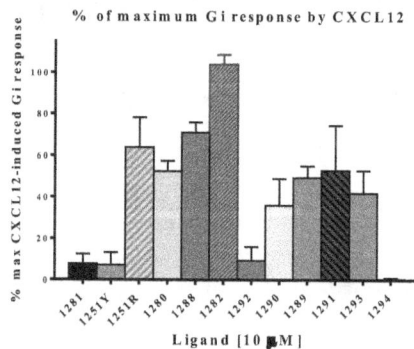

Fig. 1. BRET monitoring of cAMP levels following Gi activation, using EPAC biosensor. (N=3).

Table 1. Structures: SDF-1α derived side chain and its site of attachment on T140.

Cmp No		(EC_{50}) in nM
Graft Position 12	cyclo-Arg-Arg-Nal-Cys-Tyr-Arg-Lys-DLys-Pro-Tyr-Arg-**Lys(x)**-Cys-Arg	
1294*	Lys-Pro-Bpa-Ser-Leu-Ser-Tyr- x	---
Graft Position 14-Arg[12]	cyclo-Arg-Arg-Nal-Cys-Tyr-Arg-Lys-DLys-Pro-Tyr-Arg-Arg-Cys-**Lys(x)**	
1281*	Lys-Pro-Val-Ser-Leu-Ser-x	---
1251Y*	Lys-Pro-Val-Ser-Leu-Ser-Tyr- x	---
1251R	Lys-Pro-Val-Ser-Leu-Ser-Tyr-Arg- x	18.83 ± 1.11
1280	Lys-Pro-Val-Ser-Leu-Ser-Tyr-Arg-Ser- x	23.35 ± 3.88
1282	Lys-Pro-Bpa-Ser-Leu-Ser-Tyr-Arg-Ser- x	29.59 ± 1.39
1287	Lys-Pro-Val-Ser-Leu-Ser-Tyr-Arg-Ser-Pro- x	29.06 ± 3.45
1288	Lys-Pro-Val-Ser-Leu-Ser-Tyr-Arg-Ser-Ala- x	56.20 ± 2.08
Graft Position 2	cyclo-Arg-**Lys(x)**-Nal-Cys-Tyr-Arg-Lys-DLys-Pro-Tyr-Arg-Arg-Cys-Arg	
1292*	Lys-Pro-Val-Ser-Leu- x	---
1290*	Lys-Pro-Val-Ser-Leu-Ser- x	---
1289*	Lys-Pro-Val-Ser-Leu-Ser-Tyr- x	---
1291	Lys-Pro-Val-Ser-Leu-Ser-Tyr-Arg- x	10.33 ± 1.74
1293	Lys-Pro-Val-Ser-Leu-Ser-Tyr-Arg-Ser- x	20.49 ± 3.04

taxis *in vitro*: a longer chain appears to confer greater efficacy. As expected, the compound with the chain grafted to Arg-12 that is antagonistic of chemotaxis, **1294**, does not induce $G\alpha_i$ response. Through this study we have come to the conclusion that our synthetic CXCR4 agonists are able to induce a full $G\alpha_i$ response (**1282**), but most of the peptides are partial $G\alpha_i$ agonists. The compounds with the chain grafted onto the Arg-2 of T140 exhibit poorer chemotactic responses and affinities, but appear only slightly poorer to those with the chain grafted onto Arg-14 of T140, for $G\alpha_i$.

In conclusion, these assays have confirmed the important role of $G\alpha_i$ signaling in CXCR4 mediated chemotaxis, as $G\alpha_i$ responses are proportional to chemotactic activities. Strikingly, β-arrestin responses seemed to be dispensable for chemotactic response. Thus, these compounds, by their biased agonism towards the $G\alpha_i$ pathway, have great potential for the study of biased activities of CXCR4. Further study of these peptides will include an evaluation of the activation of the ERK signaling pathway.

Acknowledgments

Special thanks to Pr Sylvie Marleau for sharing her laboratory's expertise on the air pouch technique, and Cindy Rosa-Boisvert for her technical help and insightful advice. Thanks to Stephanie Gravel, Nicolas Montpas, Nassr Nama and François Guite-Vinet for their valuable insights. This research was supported by funds from the Canadian Institutes of Health Research and l'Institut de Pharmacologie de Sherbrooke for a graduate studies scholarship. An APS Merit Award has made possible the presentation of these results at the 23rd APS.

References

1. Pettersson, S., Perez-Nueno, V.I., Ros-Blanco, L., de La Bellacasa, R.P., Rabal, M.O., Batllori, X., Clotet, B., Clotet-Codina, I., Armand-Ugon, M., Este, J., Borrell, J.I., Teixido, J. *ChemMedChem*, **3**, 1549-1557 (2008).
2. Lefrancois, M., Lefebvre, M.R., Saint-Onge, G., Boulais, P.E., Lamothe, S., Leduc, R., Lavigne, P., Heveker, N., Escher, E. *ACS Med. Chem. Lett.* **2**, 597-602 (2011).
3. Tamamura, H., Xu, Y., Hattori, T., Zhang, X., Arakaki, R., Kanbara, K., Omagari, A., Otaka, A., Ibuka, T., Yamamoto, N., Nakashima, H., Fujii, N. *Biochem. Biophys. Res. Commun.* **253**(3), 877-882 (1998).

Proceedings of the 23rd American Peptide Symposium
Michal Lebl (Editor)
American Peptide Society, 2013

The Effects of Modulating Endothelial Nitric Oxide Synthase (eNOS) Activity and Coupling in Extracorporeal Shock Wave Lithotripsy (ESWL)

Alexandra C. Lopez, Qian Chen, Brittany L. Deiling, Edward S. Iames, Robert Barsotti, and Lindon H. Young

Department of Bio-Medical Sciences, Philadelphia College of Osteopathic Medicine, Philadelphia, PA, 19131, U.S.A.

Introduction

ESWL therapy utilizes high-energy shock waves to break down kidney and uretal stones. The shock waves produce shear stress and cavitation bubbles which synergistically ablate the stone into small fragments that can be passed through the urinary tract. ESWL is highly effective and minimally invasive compared to surgical treatments, making it the preferred treatment option by many urologists. However, repetitive shock waves may also cause damage to the renal vasculature endothelium and can lead to the development of chronic hypertension [1]. Previous studies have found that ESWL causes endothelial dysfunction which is characterized by reduced nitric oxide (NO) bioavailability and increased production of reactive oxygen species (ROS) such as superoxide (O_2^-) and hydrogen peroxide (H_2O_2) [2]. Normally, eNOS is in a coupled state which produces NO in the presence of essential cofactor tetrahydrobiopterin (BH_4). Oxidative stress, such as that caused by ESWL-induced overproduction of ROS, can cause BH_4 to be oxidized to dihydrobiopterin (BH_2). When the BH_4:BH_2 ratio is reduced, eNOS becomes uncoupled and produces O_2^- instead of NO [2,3]. O_2^- is short-lived and converted to H_2O_2 in blood by superoxide dismutase. Protein kinase C epsilon (PKCε) positively regulates eNOS activity via phosphorylation at serine-1177. Cell-permeable myristoylated (Myr) PKCε peptide activator (PKCε+) and inhibitor (PKCε-) have been shown to increase and decrease eNOS activity, respectively [2,4]. Using a combination of eNOS cofactors, BH_4 or BH_2, with eNOS activity regulators, PKCε+ or PKCε-, we can explore the role of modulating eNOS activity to reduce oxidative stress and endothelial dysfunction caused by ESWL.

Results and Discussion

We hypothesized that this study would confirm previous findings that ESWL decreased blood NO and increased H_2O_2 levels in rat renal veins compared to no-ESWL controls. We predicted that a post-ESWL i.v. bolus of PKCε+ (N-Myr-HDAPIGYD, MW: 1097 g/mol, Genemed Synthesis) with BH_4 (MW: 314 g/mol, Cayman Chemicals) will increase blood NO and decrease H_2O_2 release compared to ESWL controls. Whereas, we expect a post-ESWL infusion of PKCε+ with BH_2 (MW: 239 g/mol, Cayman Chemicals) will decrease NO and increase H_2O_2 compared to ESWL controls. Further, we hypothesized that PKCε- (N-Myr-EAVSLKPT, MW: 1054 g/mol, Genemed Synthesis) given with either BH_4 or BH_2 will increase NO and decrease H_2O_2 release compared to ESWL controls.

Left renal veins were cannulated with a 22-gauge angiocatheter which supported a NO or H_2O_2 microsensor (100 µM). The microsensor was connected to the TBR 4100 free radical analyzer (World Precision Instruments, Inc.) which recorded the real-time blood NO or H_2O_2 response in picoamps which was later converted to molar concentration using the standard calibration curve of each microsensor. ESWL was administered using a Dornier Epos Ultra high-energy lithotripter (Multimed Technical Services, Inc.) (1000 shocks total; 500 at 60 shocks/min, 500 at 120 shocks/min; 16 kV intensity). Immediately following ESWL or at the same time for the no-ESWL controls, 0.5 mL of saline or drug bolus was infused into the jugular vein followed by a 0.5 mL saline flush. Recordings of blood NO and H_2O_2 release were taken throughout the experiment (e.g. baseline, ESWL end, 30 min post-ESWL). All data were analyzed using ANOVA with Student Newman Keuls post-hoc test.

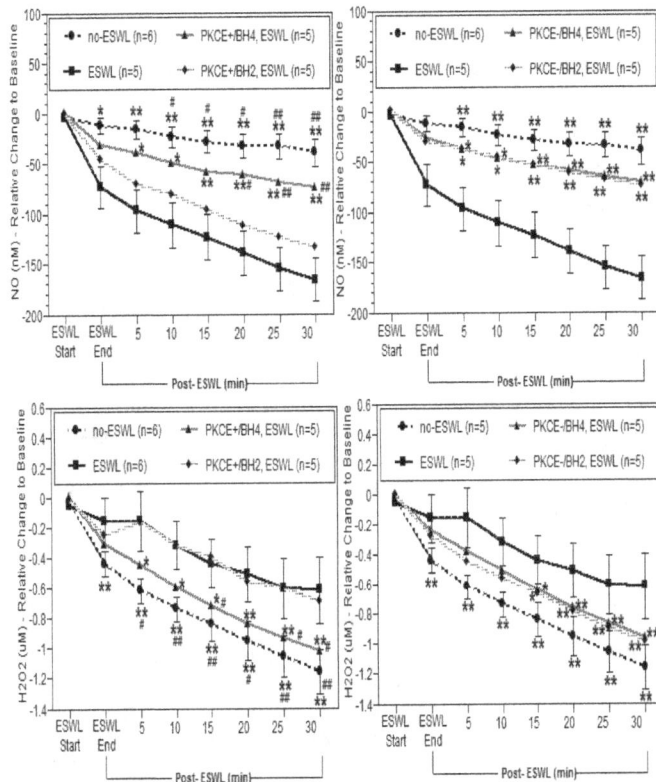

Fig. 1. Real-time blood NO and H_2O_2 levels relative to baseline. ESWL significantly reduced NO release by 126 nM and increased H_2O_2 release by 0.54 µM compared to no-ESWL controls (**$p \leq 0.01$). PKCε+ with BH_4 significantly attenuated the ESWL-induced effects by increasing NO and decreasing H_2O_2 release similar to no-ESWL controls (*$p \leq 0.05$, **$p \leq 0.01$, compared to ESWL controls). PKCε+ with BH_2 was similar to ESWL controls with a 97 nM NO decrease and 0.47 µM H_2O_2 increase compared to no-ESWL controls (#$p \leq 0.05$, ##$p \leq 0.01$). PKCε- with BH_4 or BH_2 significantly increased NO by ~93 nM and decreased H_2O_2 by ~0.35 µM compared to ESWL controls (*$p \leq 0.05$, **$p \leq 0.01$).

ESWL significantly decreased NO and increased H_2O_2 blood release compared to no-ESWL controls (Figure 1). This supports our hypothesis that ESWL causes oxidative stress and reduced NO bioavailability. Post-ESWL infusion of PKCε+/BH_4 significantly attenuated the ESWL-induced effects on NO and H_2O_2 release suggesting that this combination enhances coupled eNOS activity. Whereas, post-ESWL PKCε+/BH_2 was similar to ESWL controls, suggesting that BH_2 is nearing saturation at the eNOS binding site. In contrast, post-ESWL PKCε- with either BH_4 or BH_2 significantly attenuated ESWL-induced effects on NO and H_2O_2 which suggests that PKCε- attenuates uncoupled eNOS activity after ESWL. These results suggest that PKCε+/BH_4 or PKCε- may have clinical application by reducing endothelial dysfunction in patients undergoing ESWL treatment.

Acknowledgments

This study was supported by the Center for Chronic Disorders of Aging and the Department of Bio-Medical Sciences at Philadelphia College of Osteopathic Medicine.

References

1. McAteer, J.A., Evan, A.P. Seminars in Nephrology **28**, 200-213 (2008).
2. Iames, E.S., et al. In Lebl, M. (Ed.) Proceedings of the 22nd American Peptide Symposium, PSP, San Diego, 2011, 278-279.
3. Chen, Q., et al. Advances in Pharmacological Sciences **11**, 1-11 (2010).
4. Csukai, M., Mochly-Rosen, D. Pharmacol. Res. **39**, 253-259 (1999).

Proceedings of the 23rd American Peptide Symposium
Michal Lebl (Editor)
American Peptide Society, 2013

Structure-Activity Relationship Study of Tachykinin Peptides for the Development of Novel NK3 Receptor Agonists

Ryosuke Misu, Taro Noguchi, Hiroaki Ohno, Shinya Oishi, and Nobutaka Fujii

Graduation School of Pharmaceutical Sciences, Kyoto University, Sakyo-ku, Kyoto, 606-8501, Japan

Introduction

Pulsatile release of gonadotropin-releasing hormone (GnRH) is prerequisite for reproductive success in mammals. Recently, it was suggested that neurokinin B (NKB), an endogenous ligand for neurokinin-3 receptor (NK3R), plays a pivotal role in the central control of pulsatile GnRH secretion [1]. Thus, the NKB receptor(s) could be a pharmaceutical target to regulate pulsatile GnRH secretion. Although naturally occurring tachykinin peptides could provide appropriate lead peptides, most of them bind to all tachykinin receptors [neurokinin-1 receptor (NK1R), neurokinin-2 receptor (NK2R) and NK3R] with low selectivity. In order to identify the essential structural requirements for selective NK3R agonists, we carried out structure-activity relationship (SAR) study of [MePhe^7]-NKB and other tachykinin peptides [2].

Results and Discussion

All peptides were synthesized by Fmoc-based solid-phase peptide synthesis (SPPS). Binding affinity of the synthetic peptides to each receptor (NK1R, NK2R, or NK3R) was evaluated by a binding inhibition assay using the corresponding radiolabeled ligand. The agonistic activity of $G_q\alpha$-coupled NK3R was evaluated by Ca^{2+} flux assay.

In the SAR study of NKB and eledoisin derivatives (Table 1), dual substitutions of Phe^5 and Val^7 in NKB, and the corresponding residues in the tachykinin peptide (i.e. eledoisin) with Asp and MePhe, respectively, improved the agonistic activity and receptor selectivity for NK3R in most cases (Table 2). To understand the optimization rationale for Asp and MePhe substitution, we designed eledoisin derivatives, in which each residue at position 5 or 7 was substituted (Table 1 and 2). The MePhe substitution at position 7 of eledoisin led to increased receptor binding. On the other hand, the Asp substitution at position 5 led to an increase in agonistic activity. Of note, NK1R and NK2R binding remained for both single substituted analog at 10 µM, whereas double substituted analog did not bind to these receptors. Thus, dual Asp/MePhe substitutions provided the desired biological properties with high potency and selectivity for NK3R selective agonist.

On the basis of the favorable effects of Asp/MePhe substitutions at position 5 and 7 for selective NK3R agonists, we next designed Phe^5-substituted analogs of [MePhe^7]-NKB. Among these 19 peptides, moderate correlations were observed between the binding affinity and agonistic activity for NK3R, suggesting that the amino acid at this position is not a critical factor in determining the biological activity for NK3R. We also assessed the binding affinity of these peptides to NK1R and NK2R to evaluate the receptor selectivity. Peptides with a substitution with Lys, Arg and Gln, respectively at position 5, exhibited moderate NK1R binding, whereas the other peptides did not bind to NK1R at 10 µM. In contrast, moderate to low NK2R binding was observed for most of peptides, except for two peptides with a substitution with Asp and Glu, respectively. These observations suggest that the negatively charged residue at this position is unfavorable for NK2R binding, and that the electrostatic property at this position is more important for an NK3R selective agonist.

In conclusion, we demonstrated that both Asp and MePhe at position 5 and 7 contribute to potent receptor binding to NK3R and high NK3R selectivity. It was also revealed that an acidic amino acid at position 5 improved the NK3R selectivity. These SAR data would be helpful to develop novel NK3R selective peptide ligands for therapeutic agents that induce pulsatile GnRH secretion.

Table 1. Design of tachykinin derivatives for the structure-activity relationship study.

Peptide	Sequence
NKB	H-Asp-Met-His-Asp-Phe-Phe- Val -Gly-Leu-Met-NH$_2$
[MePhe^7]-NKB	H-Asp-Met-His-Asp-Phe-Phe-<u>MePhe</u>-Gly-Leu-Met-NH$_2$
[Asp^5, MePhe^7]-NKB	H-Asp-Met-His-Asp-<u>Asp</u>-Phe-<u>MePhe</u>-Gly-Leu-Met-NH$_2$
eledoisin	pGlu-Pro-Ser-Lys-Asp-Ala-Phe- Ile -Gly-Leu-Met-NH$_2$
[MePhe^8]-eledoisin	pGlu-Pro-Ser-Lys-Asp-Ala-Phe-<u>MePhe</u>-Gly-Leu-Met-NH$_2$
[Asp^6]-eledoisin	pGlu-Pro-Ser-Lys-Asp-<u>Asp</u>-Phe- Ile -Gly-Leu-Met-NH$_2$
[Asp^6, MePhe^8]-eledoisin	pGlu-Pro-Ser-Lys-Asp-<u>Asp</u>-Phe-<u>MePhe</u>-Gly-Leu-Met-NH$_2$

Table 2. Biological activities of tachykinin derivatives.

Peptide	NK3R IC$_{50}$ (nM)^a	EC$_{50}$ (pM)^b	NK1R % inhibition^c	NK2R
[MePhe^7]-NKB	3.4	13	<10	57
[Asp^5, MePhe^7]-NKB	36	7.4	<10	<10
eledoisin	96	241	100	99
[MePhe^8]-eledoisin	7.9	71	48	26
[Asp^6]-eledoisin	35	5.4	91	56
[Asp^6, MePhe^8]-eledoisin	5.5	78	<10	<10

^{a}IC$_{50}$ was determined by a competitive binding experiment using ([^{125}I]His^3, MePhe^7)-NKB at 0.1 nM. ^{b}Agonistic activity was determined by the Ca^{2+} flux assay. ^{c}Binding inhibition of the tachykinin peptides against NK1R and NK2R at 10 μM.

Table 3. Biological activities of [Xaa^5, MePhe^7]-NKB derivatives.

Xaa^5	NK3R IC$_{50}^a$ (nM)	EC$_{50}^b$ (pM)	NK1R % inhibition^c	NK2R	Xaa^5	NK3R IC$_{50}^a$ (nM)	EC$_{50}^b$ (pM)	NK1R % inhibition^c	NK2R
Phe	3.0	70	<10	57	Asp	38	40	<10	<10
His	5.9	17	<10	47	Glu	7.5	24	<10	<10
Trp	1.8	281	<10	79	Lys	11	26	40	35
Tyr	1.5	72	<10	51	Arg	2.5	80	60	62
Ala	4.4	12	<10	33	Met	1.7	32	<10	41
Gly	13	36	<10	22	Asn	18	36	<10	10
Ile	4.0	60	<10	35	Gln	4.6	43	24	21
Leu	6.9	51	<10	29	Ser	3.6	40	<10	31
Pro	45	41	<10	38	Thr	4.3	46	<10	19
Val	1.5	65	<10	29					

^{a}IC$_{50}$ was determined by a competitive binding experiment using ([^{125}I]His^3, MePhe^7)-NKB at 0.1 nM. ^{b}Agonistic activity was determined by the Ca^{2+} flux assay. ^{c}Binding inhibition of the tachykinin derivatives against NK1R and NK2R at 10 μM.

Acknowledgments

This work was supported by a Grant from Ministry of Agriculture, Forestry and Fisheries of Japan (Research Program on Innovative Technologies for Animal Breeding, Reproduction, and Vaccine Development, REP-2003); and Grants-in-Aid for Scientific Research from MEXT. R.M. is grateful for the JSPS Research Fellowships for Young Scientists.

References

1. Wakabayashi, Y., et al. *J. Neurosci.* **30**, 3124-3132 (2010).
2. Misu, R., Oishi, S., Fujii, N., et al. *Bioorg. Med. Chem.* **21**, 2413-2417 (2013).

Proceedings of the 23rd American Peptide Symposium
Michal Lebl (Editor)
American Peptide Society, 2013

A Minimal Labor Approach to Identifying the Optimal Anoplin Analog Using ANOVA

Jens K. Munk[1], Christian Ritz[2], Frederikke P. Fliedner[1], Niels Frimodt-Møller[3], and Paul R. Hansen[1]

[1]Department of Drug Design and Pharmacology, Faculty of Health and Medical Sciences, University of Copenhagen, Copenhagen Ø, 2100, Denmark; [2]Department of Nutrition, Exercise and Sports, Faculty of Science, University of Copenhagen, Frederiksberg, 1871, Denmark; [3]Department of Clinical Microbiology, Hvidovre Hospital, Hvidovre, 2650, Denmark

Introduction

The Gram negative bacterium *Pseudomonas aeruginosa* is, together with methicillin-resistant *Staphylococcus aureus*, a leading cause of hospital-acquired infections, and causes complications in compromised patients [1]. Some strains of *P. aeruginosa* have developed resistance against antibiotics, leading to fatalities among these patients. Antimicrobial peptides show promise as new drug leads for treatment of such resistant strains. Optimization of antimicrobial peptide specificity can be complex and labor-extensive due to the size of the chemical space. We demonstrate the use of statistical design and biological activity modeling based on analysis of variance of biological data. We investigate the effects of modifications to our lead structure, anoplin (H-GLLKRIKTLL-NH$_2$) [2]. Five of the side chains are modified, these are: position 2: leucine to cyclohexylalanine; 5: arginine to lysine; 6: isoleucine to 2-naphthylalanine; 8: threonine to lysine; and 10: leucine to cyclohexylalanine. The sixth modification is L- to D-form at position 6. Hence, there are $2^6 = 64$ amino acid combinations in total. We use a training set of 12 peptides, representative of all 64 peptides, to derive a simple mathematical model for specificity against *P. aeruginosa* relative to human red blood cells, and verify our model experimentally.

Results and Discussion

Hemolytic activities (EC$_{10}$, the concentration needed to lyse 10% of red blood cells suspended in PBS) and minimum inhibitory concentrations (MICs) against *P. aeruginosa* of the training set peptides were measured by dilution row. These data were log$_2$ transformed and fed into analysis of variance (ANOVA). Initial ANOVA-based models contained the six factor terms, one for each modification, but also several interaction terms for combinations of the modifications. These highly complex models were reduced by the Akaike Information Criterion (AIC) procedure [3]. This procedure is similar to Occam's Razor, and eliminated all interaction terms. In these simple models almost all of the variance (95-98%) was assigned to the six sequence modifications; the remaining variance was considered noise. Figure 1 is a double log plot and shows excellent correlations between the measured and the fitted biological activity values.

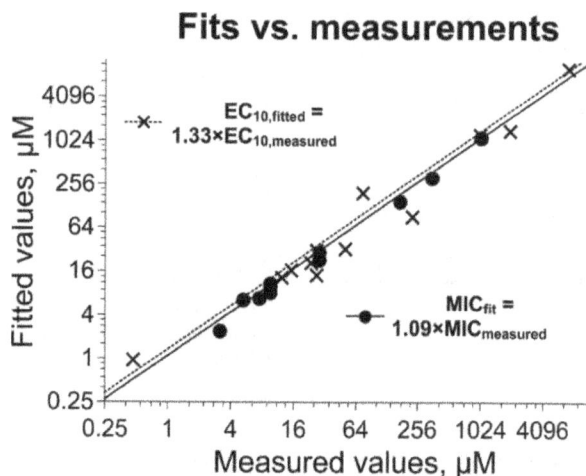

Fig. 1. Double log plot of the correlation between measured and fitted values for hemolytic (dashed line, crosses) and antipseudomonal (full line, dots) activities for the 12 training set peptides.

Table 1. Factor coefficients fitted by ANOVA of the training set. The coefficients can be applied to anoplin by multiplication.

Factor	Anoplin	2Cha	5Lys	6(2Nal)	8Lys	10Cha	D-form
EC_{10}, μM	1052	0.09	0.77	0.19	0.14	0.08	10.33
MIC, μM	148	0.25	1.50	0.20	0.28	0.16	5.09

The fitted ANOVA coefficients are measures of the effect of each modification on biological activity, and are shown in Table 1. Predicted biological activities for all 64 peptides can be calculated by multiplying the relevant coefficients. As an example, EC_{10} for the peptide with a D-isoleucine at position 6 (called 6(D-Ile)) is 1052 μM (value for anoplin) multiplied by 10.33 (the effect of L- to D-form modification at position 6), equal to 12982 μM. Likewise, MIC against *P. aeruginosa* for 6(D-Ile) is 148 μM × 5.09 = 797 μM. For each peptide, predicted hemolytic activity divided by predicted antipseudomonal activity gives a measure of antipseudomonal specificity. Among the 64 combinations, 6(D-Ile) has the highest predicted specificity at 12982 μM / 797 μM = 16.3. We therefore synthesized, purified and characterized 6(D-Ile) and other peptides with high predicted specificity. We verified that the predicted activities were within error of experimentally obtained values, and that the three best peptides by prediction were also the best peptides experimentally.

As illustrated in Table 1, anoplin specificity is 1052 μM / 148 μM = 7.1. The factors associated with sequence modifications quantify the effects of the modifications. Table 1 shows that any side chain modification decreases EC_{10}, while the L- to D-form modification increases EC_{10}. The Arg^5 to Lys and the L- to D-form modifications increase MIC against *P. aeruginosa*, while the side chain modification at positions 2, 6, 8 and 10 all decrease MIC. However, the same modifications decrease EC_{10} more than they decrease MIC, so these modifications cause loss of specificity. Only the L- to D-form modification shows a lower coefficient for MIC than for EC_{10}, so this is the only modification which, by itself, increases antipseudomonal specificity.

In conclusion we have 1) identified the optimal peptide in a confined chemical space in a labor-efficient manner, and 2) described both hemolytic and antipseudomonal activities, and hence antipseudomonal specificity, in mathematical terms.

Acknowledgments

We thank technician Sabaheta Babajic for peptide synthesis and purification, technician Jytte Mark Andersen for measuring antipseudomonal activities, and Professor, Dr. med. Kaj Winther for providing blood. The present work is performed as a part of the Danish Center for Antibiotic Research and Development (DanCARD) financed by The Danish Council for Strategic Research (grant no. 09_067075).

References

1. Driscoll, J.A., Brody, S.L., Kollef, M.H. *Drugs* **67**, 351-368 (2007).
2. Konno, K., Hisada, M., Fontana, R., Lorenzi, CCB., Naoki, H., Itagaki, Y., Miwa, A., Kawai, N., Nakata, Y., Yasuhara, T., Neto, J.R., de Azevedo, W.F., Palma, M.S., Nakajima, T. *Biochim. Biophys. Acta* **1550**(1), 70-80 (2001).
3. Hastie, T., Tibshirani, R., Friedman, J. *The Elements of Statistical Learning* Springer, New York, 2009.

Proceedings of the 23rd American Peptide Symposium
Michal Lebl (Editor)
American Peptide Society, 2013

Cardioprotective Effects of Mitochondrial-Targeted Antioxidants in Myocardial Ischemia/Reperfusion (I/R) Injury

Regina Ondrasik, Qian Chen, Katelyn Navitsky, William Chau, Issachar Devine, On S. Lau, Tyler Galbreath, Robert Barsotti, and Lindon H. Young

Department of Bio-Medical Sciences, Philadelphia College of Osteopathic Medicine, Philadelphia, PA, 19131, U.S.A.

Introduction

During myocardial ischemia cardiomyocytes are deprived of oxygen, glucose, and fatty acids. Prompt reperfusion limits the extent of irreversible ischemic damage. However, this damage is exacerbated at reperfusion by a burst of reactive oxygen species (ROS); much of which is produced by the interaction of oxygen with damaged mitochondrial electron transport chain complexes I and III [1]. Excess ROS production increases oxidative stress and, thereby, directly damages nucleic acids, proteins, and lipids; thus, as a result, disrupting membrane integrity, cell function, and activating apoptosis signaling cascades [1,2]. Conventional antioxidants have limited efficacy in alleviating myocardial I/R injury since they are not selectively targeted to mitochondria, major sites of ROS overproduction [1,3-5]. It has been suggested that mitoquinone (mitoQ, MW=600g/mol; the mitoQ referred to in this study was complexed with cyclodextrin to improve water solubility, total MW=1714g/mol) and the SS-31 (Szeto-Schiller) peptide ((D-Arg)-Dmt-Lys-Phe-Amide, MW=640g/mol, Genemed Synthesis, Inc., San Antonio, TX), mitochondrial-targeted antioxidants, can exert cardioprotective effects in I/R [3-5]. Although both selectively concentrate within mitochondria, mitoQ is dependent on the electrochemical membrane gradient; whereas, an alternating cationic-aromatic amino acid sequence allows SS-31 to diffuse independently from membrane potential [4,5]. Prior studies have reported that pretreatment with mitoQ or SS-31 before the onset of ischemia effectively limits I/R injury, but pretreatment of acute myocardial infarction is not always possible [3,4]. Therefore, we evaluated the cardioprotective efficacy of both compounds when given only at reperfusion. We hypothesized that mitoQ or SS-31 given at reperfusion will attenuate myocardial I(30min)/R(45min) injury by limiting ROS production and thereby improve cardiac contractile function and limit infarct size in isolated perfused rat hearts subjected to I/R compared to untreated I/R hearts.

Results and Discussion

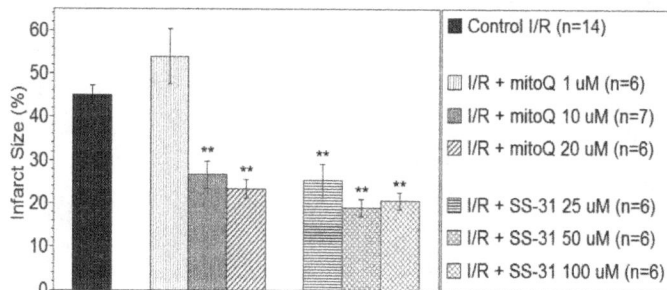

Fig. 1. Ratio of infarct to area at risk as determined by 1% 2,3,5-triphenyltetrazolium chloride staining (**$p < 0.01$ compared to control I/R). All data were analyzed using ANOVA with the Student Newman Keuls test.

MitoQ (10, 20µM) and SS-31 (25, 50, 100µM) given at reperfusion significantly reduced infarct size compared to untreated I/R hearts (Figure 1, Table 1). MitoQ (10, 20µM) reduced infarct size, whereas all SS-31 doses reduced infarct size (Figure 1, Table 1). Left ventricular developed pressure (LVDP, systolic-diastolic pressure) and dP/dtmax, the derivative of the maximum rate of pressure rise, were measured as indicies of cardiac function (Figure 2, Table 1). In addition to reducing infarct size, mitoQ (10, 20µM) and SS-31 (50µM) restored post-reperfused cardiac function (LVDP and dP/dt$_{max}$) significantly compared to untreated I/R hearts (Figures 1,2; Table 1). I/R hearts treated with 25 and 100µM doses of SS-31 did not exhibit significant improvement in cardiac function, however both doses caused a significant reduction in infarct size (Figures 1,2; Table 1).

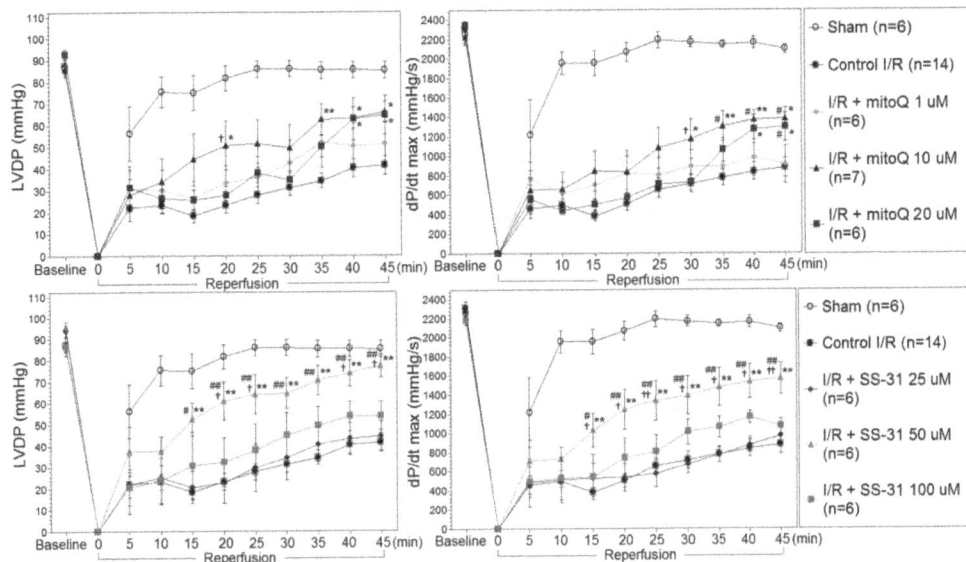

Fig. 2. LVDP (upper left) and dP/dtmax (upper right) for mitoQ and LVDP (lower left) and dP/dtmax (lower right) for SS-31 I/R hearts (*p<0.05, **p<0.01 compared to control I/R; #p<0.05, ##p<0.01 compared to low dose I/R; †p<0.05, ††p<0.01 compared to high dose I/R). All data were analyzed using ANOVA with the Student Newman Keuls test.

Table 1. Infarct Size and Percent Recovery of LVDP and dP/dt_{max}.

| | Control I/R | mitoQ | | | SS-31 | | |
		1μM	10μM	20μM	25μM	50μM	100μM
Infarct Size (%)	45.0±2.1	53.9±6.3	26.7±3.2**	23.3±2.1**	25.2±3.8**	18.9±2.0**	20.6±1.9**
LVDP (%)	47.9±4.6	58.7±11.0	77.0±5.8*	70.9±11.5*	47.7±7.6	81.1±6.1**	62.3±7.4
dP/dt_{max} (%)	37.9±3.8	41.4±9.0	62.8±6.0*	55.7±7.1*	43.3±5.0	70.5±6.9**	49.0±3.4

All data were analyzed using ANOVA with the Student Newman Keuls test (*p<0.05, **p<0.01 compared to control I/R).

These results suggest mitochondrial-derived ROS are important contributors to I/R injury. Additionally, the data suggest mitoQ and SS-31 can work expeditiously and effectively when administered at reperfusion in I/R to limit infarct size. Therefore, mitoQ or SS-31 may potentially augment the benefit of angioplasty or thrombolytic treatment in the clinical setting for myocardial infarction, where pretreatment may not be a practical option.

Acknowledgments

This study was supported by the Center for Chronic Disorders of Aging and the Department of Bio-Medical Sciences at the Philadelphia College of Osteopathic Medicine.

References

1. Szeto, H.H. AAPS Journal 8, 521-531 (2006).
2. Ide, T., Tsutsui, H., Kinugawa, S., et al. Circ Res. 85, 357-363 (1999).
3. Bayeva, M., Gheorghiade, M., Ardehali, H. J. Am. Coll. Cardiol. 61, 599-610 (2013).
4. Szeto, H.H. Antioxidants and Redox Signaling 10, 601-619 (2008).
5. Adlam, V.J., Harrison, J.C., Porteous, C.M., et al. FASEB J. 19, 1088-1095 (2005).

Proceedings of the 23rd American Peptide Symposium
Michal Lebl (Editor)
American Peptide Society, 2013

The Role of NADPH Oxidase on L-NAME-Induced Leukocyte-Endothelial Interactions in Rat Mesenteric Postcapillary Venules

Hung T. Pham, Robert Barsotti, Amber N. Koon, Brian Rueter, Lindon H. Young, and Qian Chen

Department of Bio-Medical Sciences, Philadelphia College of Osteopathic Medicine PCOM,
4170 City Avenue, Philadelphia, PA, 19131, U.S.A.

Introduction

Chronic inflammation can lead to endothelial dysfunction causing further inflammation, setting up a vicious cycle. Inflammatory responses are thought to underlie the pathogenesis of many vascular-related diseases and are characterized by decreased production of endothelium-derived nitric oxide (NO) by endothelial NO synthase (eNOS) and increased reactive oxygen species (ROS) particularly, superoxide (SO) [1,2]. The inflammatory responses involve increased leukocyte-endothelial interactions and activation of NADPH oxidase causing ROS release. We hypothesized that the selective NADPH oxidase inhibitors, apocynin (mol.wt.=166 g/mol, Sigma Chemicals) [3] and gp91 ds-tat (RKKRRQRRR-CSTRIRRQL-amide, mol.wt.=2452 g/mol, Genemed Synthesis, San Antonio, TX) [4], will attenuate leukocyte-endothelial interactions induced by inhibition of eNOS activity *via* N^G-nitro-L-arginine-methyl-ester (L-NAME).

Results and Discussion

Intravital microscopy was conducted on male Sprague-Dawley rats (275-325 g) after anesthesia as previously published [1]. The mean arterial blood pressure was maintained (mean=90 $\pm$ 3 mmHg) throughout the experiment in all groups. We examined leukocyte-endothelial interactions in rat mesenteric postcapillary venules (mean diameter=20.91 $\pm$ .23μm) at baseline and every 30 min throughout the 120 min of superfusion of Krebs' buffer or Krebs' with test compounds. The superfused mesenteric tissue was harvested, fixed, embedded, and subjected to hematoxylin and eosin (H&E) staining to evaluate leukocyte adherence and transmigration. We found that L-NAME (n=5, p<0.05) significantly increased leukocyte-endothelial interactions compared to Krebs's control group (n=6). The administration of 400 μM apocynin (n=7) did not have any effects on basal levels of leukocyte-endothelial interactions. However, 1000 μM apocynin exerted inflammatory effects suggesting non-specificity of apocynin at this higher dose. Leukocyte-endothelial interactions induced by L-NAME were significantly attenuated by 40 and 400 μM apocynin (n= 6-7, p<0.05, Figure 1). 20 μM gp91 ds-tat also significantly attenuated leukocyte-endothelial interactions induced by L-NAME (n=6, p<0.05, Figure 2). Additionally, leukocyte vascular adherence and transmigration evaluated by H&E staining are consistent with intravital microscopy results (Table 1.).

Table 1. *H&E summary of leukocyte vascular adherence/transmigration.*

Experimental groups	Leukocyte/mm^2	
	Adherence	*Transmigration*
Krebs' buffer control (n=3)	73±33	62±30
L-NAME (n=3)	299±23**	271±26**
L-NAME+ 40 μM apocynin (n=3)	85±7##	74 ±12##
L-NAME + 400 μM apocynin (n=3)	83 ±20##	71±14##
L-NAME + 20 μM gp91 ds-tat (n=3)	83±9##	55±9##

*p<0.05, **p<0.01 from Krebs'; #p<0.05, ##p<0.01 from L-NAME by ANOVA using Bonferroni/Dunn test.

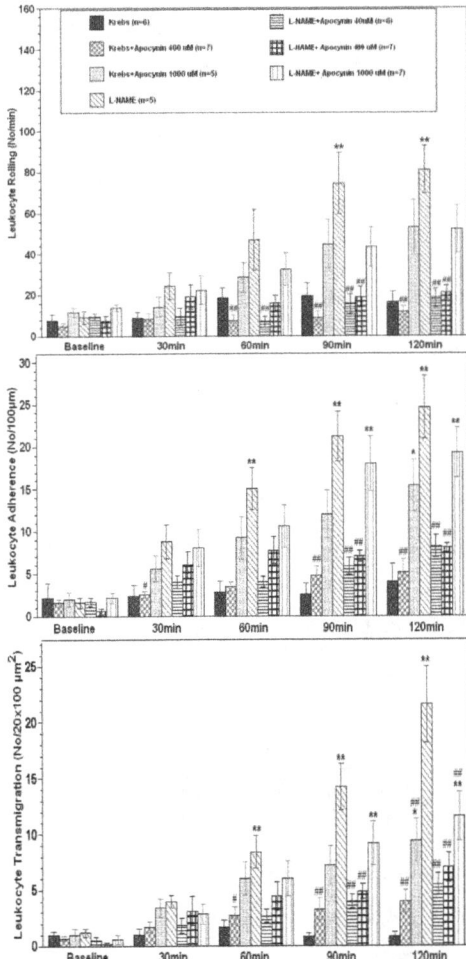

*Fig. 1. Leukocyte rolling (top), adherence (middle), and transmigration (bottom), among Krebs', L-NAME, and apocynin groups. *p<0.05, **p<0.01 from Krebs'; #p<0.05, ##p<0.01 from L-NAME by ANOVA using Bonferroni/Dunn test.*

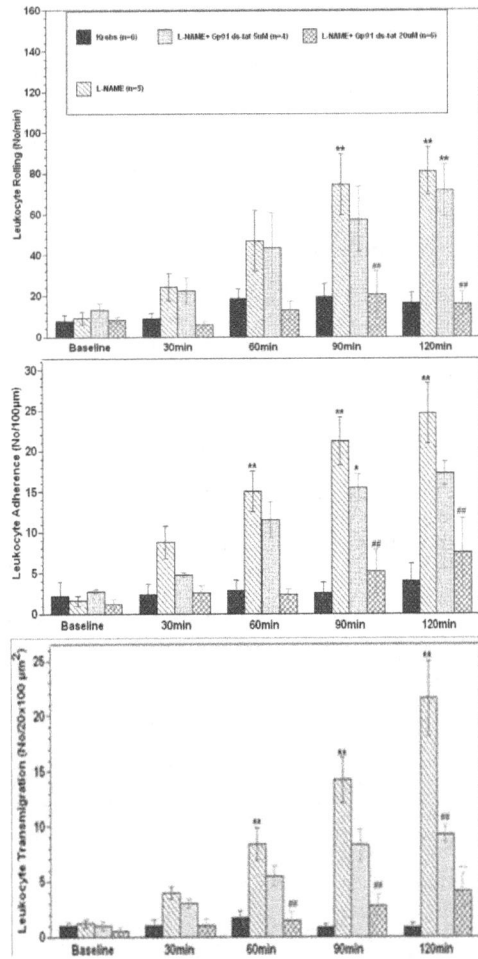

*Fig. 2. Leukocyte rolling (top), adherence (middle), and transmigration (bottom), among Krebs', L-NAME, and gp91 ds-tat groups. *p<0.05, **p<0.01 from Krebs'; #p<0.05, ##p<0.01 from L-NAME by ANOVA using Bonferroni/Dunn test.*

The results suggest that inhibiting NAPDH oxidase activity is an effective method to attenuate leukocyte-endothelial interactions induced by endothelial dysfunction. Therefore, NADPH oxidase inhibitors may be beneficial in mitigating the pathogenesis of inflammatory-mediated vascular diseases.

Acknowledgments

This study was supported by the Center for the Chronic Disorders of Aging and the Department of Bio-Medical Sciences at PCOM.

References

1. Chen, Q., et al. *Current Topics in Pharmacology* **14**, 11-24 (2010).
2. Brandes, R.P. *Circulation Research* **92**, 583-585 (2003).
3. Mora-Pale, M., et al. *Free Radical Biology and Medicine* **52**, 962-969 (2012).
4. Rey, F.E., et al. *Circulation Research* **89**, 408-414 (2001).

Proceedings of the 23rd American Peptide Symposium
Michal Lebl (Editor)
American Peptide Society, 2013

Novel Camptothecin Somatostatin Conjugate in Treatment of Tumor Cells Expressing sst2 Receptors

Philip G. Kasprzyk, Ann V. Fiore, Yeelana Shen, Jundong Zhang, Jennifer Morgan, Lynda Cooper, Kevin L. Zhou, and Jesse Z. Dong

IPSEN, 27 Maple Street, Milford, MA, U.S.A.

Introduction

Many G protein-coupled receptors internalize into cells together with their ligands, usually agonists. Somatostatin receptors have 5 subtypes, sstr1-5 and the sstr2 is over-expressed in a number of human tumors. IPSEN has a targeted peptide technology. In its cytotoxic conjugate program, a novel camptothecin (CPT) somatostatin (sst) conjugate X has been developed in treatment of tumor cells expressing sst2 receptors.

Results and Discussion

IPSEN's cytotoxic conjugate program using targeted peptide technology (Figure 1) to deliver cytotoxic "warhead" led to development of conjugate X, a CPT-sst2 complex with an innovative design of linkers. Combination of the solution phase preparation of cytotoxic

Peptide-cytotoxic conjugates target cancer cells

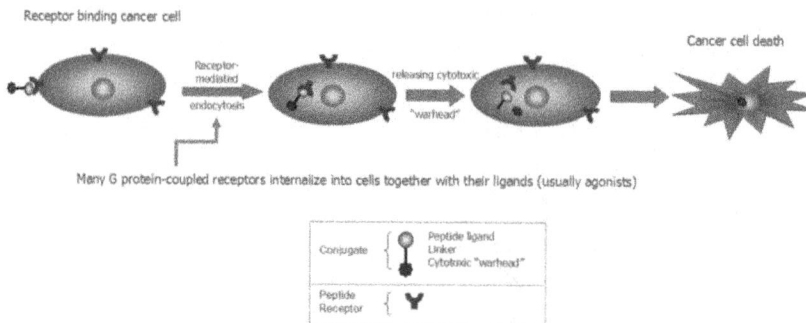

Fig. 1. Schematic diagram of targeted delivery of conjugates.

precursors and the solid support chemistry of peptide/linker synthesis was an easy process to produce the conjugate X in a moderate yield and high chemical purify (97.8-99.9%) after purification by RP prep HPLC. It's worthwhile to point out that the linker may be either L- or D-isomer, with the later one as desired. Under the optimized reaction conditions, D-isomer was obtained as a major component, in a ratio of D/L-isomers as 96.8:3.2. The linker used conferred a long plasma half-life ($T_{1/2}$) in mouse similar to human. $T_{1/2}$ of 1.5 hours was also observed in live animals. In addition, the conjugate X showed a very good selectivity on sst2 against other sst receptors 1 and 3-5 (Table 1).

In terms of its biological activity, the conjugate X was better than the cytotoxic portion itself in various somatostatin receptor positive tumor models. This conjugate demonstrated

Table 1. Half-life and receptor binding.

In vitro plasma $T_{1/2}$ (h)		In vivo $T_{1/2}$ (h)	Receptor binding Ki (nM)				
Mouse	Human	Mouse	sst1	sst2	sst3	sst4	sst5
15.5	13.9	1.5	>1000	2.94	1029	>1000	843

potent *in vitro* cytotoxicity only in the cells expressing sst2 receptors. The effects of this new conjugate were evaluated in several different tumor models with varying levels of sst2 receptors. The conjugate was administered at doses equivalent to 1 to 3-fold of the MTD of the parent cytotoxic CPT. In three of the models, the highest concentrations of the conjugate tested so far exhibited superior efficacy to that observed with the warhead, CPT alone at its MTD on the schedule used (qwk x 3, i.v.). For example, in the human chronic myelogenous leukemia K562 model four of six and five of six tumor free survivals were recorded with, respectively, 2.5 and 3 fold equivalent doses of the conjugate with no toxicity and a slight weight loss similar to that observed with CPT.

In summary, the conjugate X showed increased efficacy compared to the parent cytotoxic agent, camptothecin (CPT), in models with medium to high levels of sst2 somatostatin receptor expression. Even in a model with low levels of receptor expression, a slight increase in efficacy was demonstrated. In addition, no significant weight loss was observed with the conjugate X in any of the models tested, therefore, higher doses could be used that might result in increased efficacy.

References

1. Partial work has been presented at AACR, April 2007, #676.

Proceedings of the 23rd American Peptide Symposium
Michal Lebl (Editor)
American Peptide Society, 2013

Targeted Cytotoxic Somatostatin Conjugates Specifically Inhibit the Growth of SCLC

**Philip G. Kasprzyk[1], Ann V. Fiore[1], Mark Carlson[1], Yeelana Shen[1],
Jundong Zhang[1], Jeanne M. Comstock[1], Xiaojun Zou[1], Marc Teillot[2],
Caroline Touvay[2], Kevin L. Zhou[1], and Jesse Z. Dong[1]**

[1]IPSEN, 27 Maple Street, Milford, MA, U.S.A.; [2]IPSEN, Avenue du Canada 5, Les Ulis, France

Introduction

Cancer chemotherapy is limited by intrinsic or acquired multi-drug resistance of tumor cells and toxicity to normal cells. A more selective delivery of the cytotoxic agent to the primary tumors by targeting receptors expressed on tumor cells would allow a dose escalation and reduce toxicity. Certain human tumor cells have been shown to express high levels of somatostatin. Therefore, the cytotoxic agent, Camptothecin (CPT) or Homocamptothecin (hCPT) could be linked to somatostatin analogs *via* a linker to make conjugates. Ideally the conjugate should be stable and inactive in the circulation and release the cytotoxic agent in the target tumor tissues. Conjugates, A and B, composed of cytotoxic "warhead" CPT or hCPT, a linker and targeted moiety somatostatin peptide have been developed to target the tumor cells, especially SCLC, that express sst2 receptors (Figure 1).

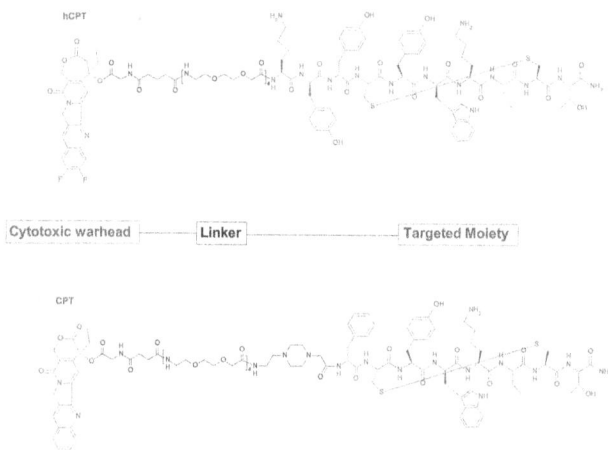

Fig. 1. Structures of tested conjugates A and B.

Results and Discussion

A synthesis started with the preparation of the cytotoxic precursors employing general organic chemistry. The peptide/linker part was assembled on a peptide synthesizer using Fmoc chemistry on a resin. The conjugation followed by purification gave the product with >95% chemical purity in 20-30% yield.

Both conjugates A and B showed a good stability in either mouse or human plasma, with $T_{1/2}$ in a range of 1.5-5.2 hours. In the *in vitro* binding assays, they retained high affinity to sst2 receptor (sub nM) and selectivity for the same receptor against other sst receptor subtypes (1 and 3-5).

In vitro internalization was both time and temperature dependent and was observed only in the cells expressing sst2 receptors. Internalization was blocked by a specific sst2 receptor antagonist. Potent *in vitro* cytotoxicity was observed only in the cells expressing sst2 receptors. Cytotoxicity was blocked by specific sst2 receptor antagonists suggesting that binding and internalization were required to bring the cytotoxic agents into the cells. These demonstrated the expected sequence of events: (sst2 binding – receptor activation – receptor/conjugate internalization – release of the free cytotoxic ligand inside of the cell).

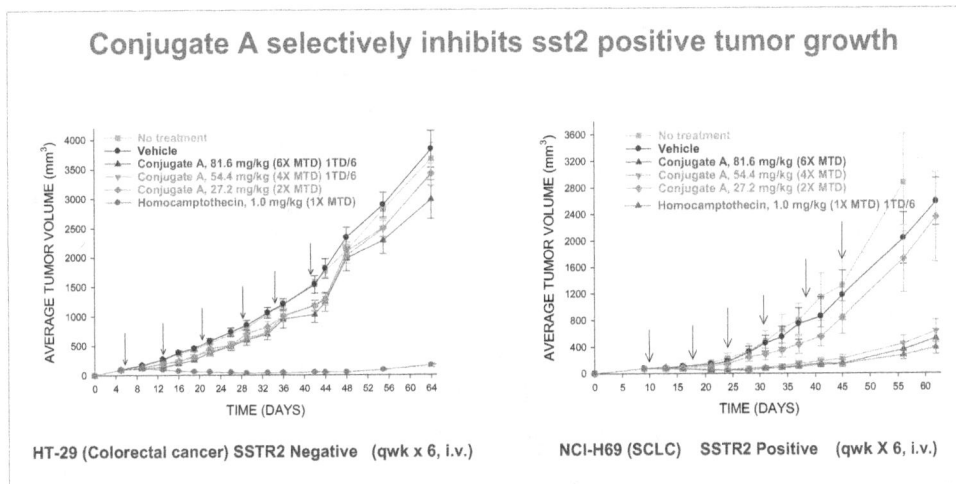

Conjugate A selectively inhibits sst2 positive tumor growth

HT-29 (Colorectal cancer) SSTR2 Negative   (qwk x 6, i.v.)   NCI-H69 (SCLC)   SSTR2 Positive   (qwk X 6, i.v.)

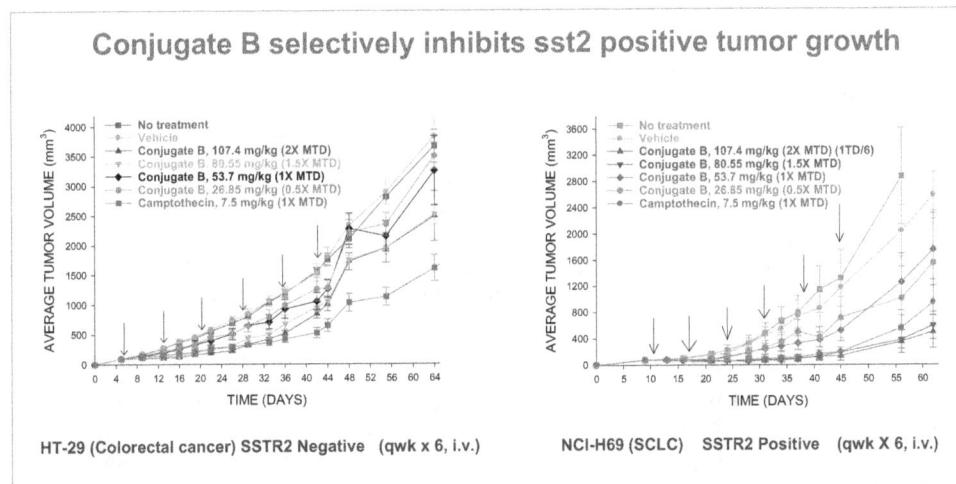

Conjugate B selectively inhibits sst2 positive tumor growth

HT-29 (Colorectal cancer) SSTR2 Negative   (qwk x 6, i.v.)   NCI-H69 (SCLC)   SSTR2 Positive   (qwk X 6, i.v.)

Fig. 2. In vivo results of tested conjugates A and B.

We evaluated the effects of the two conjugates in the NCI-H69, SCLC xenograft model (sst2 receptor positive) and in the negative control HT-29 colon xenograft model (sst2 receptor negative) (Figure 2). Nude mice bearing the tumors were administered the analogs on a qwk x 6, i.v. schedule at doses equivalent to 1 to 6-fold of the MTD of the parent cytotoxic agents. The results of the *in vivo* experiments suggested that specific cytotoxic efficacy could be observed at doses up to 6X the parent cytotoxic with no added toxicity. Preliminary acute toxicity analysis showed that the analogs did not decrease white blood count, neutrophils, and platelets while CPT and hCPT did.

The *in vivo* studies verified the *in vitro* results by demonstrating efficacy in sst2 receptor containing model versus non-sst2 containing models. Also, toxicity observed with the parent cytotoxic agent was absent with the cytotoxic conjugates whereas the efficacy was the similar. It seems that the somatostatin-directed cytotoxic conjugates selectively target the cancer cells expressing sst2 receptors *in vitro* and *in vivo*.

References

1. Partial work has been presented at AACR, April 2005, #133.

Proceedings of the 23rd American Peptide Symposium
Michal Lebl (Editor)
American Peptide Society, 2013

Novel TIPP (Tyr-Tic-Phe-Phe) Analogues with Unexpected Opioid Actvity Profiles

Grazyna Weltrowska[1], Thi M.-D Nguyen[1], Nga N. Chung[1], Brian C. Wilkes[1], and Peter W. Schiller[1,2]

[1]Laboratory of Chemical Biology and Peptide Research, Clinical Research Institute of Montreal, Montreal, QC, H2W 1R7 Canada; [2]Department of Pharmacology, Université de Montréal, Montreal, QC, H3C 3J7 Canada

Introduction

The TIPP peptides H-Tyr-Tic-Phe-Phe-OH (TIPP, Tic = 1,2,3,4-tetrahydroisoquinoline-3-carboxylic acid), H-Dmt-Tic-Phe-Phe-OH (DIPP, Dmt = 2',6'-dimethyltyrosine) and H-Tyr-TicΨ[CH$_2$NH]Phe-Phe-OH (TIPP[Ψ]) are highly selective δ opioid antagonists, whereas the TIPP-derived tetrapeptide amides H-Dmt-Tic-Phe-Phe-NH$_2$ (DIPP-NH$_2$) and H-Dmt-TicΨ[CH$_2$NH]Phe-Phe-NH$_2$ (DIPP-NH$_2$[Ψ]) act as agonists at the μ opioid receptor and as antagonists at the δ opioid receptor [1]. Such mixed μ agonist/δ antagonists have therapeutic potential as analgesics with low propensity to induce analgesic tolerance and dependence [2]. In an effort to improve the ability of TIPP peptides to cross the blood-brain barrier, we prepared N-terminally guanidinylated analogues (Gu-peptides). Compounds were prepared by solid-phase and solution techniques and were tested in opioid receptor binding assays and in the functional guinea pig ileum (GPI) and mouse vas deferens (MVD) assays.

Results and Discussion

The Gu-analogues of the δ antagonists TIPP, DIPP and TIPP[Ψ] showed δ receptor binding affinities similar to those of their respective parent peptides and retained high δ receptor selectivity with weak binding affinities for μ receptors and very weak affinities for κ receptors (Table 1). However, in the functional MVD assay they displayed δ partial agonist behavior. Amidation of the dipeptide δ antagonist H-Dmt-Tic-OH had no effect on δ receptor affinity, increased μ receptor affinity about 100-fold and resulted in partial agonism in the MVD and GPI assays. Thus, Gu-Dmt-Tic-OH is a partial μ agonist/partial δ agonist with low nanomolar binding affinity for μ and δ receptors.

Guanidinylation of the mixed μ agonist/δ antagonist DIPP-NH$_2$ had unexpected effects on the *in vitro* opioid activity profile. While Gu-DIPP-NH$_2$ retained subnanomolar δ and μ receptor binding affinities, it showed significant κ receptor affinity K$_i^\kappa$ = 35.8 nM), in contrast to the very low κ affinity of the non-guanidinylated parent peptide. Surprisingly, this compound turned out to be a potent δ full agonist in the MVD assay (IC$_{50}$ = 1.72 nM). It showed high μ agonist potency in the GPI assay and thus represents a potent, balanced μ agonist/δ *agonist*. In a flexible docking study, the mixed μ agonist/δ antagonist DIPP-NH$_2$ and the mixed μ agonist/δ agonist Gu-DIPP-NH$_2$ were docked to models of the inactive and activated form of the δ receptor, respectively. Both the N-terminal amino group of DIPP-NH$_2$ and the N-terminal guanidino group of Gu-DIPP-NH$_2$ were engaged in an electrostatic interaction with Asp^{128} in the third transmembrane helix of the receptor. However, due to the steric bulk of the guanidino group, the position of Gu-DIPP-NH$_2$ was slightly shifted relative to that of DIPP-NH$_2$ (average RMS deviation = 1.06 Å). This resulted in somewhat different interactions with receptor binding site residues, thus explaining the δ antagonist vs. δ agonist behavior of DIPP-NH$_2$ and Gu-DIPP-NH$_2$. Compared to Gu-DIPP-NH$_2$, the pseudopeptide Gu-DIPP-NH$_2$[Ψ] displayed a similar opiod receptor selectivity profile and similar balanced μ agonist/δ *agonist* properties. Gu-DIPP-NH$_2$ and Gu-DIPP-NH$_2$[Ψ] with a mixed μ agonist/δ *agonist* profile are of interest because there is evidence to indicate that compounds with this profile may also be effective for the treatment of pain with reduced side effects [3].

In comparison with DIPP-NH$_2$, its Lys^3-analogue, H-Dmt-Tic-Lys-Phe-NH$_2$, showed similar subnanomolar δ receptor binding affinity, 16-fold lower μ receptor affinity and >1000-fold higher κ receptor affinity. In the functional assays it displayed a mixed μ agonist/δ antagonist profile similar to that of DIPP-NH$_2$. Its amidation produced a compound,

Table 1. *In vitro opioid activity profiles of guanidinylated TIPP and TIPP-NH$_2$ analogues.*

Compound	Binding K_is, nM			MVD		GPI
	δ	μ	κ	K_e, nM	IC_{50},nM	IC_{50}, nM
Gu-Tyr-Tic-Phe-Phe-OH	2.29	875	>5000	PA (33%)[a]		inactive
H-Tyr-Tic-Phe-Phe-OH[b]	1.22	1720	>1000	4.80		inactive
Gu-Dmt-Tic-Phe-Phe-OH	0.146	126	2260		1.57[c]	1050
H-Dmt-Tic-Phe-Phe-OH[b]	0.248	141	>1000	0.196		inactive
Gu-Tyr-TicΨ[CH$_2$NH]Phe-Phe-OH	0.968	704	>1000	PA (50%)[a]		inactive
H-Tyr-TicΨ[CH$_2$NH]Phe-Phe-OH[b]	0.308	3230	>1000	2.89		inactive
Gu-Dmt-Tic-OH	2.66	15.0	>1000	PA (50%)[a]		PA (50%)[a]
H-Dmt-Tic-OH[b]	1.64	1360	>1000	6.55		inactive
Gu-Dmt-Tic-Phe-Phe-NH$_2$	0.146	0.518	35.8		1.72	8.09
H-Dmt-Tic-Phe-Phe-NH$_2$[b]	0.118	1.19	>1000	0.209		18.2
Gu-Dmt-TicΨ[CH$_2$NH]Phe-Phe-NH$_2$	0.789	1.02	40.1		0.750	1.22
H-Dmt-TicΨ[CH$_2$NH]Phe-Phe-NH$_2$[b]	0.447	0.94	>1000	0.537		7.71
Gu-Dmt-Tic-Lys-Phe-NH$_2$	2.20	0.354	1.69		4.69	0.289
H-Dmt-Tic-Lys-Phe-NH$_2$	0.306	19.4	2.23	2.18		16.5
Gu-Dmt-TicΨ[CH$_2$NH]Lys-Phe-NH$_2$	1.83	18.1	24.7		7.35[c]	9.46
H-Dmt-TicΨ[CH$_2$NH]Lys-Phe-NH$_2$	2.73	9.11	2.06		23.2[c]	5.69

[a]*Partial agonist (% of maximal inhibition of contractions).* [b]*Data taken from ref. [1].* [c]*Partial agonist (IC$_{35}$).*

Gu-Dmt-Tic-Lys-Phe-NH$_2$, showing 7-fold lower δ receptor affinity, 55-fold higher μ receptor affinity and similar κ receptor affinity, but again agonist activity in the MVD assay. In comparison with H-Dmt-Tic-Lys-Phe-NH$_2$, the pseudopeptide H-Dmt-TicΨ[CH$_2$NH]Lys-Phe-NH$_2$ had a similar receptor binding profile, except for its 9-fold lower δ receptor affinity, but it behaved as a δ partial agonist in the MVD assay. Its guanidinylated derivative, Gu-Dmt-Tic[CH$_2$NH]Lys-Phe-NH$_2$, showed a similar profile in the receptor binding and functional assays. The Lys^3-analogues essentially are mixed μ/κ agonists with varying efficacy at the δ receptor. Compounds with mixed μ/κ opioid activity have therapeutic potential for treatment of cocaine abuse [4].

In conclusion, guanidinylation of various TIPP peptides produced major changes in the opioid activity profiles with particularly drastic effects on δ receptor efficacy.

Acknowledgements

Work supported by grants from the CIHR (MOP-89716) and the NIH (DA004443).

References

1. Schiller, P.W., et al. *Biopolymers (Peptide Sci.)* **51**, 411-425 (1999).
2. Schiller, P.W., et al. *J. Med. Chem.* **42**, 3520-3526 (1999).
3. Coop, A. and Rice, K.C. *Drug News Perspect.* **13**, 81-487 (2000).
4. Mello, N.K., Negus, S.S. *Ann. N.Y. Acad. Sci.* **909**, 104-132 (2000).

Proceedings of the 23rd American Peptide Symposium
Michal Lebl (Editor)
American Peptide Society, 2013

Influence of the Isoleucine and Histidine Residues in Anti-Plasmodium Activity of the Angiotensin II Analogs

Adriana F. Silva[1], Luiz H. F. Ferreira[1], Margareth L. Capurro[2], and Vani X. Oliveira Jr[1]

[1]*Universidade Federal do ABC, Santo André,09210-170, Brazil;* [2] *Departamento de Parasitologia, Universidade de São Paulo, São Paulo,05508-000, Brazil*

Introduction

Angiotensin II (AII, Asp-Arg-Val-Tyr-Ile-His-Pro-Phe) is a natural peptide with pressor activity, which also features anti-plasmodium action in *Plasmodium gallinaceum* [1]. In order to find an effective antiplasmodium compound, this study refers to four AII analogs that presented scan study with Ile and His residues, removing selectively or inverting these amino acids to understand the hydrophobic cluster influence, explained by Tzakos, et al. [2]. The peptides were synthesized by the Fmoc solid phase strategy, purified by RP-HPLC and characterized by LC/ESI-MS. The conformational studies were performed by circular dichroism (CD – data not shown).

Results and Discussion

Lytic activity assays were performed using mature sporozoites, collected from salivary glands of *A. aegypti* infected. The sporozoites were incubated with each peptide for 1 hour at 37°C, and the cell membrane integrity was monitored by fluorescence microscopy. The results are shown in Figure 1. Purity percent of the AII analogues determined by HPLC and mass spectrometry are shown in Table 1.

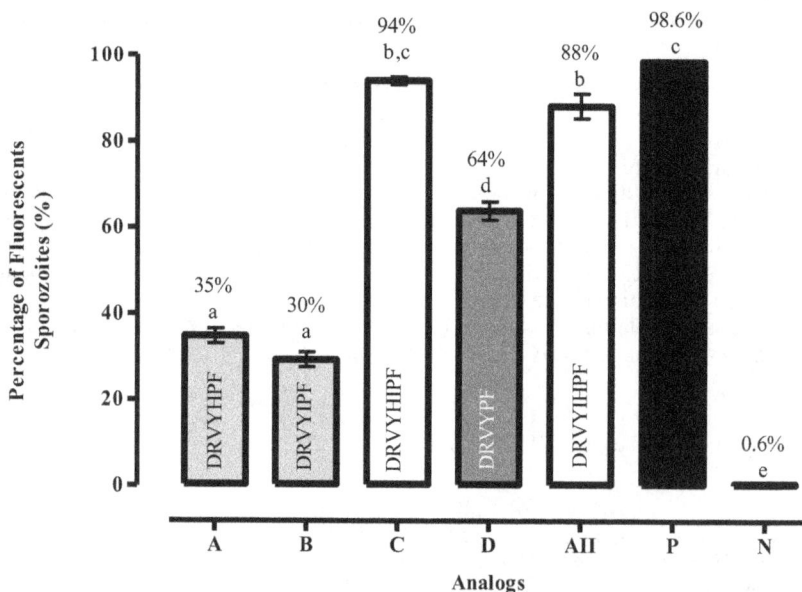

Fig. 1. Effects of the AII analogs on membrane permeability of the mature sporozoites. Data are presented as mean ± standard deviation of fluorescent sporozoites percentage/blade (n=9). Different letters indicate significant difference between the groups treated with AII, peptide analogues, positive control group (P, treated with digitonin and PBS) and negative (N, treated with PBS), respectively (ANOVA followed by Tuckey test, p<0,05).

Table 1. Purity percent of the AII analogues determined by HPLC and mass spectrometry.

Letter & Name	Sequence	HPLC Purity^a (%)	Calcd mass^b (Da)	Obsd mass^b (Da)
A – des-Ile^6-AII	DRVYHPF	99	933	934
B – des-His^5-AII	DRVYIPF	99	909	910
C – [Ile^6,His^5]-AII	DRVYHIPF	98	1045	1046
D – des-Ile^6,His^5-AII	DRVYPF	98	795	796

^{a}HPLC profiles were obtained under the following conditions: Column Supelcosil C18 (4.6 x 150 mm), 60 Å, 5 μm; Solvent System: A (0.1% TFA/H$_2$O) and B (0.1% TFA in 60% ACN/H$_2$O); Gradient: 5-95% B in 30 minutes, Flow: 1.0 mL/min; λ=220 nm; Injection Volume: 50 μL and Sample Concentration: 1.0 mg/mL. ^{b}The observed mass were determined by LC/ESI-MS using a Micromass instrument, model ZMD coupled on a Waters Alliance, model 2690 system. Mass measurements were performed in a positive mode in the following conditions: mass range between 500 and 2000 m/z; nitrogen gas flow: 4.1 L/h; capillary: 2.3 kV; cone voltage: 32 V; extractor: 8 V; source heater: 100°C; solvent heater: 400°C; ion energy: 1.0 V and multiplier: 800 V.

We observed that in analog C (94% antiplasmodium activity), Tyr is close to imidazole group of His that could promote a hydrogen bond formation. Besides that, van der Waals interactions could occur between Ile and Phe residues due to its proximity and non-polar characteristic and these simultaneous interactions may be responsible for increase of the bioactivity of this analog. These interactions could not be effective in native AII (88%) [1] nor in other analogs studied here, because Ile and His promote a steric influence on the organization of Phe and Tyr residues [3]. CD data showed that only analog C adopted a β-turn conformation in aqueous solvent - the same conformation as the AII. We conclude that hydrophobic cluster modifications and interactions of amino acid side-chains influences the biological activity against *Plasmodium gallinaceum*.

Acknowledgments

This research was supported by FAPESP.

References

1. Maciel, C., Oliveira, V.X., Fázio, M.A., Nacif-Pimenta, R., Miranda, A., Pimenta, P.F., Capurro, M.L. *PLoS ONE* **3**, e3296 (2008).
2. Tzakos, A.G., Bonvin, A.M.J.J., Troganis, A., Cordopatis, P., Amzel, M.L., Gerothanassis, I.P., van Nuland, N.A.J. *Eur. J. Biochem.* **270**, 849-860 (2003).
3. Fermandjian, S., Sakarellos, C., Piriou, F., Juy, M., Toma, F., Thanh, H.L., Lintner, K., Khosla, M.C., Smeby, R.R., Bumpus, F.M. *Biopolymers* **22**, 227-231 (1983).

Proceedings of the 23rd American Peptide Symposium
Michal Lebl (Editor)
American Peptide Society, 2013

Cell Penetrating Peptides Mediate Internalization and Accumulation of Bioactive Cargoes into Lipid Vesicles

Jean-Marie Swiecicki, Annika Bartsch, Julien Tailhades, Christelle Mansuy, Fabienne Burlina, Gérard Chassaing, and Solange Lavielle

Laboratoire des BioMolécules, UMR 7203, Université Pierre et Marie Curie – CNRS – Ecole Normale Supérieure, 24 rue Lhomond, 75005, Paris, France

Introduction

Cell penetrating peptides (CPPs) correspond to small cationic sequences, which have the unique capacity to cross cell membranes and deliver a cargo into cells. Since their discovery in the 90's, CPPs have been successfully used to transport *in vitro* and *in vivo* nucleic acids, proteins and nanoparticles into different cells. Nevertheless, the exact mechanism(s) by which CPPs cross biological membranes remain(s) to be elucidated. In living cells, there is some evidence for both energy-independent penetration ("direct translocation") and energy-dependent penetration (endocytosis) [1]. The direct translocation pathway is *a priori* the uptake mechanism that needs to be increased because it avoids endosomal trapping of the cargo associated to the CPP. Moreover, understanding direct translocation may help understanding an eventual escape from endosomes of the CPP-cargo conjugates.

To date, studies on the ability of CPPs to translocate across model lipid bilayers have given divergent results. Drin, et al. [2] found that the CPP Penetratin did not translocate significantly into small unilamellar vesicles. In contrast, fluorescence-microscopic studies have been reported, which suggest that Penetratin may cross the membranes of giant unilamellar vesicles [3]. Silvius, et al. demonstrated that Penetratin can translocate at significant rates into large unilamellar lipid vesicles (LUVs) only in the presence of a transbilayer potential [4]. More recently, Wimley, et al. with a high-throughput screening protocol found a new family of peptides that translocate across synthetic lipid bilayer, called spontaneous membrane-translocating peptides (SMTPs) [5].

In the present study, we have investigated the spontaneous translocation of different widely used CPPs in anionic (LUVs) (Figure 1) and compared their behaviour to the one of the antimicrobial peptide (AMP) Magainin. In addition, we have evaluated the capacity of CPPs to mediate the delivery of the bioactive peptide cargo PKCi into vesicles (Figure 1). This cationic peptide has been shown before to be unable to enter cells by itself.

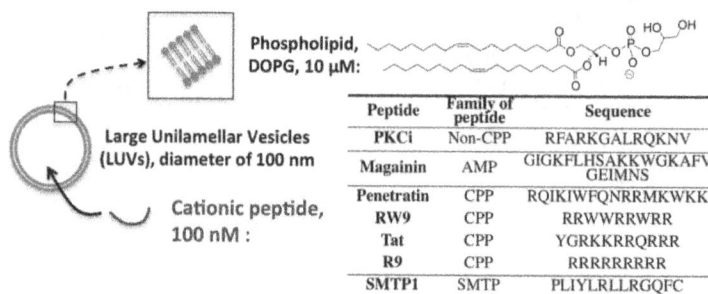

Phospholipid, DOPG, 10 µM:

Large Unilamellar Vesicles (LUVs), diameter of 100 nm

Cationic peptide, 100 nM :

Peptide	Family of peptide	Sequence
PKCi	Non-CPP	RFARKGALRQKNV
Magainin	AMP	GIGKFLHSAKKWGKAFV GEIMNS
Penetratin	CPP	RQIKIWFQNRRMKWKK
RW9	CPP	RRWWRRWRR
Tat	CPP	YGRKKRRQRRR
R9	CPP	RRRRRRRRR
SMTP1	SMTP	PLIYLRLLRGQFC

Fig. 1. Internalization studies have been performed using LUVs (100 nm diameter) produced by extrusion and composed exclusively by DOPG.

Results and Discussion

The peptides have been functionalized on their *N*-terminus by 7-chloro-4-nitro-benzofurazan (NBD-Cl) to assess their internalization inside LUVs. This fluorophore can be chemically reduced by dithionite, which does not enter vesicles. As a consequence, after incubation of the NBD-functionalized peptides with the vesicles, the extravesicular fluorescence can be totally quenched and the quantity of internalized peptide accurately measured (Figure 2a) [6].

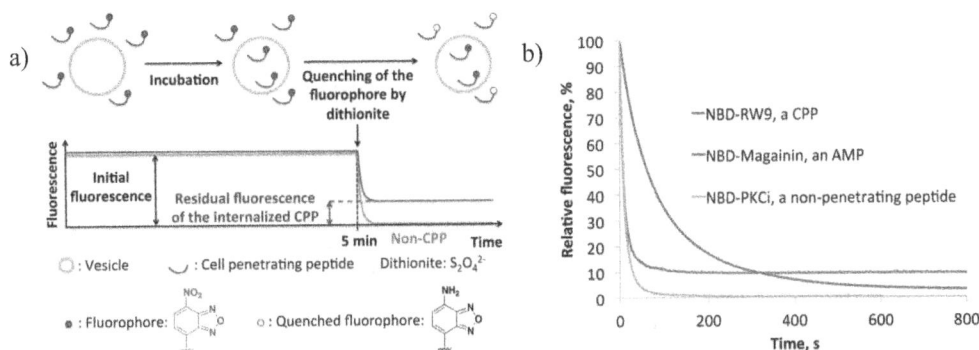

Fig. 2. a) General methodology for the determination of the quantity of internalized peptides. After 5 min incubation of the peptides (100 nM) with LUVs (DOPG, 10 µM) at 20°C, in PBS buffer, the fluorescence of the non-internalized peptide is quenched by addition of dithionite (to a final concentration of 10 mM). The fluorescence decay is then monitored over 800 s. The quantity of internalized peptide is deduced from the ratio between the fluorescence after quenching and the initial fluorescence. b) Experimental time-course measurement of the reduction by dithionite of various NBD-peptides incubated with vesicles.

The various cationic peptides show very distinct behaviors. The fluorescence of NBD-PKCi is quantitatively quenched showing that this peptide does not enter inside LUVs. In contrast, the fluorescence of the CPP NBD-RW9 is only partially reduced and reaches a plateau, meaning that 10% of the NBD-peptide is internalized (Figure 2b). All the investigated CPPs (and SMTP1) give similar results showing that they are internalized in the LUVs. The behavior of the AMP Magainin is different: the fluorescence decrease is much slower, probably due to the fact that this peptide is deeply buried into the membrane, and finally reaches zero. The quantitative reduction of this peptide has been explained by the formation of transmembrane pores, through which dithionite can passively diffuse [7]. As a consequence, it is unlikely that CPPs forms permanent transmembrane pores.

CPPs have been described to be efficient drug carriers. To characterize this process, we have conjugated the non-permeable peptide PKCi to the CPP RW9 through a disulfide bridge (NBD-RFARKGALRQKNVC-CRRWWRRWRR-NH₂). Interestingly, this drug-CPP conjugate is internalized in a very similar quantity as the free CPP: after 5 min incubation, 9% of the conjugate is located inside vesicles, corresponding to an intravesicular concentration of about 300 µM (using an initial extravesicular concentration of 100 nM). This demonstrates that CPPs are not only able to facilitate the penetration of non-penetrating biomolecules; they also induce their accumulation. The driving force of such an accumulation is most probably the high affinity of CPPs for negatively charged membranes.

In summary, we have shown that CPPs are able to penetrate inside LUV made of DOPG. Under the herein described conditions, CPPs exhibit a behavior different from the one of AMPs. In particular, CPPs are unable to form pores, in agreement with their low cell toxicity. Moreover, we have demonstrated that CPPs are able to promote the direct translocation and the accumulation of non-penetrating peptide inside vesicles.

Acknowledgments

Jean-Marie Swiecicki gratefully acknowledges the APS for financial support (travel grant).

References

1. Jones, A.T. and Sayers, E.J. *J. Control. Release* **161**, 582-591 (2012).
2. Drin, G., Cottin, S., Blanc, E., Rees, A.R., Temsamani, J. *J. Biol. Chem.* **278**, 31192-31201 (2003).
3. Thore´n, P.E., Persson, D., Karlsson, M., Norde´n, B. *FEBS Lett.* **482**, 265-268 (2000).
4. Terrone, D., Leung Wai Sang, S., Roudaia, L., Silvius, J.R. *Biochemistry* **43**, 13787-13799 (2003).
5. Marks, J.R., Placone, J., Hristova, K., Wimley, W.C. *J. Am. Chem. Soc.* **133**, 8995-9004 (2011).
6. Henriques, S.T., Melo, M.N., Castanho, M.A.R.B. *Mol. Membr. Biol.* **24**, 173-184 (2007).
7. Matsuzaki, K., Murase, O., Fujii, N., Miyajima, K. *Biochemistry* **35**, 11361-11368 (1996).

Proceedings of the 23rd American Peptide Symposium
Michal Lebl (Editor)
American Peptide Society, 2013

Urotensin-II Receptor Regulates Cell Mobility/Invasion and Determines Prognosis of Bladder Cancer

Paolo Grieco[1], Michele Caraglia[2], Silvia Zappavigna[2], Francesco Merlino[1], Ettore Novellino[1], Amalia Luce[2], Gaetano Facchini[3], Luigi Marra[3], Vincenzo Gigantini[4], and Renato Franco[4]

[1]Dept. of Pharmacy, University of Naples Federico II, Naples, 80131, Italy; [2]Dept. of Biochemistry, Biophysics and General Pathology, Second University of Naples, Naples, 80131, Italy; [3]Urogynecological Unit, Istituto nazionale Tumori "Fondazione G. Pascale"-IRCCS, Naples, 80131, Italy; [4]Pathology Unit, Istituto nazionale Tumori "Fondazione G. Pascale"-RCCS, Naples, 80131, Italy

Introduction

Non Muscle Invasive Bladder Transitional Cancer (NMIBC)/superficial and Muscle Invasive Bladder Transitional Cancer (MIBC)/invasive have different genetic profile and clinical course. Among NMIBC the prognosis is not completely predictable, since 20% of the cases experience a relapse, even in the form of MIBC and, therefore, the search for new molecular prognostic markers is urgently needed. In this light, we have recently reported that expression of Urotensin II (U-II) Receptor (UTR) is correlated to the prognosis of prostate adenocarcinoma [1]. U-II receptor (UTR), a GcRP protein binding the vaso-active endecapeptide urotensin II (U-II, H-Glu-Thr-Pro-Asp-c[Cys-Phe-Trp-Lys-Tyr-Cys]-Val-OH), is a seven-fold trans-membrane receptor. Moreover, it is involved in the differentiation and regulation of prostate cancer cells motility and invasion [2,3]. In the present study, we evaluated UTR expression in a series of bladder cancer cell lines and its involvement in the regulation of biological functions. Subsequently, we have assessed the role of UTR in bladder tumorigenesis and progression, considering NMIBC and MIBC phenotypes.

Results and Discussion

UTR expression was evaluated in four different cell lines of human bladder cancer RT112, T24, MCR and HT1376 by Western blotting (Figure 1). UTR protein was highly expressed in RT112 and, in particular, in T24 cell line whereas its expression was low in HT1376 and MCR (Figure 2).

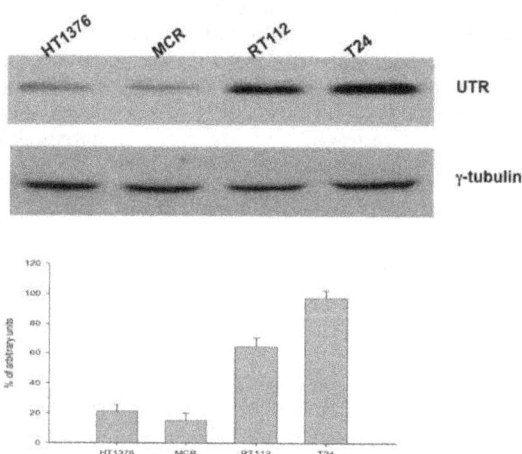

Fig. 1. UTR expression in bladder cancer cell lines.

To investigate the role of UTR in bladder cancer we also studied the biological functions of UTR on *in vitro* bladder cancer cells using either the UT-II antagonist urantide or knocking down UTR expression through the use of a specific shRNA in RT112 and T24 bladder cancer cells. We found that the down regulation of either the function or expression of UTR significantly blocked the motility and invasion of cancer cells. Interestingly, the addition of urantide to transfected cells potentiated the effects of shRNA-induced UTR downregulation in RT112.

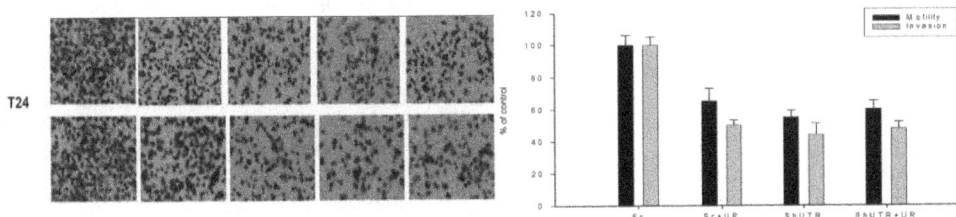

Fig. 2. *Effects of Urantide and UTR knock-down on motility and invasion of bladder cancer cells.*

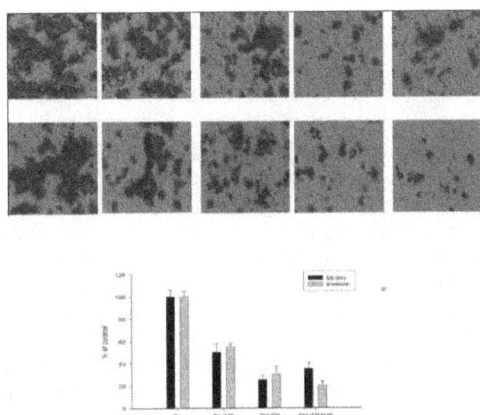

Fig. 3. *UTR expression was directly correlated with disease free survival (DFS).*

Successively, we have evaluated in a series of NMIBC a positive and significant correlation between low UTR expression and shorter DFS (Figure 3). In a multivariate analysis including UTR expression and grade, only UTR appeared to be an independent prognostic factor ($p= 0.025$).

In conclusion, we have studied the biological role of UTR in bladder cancer and obtained data suggesting its involvement in the regulation of cell motility and invasion.

These data suggest that U-II/UTR mediated pathway may play a role in bladder cancer progression. Moreover, UTR expression can be an independent predictive factor of progression in NMIBC patients. The evaluation of UTR expression can discriminate between NMIBC at high and low risk of relapse. The identification of new molecular markers in NMIBC, such as UTR, represents an important element to its prognostic definition.

Acknowledgments

PRIN 2010-11 (2010MCLBCZ_002) and FIRB (accordi di programma 2011, RBAP11884M_003). Regione Campania (Hauteville-Project C UP:B65EJIOOOI 80002).

References

1. Jemal, A., Siegel, R., Ward, E., Hao, Y., Xu, J., Thun, M.J. *Cancer Statistics, CA Cancer J. Clin.* **59**, 225-249 (2009).
2. Grieco, P., Carotenuto, A., Campiglia, P., Marinelli, L., Lama, T., Petacchini, R., Santicioli, P., Maggi, C.A., Rovero, P., Novellino, E. *J. Med. Chem.* **48**, 7290-7297 (2005).
3. Grieco, P., Franco, R., Bozzuto, G., Toccacieli, L., Sgambato, A., Marra, M., Zappavigna, S., Migaldi, M., Rossi, G., Striano, S., Marra, L., Gallo, L., Cittadini, A., Botti, G., Novellino, E., Molinari, A., Budillon, A., Caraglia, M. *J. Cell Biochem.* **112**, 341-353 (2011).

Proceedings of the 23rd American Peptide Symposium
Michal Lebl (Editor)
American Peptide Society, 2013

Importance of N-Terminal Extremity Restriction in the Antiplasmodial Activity of Angiotensin II

Marcelo D.T. Torres[1], Adriana F. Silva[1], Antonio Miranda[2], Margareth L. Capurro[3], and Vani X. Oliveira Jr.[1]

[1]Universidade Federal do ABC, Santo André, 09210-580, Brazil; [2]Universidade Federal de São Paulo, São Paulo, 04021-001, Brazil; [3]Universidade de São Paulo, São Paulo, 05508-000, Brazil

Introduction

Malaria affects around 500 million people annually causing 0.5-1 million deaths annually. Based on our previous studies [1], some angiotensin II (AII – sequence: DRVYIHPF) analogs were designed with a N-terminal extremity restriction, presenting antiplasmodial activity against *Plasmodium gallinaceum*. The introduction of conformational restrictors can potentiate the pharmacological activity, reducing the enzymatic degradation by elimination of the metabolized forms. Moreover, it can enhance selectivity by lowering the number of bioactive conformers. The formation of these bridges leads to a stabilized conformation, providing a turn in the molecule that may increase the action of antimicrobial peptides in the disruption of lipid membranes [2]. Lactam bridged analogs is a well-known strategy to study the conformation of peptides and its importance in biological activities [2]. In an attempt to increase this activity, we synthesized by solid phase method, cyclic analogs of AII with i-(i+2), i-(i+3), and i-(i+4) lactam bridge scaffold, using (D/E) and (K/O - ornithine) residues.

Results and Discussion

The cyclic analogs that contained i-(i+2) and i-(i+3) lactam bridge scaffold using (D) and (K) residues, presented higher antiplasmodial activity when the bridge was inserted next to N-terminal extremity, among them, the most active was DRDVKYIHPF (76%) [1,3]. Therefore analogs containing i-(i+4) lactam bridge presenting (E) as the acidic residue of the lactam bridge showed higher activity (81%) compared to those which contained (D), suggesting that the scaffold size affects directly the biological action, as well as the component of the lactam ring. It led us to the design of the other constrained analogs containing (E) and a basic residue (K/O) as bridgehead elements of i-(i+2) and i-(i+3) lactam bridge (Table 1). The insertion probably provided conformational changes by restricting the molecule in the N-terminal extremity and modifying its hydrophobic cluster formed by Tyr, Ile and His, which may have influence in the peptide-membrane interaction.

The restricted analogs can be separated in groups, according to their activity and conformation (Figure 1). The first one contains the more active peptides (groups a and e from Tukey's Test), which presented (E) as acidic bridgehead element in the N-terminal extremity of the molecule, bigger lactam rings and, preferentially, adopted β-turn conformation in MeOH, SDS, PBS, and TFE. The second group (analogs of groups d and f from Tukey's Test) showed lower antiplasmodial activity, smaller rings containing (E) or rings with (D) as acidic residue of the lactam bridge presented random conformations in the solvents tested.

Therefore we can highlight the importance of the lactam bridge position in these molecules, as well as its component residues to their antimalarial activity.

Table 1. Constrained peptides studied.

Analogs	Peptide Sequences	Fluorescent Sporozoites (%)
1	EDRVOYIHPF	81
2	DDRVKYIHPF	54
3	EDRVKYIHPF	77
4	DDRVOYIHPF	14
5	DREVKYIHPF	24
6	DREVOYIHPF	47
7	EDROVYIHPF	76
8	EDRKVYIHPF	75
9	DERVKYIHPF	80

Ornithine = O.

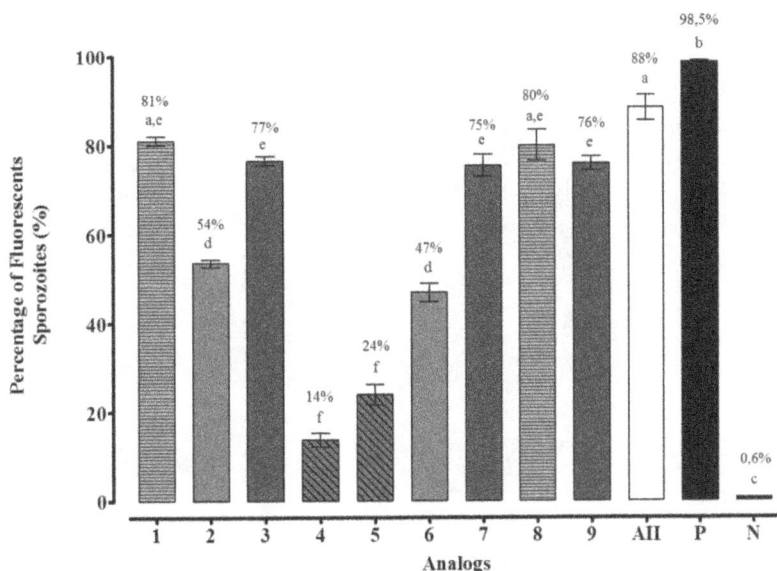

Fig. 1. Antiplasmodial activity of restricted analogs compared to AII in PBS media. Data generated by ANOVA (n = 9) and Tukey's Test (p < 0,0001). Fluorescence microscopy used to monitor the cell membrane integrity of the sporozoites, when incubated with the peptides for 1 hour at 37°C.

Acknowledgments

Supported by FAPESP.

References

1. Maciel, C., Oliveira, V.X., Fázio, M.A., Nacif-Pimenta, R., Miranda, A., Pimenta, P.F., Capurro, M.L. *PLoS ONE* **3**, 3296 (2008).
2. Taylor, J.W. *Biopolymers* **66**, 49-75 (2002).
3. Oliveira, V.X., Fázio, M.A., Silva, A.F., Campana, P.T., Pesquero, J.B., Santos, E.L., Costa-Neto, C.M., Miranda, A. *Regulatory Peptides* **172**, 1-7 (2011).

Proceedings of the 23rd American Peptide Symposium
Michal Lebl (Editor)
American Peptide Society, 2013

An Improved Synthesis of Hemiasterlin

Yu Luo, Haojun Jia, Qi Xu, Jinfeng Xu, Shawn Lee, and Xiaohe Tong

CPC Scientific Inc., 1245 Reamwood Ave, Sunnyvale, CA, 94089, U.S.A.

Introduction

Hemiasterlin is a special cytotoxic tripeptide isolated from marine sponges and composed of three special amino acids [1]. Hemiasterlin is a more potent *in vitro* cytotoxin and antimitotic agent than Taxol or vincristine, and preliminary experiments have demonstrated promising *in vivo* activity [2,3]. The total synthesis of Hemiasterlin was previously reported by several papers using a patented process requiring more than 18 tedious step reactions and generating very low total yield (<5%) [1-3]. Here we report an improved total synthesis route with 13 step reactions and an increased total yield (17%), which allows hundreds of mg of Hemiasterlin to be synthesized at one time.

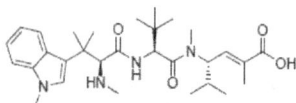

Fig. 1. The structure of Hemiasterlin.

Results and Discussion

In the synthesis of Hemiasterlin (Figure 2), **1** was treated with KHMDS and MeI in THF to obtain **2**, which was re-treated at the same condition for longer time to yield **3**. **3** was reduced with LiAlH$_4$ in THF to supply hydroxyl compound **4**, which was oxidized with Dess Martin

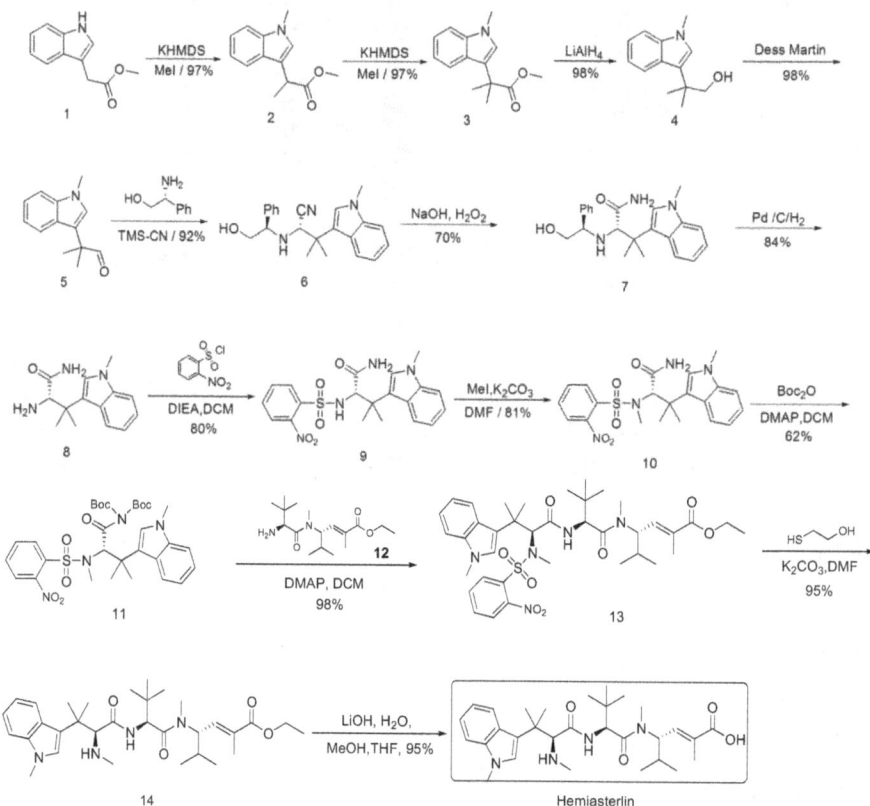

Fig. 2. Synthesis of Hemiasterlin.

reagent to form aldehyde compound **5**. Reaction of (R)-2-amino-2-phenylethanol with aldehyde **5** gave imine compound, which was directly transformed to cyanide **6** by TMS-CN addition. The hydrolysis of **6** with NaOH in the presence of H_2O_2 at room temperature obtained amide compound **7**. The benzyl group of **7** was removed by Pd/C hydrogenation to get **8**, which was protected with 2-nitrobenzene-1-sulfonyl group to supply compound **9**. N-methylation of **9** was carried out in MeI/K_2CO_3/DMF to obtain compound **10**, which was reacted with Boc_2O in presence of DMAP to supply bis-Boc compound **11**. Dipeptide **12** was made via a previously published process [1]. Compound **11** was reacted with **12** in presence of DMAP to give **13**, whose 2-nitrobenzene-1-sulfonyl group was de-protected by 2-mercaptoethanol/K_2CO_3/DMF to obtain **14**. **14** was then treated with base to receive Hemiasterlin with total yield of 17% and purity of 98% (Figure 3). ^{1}H NMR (400 MHz, CDCl$_3$): δ 7.93-7.88(m, 2H), 7.35-7.20(m,1H), 7.25-7.20(m,1H), 7.11-7.07(m,1H), 6.86(s,1H), 6.73(d,1H), 5.10(m,1H), 4.88 (d,1H), 3.76 (s,3H), 3.58(s,1H), 3.06(s,3H), 2.01(s,3H), 1.91(m,4H), 1.59(s,3H), 1.44(s,3H), 1.01(s,9H), 0.86(d,3H), 0.81(d,3H). MS(ESI): M/z for $C_{30}H_{46}N_4O_4$ 527.3 $(M+H)^+$ (Figure 4).

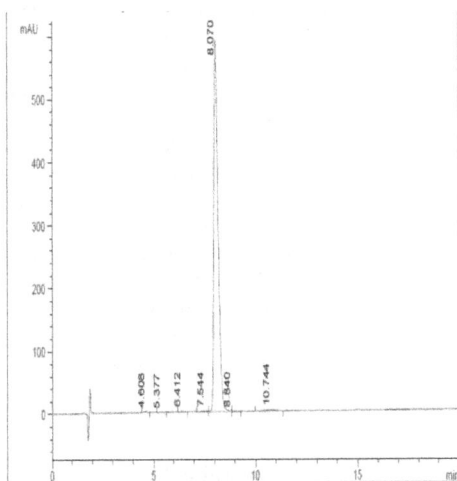

Fig. 3. RP-HPLC of Hamisterlin, A: 0.1%TFA in water, B: 0.09%TFA in (80% ACN+20% water), linear gradient 49-59% B in 20 min.

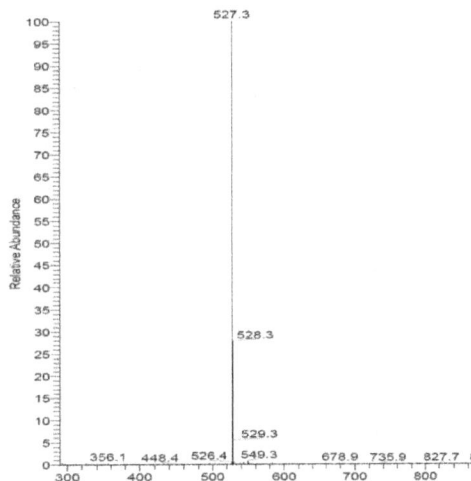

Fig. 4. Electrospray MS of Hamisterlin.

References

1. Andersen, R.J., Coleman, J.E. *Tetrahedron Letters* **38**, 317-320 (1997).
2. Vedejs, E., Kongkittingam, C. *J. Org. Chem.* **66**, 7355-7364 (2001).
3. Nieman, J.A., Coleman, J.E., Wallace, D.J., Piers, E., Lim, L.Y., Roberge, M., Andersen, R.J., *J. Nat. Prod.* **66**, 183-199 (2003).

Proceedings of the 23rd American Peptide Symposium
Michal Lebl (Editor)
American Peptide Society, 2013

HIV Viral Protein R (Vpr)-Derived Peptides Designed as HIV-1 Integrase Photoaffinity Ligands

Xue Zhi Zhao[1*], Mathieu Métifiot[2], Kasthuraiah Maddali[2], Steven J. Smith[3], Christophe Marchand[2], Stephen H. Hughes[3], Yves Pommier[2], and Terrence R. Burke, Jr.[1*]

[1]Chemical Biology Laboratory, National Cancer Institute-Frederick, Frederick, MD, 21702, U.S.A.;
[2]Laboratory of Molecular Pharmacology, Center for Cancer Research, National Cancer Institute,
National Institutes of Health, Bethesda, MD, 20892, U.S.A.; [3]HIV Drug Resistance Program,
National Cancer Institute-Frederick, Frederick, MD, 21702, U.S.A.

Introduction

HIV-1 integrase (IN) is a virally encoded polynucleotidyl transferase that inserts viral cDNA into the host genome through a process involving two sequential enzymatic steps, termed 3'-processing (3'-P) and strand transfer (ST) [1]. With FDA approval of Merck's raltegravir [2] and Gilead's elvitegravir [3], IN is a clinically validated target for anti-HIV therapy. Because drug resistant strains of HIV-1 emerge in raltegravir- and elvitegravir-treated patients, 'second-generation' inhibitors are being developed to treat patients failing regimens that involve the approved IN inhibitors. However, in experiments done in cultured cells, exposure to these second-generation inhibitors also leads to the selection of IN mutants with reduced susceptibility [4,5]. New compounds are needed that interact with IN in ways that are not affected by the extant resistance mutations. The HIV-1 viral protein R (Vpr) is a 96-residue multifunctional accessory protein that is important for viral replication in vivo. Recently, we developed a series of 7-mer peptides derived from Vpr that show low micromolar IN inhibitory potencies in vitro. Our current report details the development of biotinylated benzophenylalanine (Bpa)-containing Vpr-derived peptides designed as HIV-1 IN photoaffinity ligands.

Results and Discussion

Previous reports have detailed the in vitro screening of Vpr-derived peptide libraries for their ability to bind IN [6], and several IN-inhibitory peptides have been developed from Vpr that can block protein-protein interactions (PPIs) [7]. Recently, we identified the 7-mer Vpr (69-75)-derived peptide "FIHFRIG" (**1**), which has low micromolar IN inhibitory potency in vitro (Table 1). Identifying the binding site of these peptides on IN would be an important step that should facilitate the development of novel non-peptides inhibitors. Based on the previously reported use of benzophenone photoaffinity labeling to identify the binding site of another IN inhibitor [8], we are using a similar approach to identify the binding sites for the Vpr-derived peptides. We began by conducting a positional scan of **1** with photo-reactive Bpa residues. This resulted in the identification of several Bpa-containing peptides that retained low micromolar IN inhibitory potencies (**2 – 7**, Table 1). We then appended a biotin tag onto the most potent Bpa-containing peptide (**7**) to allow detection through cross-linking with avidin coupled with horseradish peroxidase (HRP) based immunovisualization (peptides

Table 1. Inhibitory Potencies of Benzophenylalanine (Bpa) scan peptide analogues based on Vpr-derived peptide **1** calculated from in vitro HIV-1 IN assays.

No.	Sequence	IC_{50} (μM, 3'-P)	IC_{50} (μM, ST)
1	FIHFRIG	17.6 ± 1.2	1.3 ± 0.3
2	FIHFR*Bpa*G	20 ± 3	4.3 ± 1.3
3	FIHF*Bpa*IG	> 333	43 ± 8
4	FIH*Bpa*RIG	34 ± 6	4.2 ± 0.7
5	FI*Bpa*FRIG	6.5 ± 0.9	2.2 ± 0.6
6	F*Bpa*HFRIG	37 ± 5	20 ± 3
7	*Bpa*IHFRIG	7 ± 1.2	2.1 ± 0.3

*Table 2. Inhibitory Potencies of biotin-tagged Vpr-derived Photoaffinity Ligands **8** and **9** calculated from in vitro HIV-1 IN assays.*

Benzophenone Photophore

Vpr-Derived FIHFRIG

Biotin-Tag

8. Linker a =

9. Linker b =

No.	Sequence	IC_{50} (μM, 3'-P)	IC_{50} (μM, ST)
8	Biotin-*Linker a-Bpa*-IHFRIG	32 ± 7	2.3 ± 0.3
9	Biotin-*Lnker b-Bpa*-IHFRIG	13 ± 2	1.8 ± 0.2

8 and **9**, Table 2). Both peptides **8** and **9** retain low micromolar IN inhibitory potencies, similar to the original 7-mer Vpr-derived peptide **1** and the Bpa-containing **7** (Table 1). Work is in progress to use these photoaffinity probes to identify the sites on IN where the modified peptides bind.

Acknowledgments

This work was supported in part by the Intramural Research Program of the NIH, Center for Cancer Research, Frederick National Laboratory for Cancer Research and the National Cancer Institute, National Institutes of Health.

References

1. Lewinski, M.K. and Bushman, F.D. In Jeffrey C. Hall, J.C.D.T.F. and Veronica van, H. (Eds.) *Advances in Genetics* Academic Press, 2005, Vol. 55, pp 147-181.
2. Summa, V., Petrocchi, A., Bonelli, F., Crescenzi, B., Donghi, M., Ferrara, M., Fiore, F., Gardelli, C., Gonzalez Paz, O., Hazuda, D.J., Jones, P., Kinzel, O., Laufer, R., Monteagudo, E., Muraglia, E., Nizi, E., Orvieto, F., Pace, P., Pescatore, G., Scarpelli, R., Stillmock, K., Witmer, M.V., Rowley, M. *J. Med. Chem.* **51**, 5843-5855 (2008).
3. Shimura, K., Kodama, E., Sakagami, Y., Matsuzaki, Y., Watanabe, W., Yamataka, K., Watanabe, Y., Ohata, Y., Doi, S., Sato, M., Kano, M., Ikeda, S., Matsuoka, M. *J. Virol.* **82**, 764-774 (2008).
4. Metifiot, M., Marchand, C., Maddali, K., Pommier, Y. *Viruses* **2**, 1347-1366 (2010).
5. Katlama, C., Murphy, R. *Expert Opin. Inv. Drug.* **21**, 523-530 (2012).
6. Gleenberg, I.O., Herschhorn, A., Hizi, A. *J. Mol. Biol.* **369**, 1230-1243 (2007).
7. Maes, M., Loyter, A., Friedler, A. *FEBS J.* **279**, 2795-2809 (2012).
8. Al-Mawsawi, L.Q., Fikkert, V., Dayam, R., Witvrouw, M., Burke, T.R., Jr., Borchers, C., Neamati, N. *Proc. Nat. Acad. Sci. USA* **103**, 10080-10085 (2006).

Proceedings of the 23rd American Peptide Symposium
Michal Lebl (Editor)
American Peptide Society, 2013

Identification of the Novel Bioactive Peptides for Drosophila Orphan GPCRs

T. Ida[1], H. Tominaga[1], E. Iwamoto[1], T. Sato[2], M. Miyazato[3], and M. Kojima[2]

[1]Interdisciplinary Research Organization, University of Miyazaki, 8891692 Miyazaki, Japan; [2]Division of Molecular Genetics, Institute of Life Science, Kurume University, 8390864 Fukuoka, Japan; [3]Department of Biochemistry, National Cerebral and Cardiovascular Center Research Institute, 5658565 Osaka, Japan

Introduction

There are many orphan G protein-coupled receptors (GPCRs), for which ligands have not yet been identified, in both vertebrates and invertebrates, such as *Drosophila melanogaster*. Identification of their cognate ligands is critical for understanding the function and regulation of such GPCRs. Indeed, the discovery of bioactive peptides that bind GPCRs has enhanced our understanding of mechanisms underlying many physiological processes. Here, we identified five endogenous ligands of the *Drosophila* orphan GPCR.

Results and Discussion

G-protein-coupled receptors (GPCRs) constitute a large protein superfamily that shares a 7-transmembrane motif as a common structure. Human genome sequencing has identified several hundred orphan GPCRs for which ligands have not yet been identified. GPCRs play crucial roles in cell-to-cell communication involved in a variety of physiological phenomena and are the most common target of pharmaceutical drugs. Therefore, the identification of endogenous ligands for orphan GPCRs will lead to clarification of novel physiological regulatory mechanisms and potentially facilitate the development of new GPCR-targeted therapeutics. The discovery of novel endogenous ligands for orphan GPCRs in mammals is currently challenging, possibly because of the restricted timing of expression or distribution of GPCR ligands. The recent sequencing of the *Drosophila melanogaster* genome has enabled the identification of at least 160 fly GPCRs. *Drosophila* is an excellent animal model for genetic analysis of developmental and behavioral processes, as it is a small, genetically modifiable organism with a relatively short lifecycle and can be bred easily under laboratory conditions. Structural or sequence comparison of newly discovered peptides in *Drosophila* with candidate molecules in mammals may lead to the discovery of new peptide signaling modules. Here, we succeeded to biochemically purify the endogenous ligands of *Drosophila* CG30106, CG14593, CG5811 and CG34381 from whole *Drosophila* homogenates using functional assays with the reverse pharmacological technique, and identified their primary amino acid

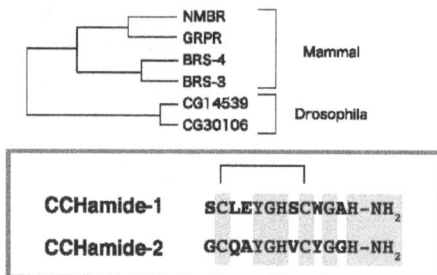

Fig. 1. Phylogenic tree of the Drosophila and mammal GPCRs and the amino acid sequences of CCHamide-1 and CCHamide-2.

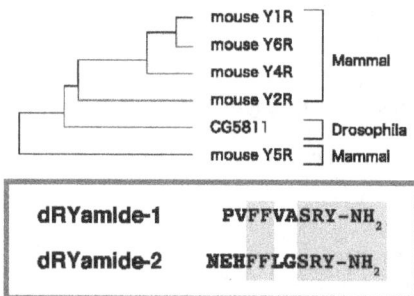

Fig. 2. Phylogenic tree of the Drosophila and mammal GPCRs and the amino acid sequences of dRYamide-1 and dRYamide-2.

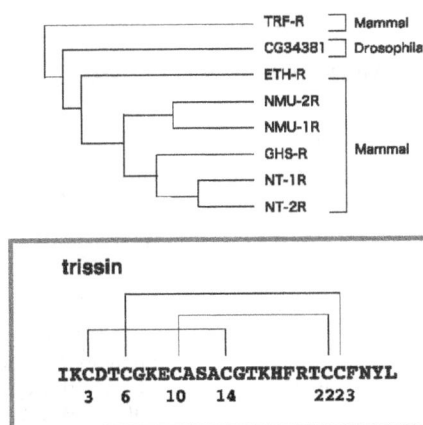

Fig. 3. Phylogenic tree of the Drosophila and mammal GPCRs and the amino acid sequence of trissin.

sequences. The purified ligands had been designated CCHamide-1, CCHamide-2 (Figure 1), dRYamide-1, dRYamide-2 (Figure 2) and trissin (Figure 3) [1,2,3].

CCHamides

Two *Drosophila* GPCRs CG14593 and CG30106 belong to the bombesin receptor subtype 3 (BRS-3) phylogenetic subgroup. BRS-3 is the orphan receptor in mammals and primarily expressed in the hypo-thalamus and plays a role in the onset of diabetes and obesity. Although the ligand for this receptor is widely sought, it has not yet been found. Here, we have identified CCHamide-1 and CCHamide-2, which are ligands for GPCRs CG30106 and CG14593, respectively, in *D. melano-gaster*. Injection of CCHamide-2 resulted in the stimulation of feeding motivation in blowflies. These bioactive peptides may provide new insights in the search for BRS-3 ligands and the elucidation of *D. melanogaster* feeding mechanisms.

dRYamides

The neuropeptide Y (NPY) family of peptides is widely conserved among vertebrates and has been implicated in feeding behavior, circadian rhythm, anxiety, and other physiological processes. *Drosophila* orphan receptor CG5811 shows some sequence similarities to NPY receptors. We identified dRYamide-1 and dRYamide-2, the ligands for CG5811 in *D. melanogaster*. dRYamides are homologs of vertebrate NPY family peptides. Our data suggest that dRYamides candidate factors for NPY family and the factor of feeding control in *Drosophila*.

trissin

We identified an endogenous ligand of the *Drosophila*. orphan GPCR, CG34381. The purified ligand is a peptide comprised of 28 amino acids with 3 intrachain disulfide bonds. We characterized the structure of intrachain disulfide bonds formation in a synthetic trissin peptide. Because the expression of trissin and its receptor is reported to predominantly localize to the brain and thoracicoabdominal ganglion, trissin is expected to behave as a neuropeptide. Cysteine-rich peptides are known to have antimicrobial or toxicant activities, although frequently their mechanism of action is poorly understood. The discovery of trissin provides an important lead to aid our understanding of cysteine-rich peptides and their functional interaction with GPCRs.

Acknowledgments

This work was financially supported in part by the Improvement of Research Environment for Young Researchers program of the Ministry of Education, Culture, Sports, Science and Technology; a grant for Scientific Research on Priority Areas from the University of Miyazaki; grants-in-aid from the Ministry of Education, Culture, Sports, Science and Technology, Japan.

References

1. Ida, T., et al. *Frontiers in Endocrinol. in press.*
2. Ida, T., et al. *Biochem. Biophys. Res. Commun.* **414**, 44-48 (2011).
3. Ida, T., et al. *Biochem. Biophys. Res. Commun.* **410**, 872-877 (2011).

Proceedings of the 23rd American Peptide Symposium
Michal Lebl (Editor)
American Peptide Society, 2013

Cleaved Intracellular SNARE Peptides are Implicated in a Novel Cytotoxicity Mechanism of Botulinum Serotype C

Jason Arsenault[1], Sabine A.G. Cuijpers[1], Enrico Ferrari[1,2], Dhevahi Niranjan[1], John A. O'Brien[1], and Bazbek Davletov[1,3]

[1]*MRC Laboratory of Molecular Biology, Cambridge, CB2 0QH, UK;* [2]*School of Life Sciences, University of Lincoln, Lincoln, LN6 7TS, UK;* [3]*Department of Biomedical Science, University of Sheffield, Sheffield, S10 2TN, UK*

Introduction

Recent advances in intracellular protein delivery have enabled more in-depth analyses of cellular functions. A specialized family of SNARE proteases, known as Botulinum Neurotoxins, blocks neurotransmitter exocytosis, which leads to systemic toxicity caused by flaccid paralysis. These pharmaceutically valuable enzymes have also been helpful in the

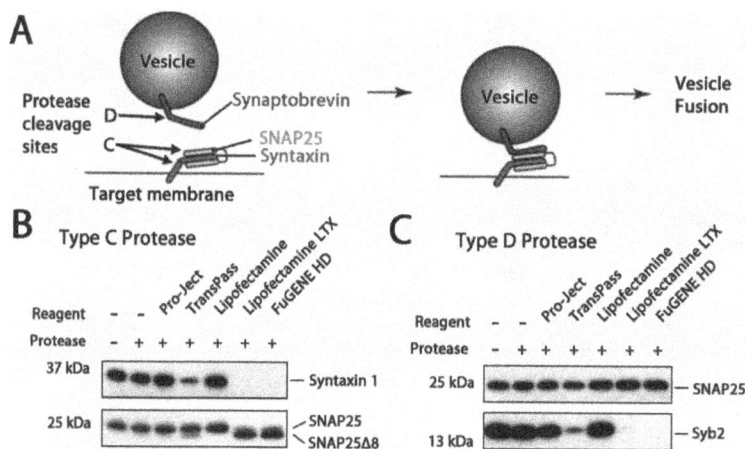

Fig. 1. (a) Schematic SNARE bundle formation. Western immunoblotting of syntaxin, SNAP25, and synaptobrevin for the type C protease; (b) and type D protease; (c) delivered with an array of transfection reagent.

study of SNARE functions. As can be seen in Figure 1A, SNARE bundle formation causes vesicle docking at the presynapse. Although these toxins are systemically toxic, no known cytotoxic effects have been reported with the curious exception of the Botulinum serotype C [1]. This enzyme cleaves intracellular SNAP25, as does serotype A and E, but also, exceptionally, cleaves Syntaxin 1. Using an array of lipid and polymer transfection reagents we were able to deliver different combinations of Botulinum holoenzymes into the normally unaffected, Neuro2A, SH-SY5Y, PC12, and Min6 cells to analyze the individual contribution of each SNARE protein and their cleaved peptide products.

Results and Discussion

Transfection reagents were able to abundantly and rapidly transduce SNARE proteases and peptides into cells [2]. Figure 1B shows the serotype C protease delivered into Neuro2A cells where it can cleave intracellular Syntaxin 1 and SNAP25 while Figure 1C shows the type D protease able to cleave synatobrevin. Using cell survival assays we also determined that the serotype C as well as the combination of type C and D caused profound cytotoxicity in the above-mentioned non-neuronal cell types that expands upon the observations seen on neuronal cells [1]. We observed that both apoptosis and necrosis mechanisms could be at play due to the appearance of morphonuclear abnormalities and increased propidium iodide staining respectively. Our results show that the freely released syntaxin and syntaxin plus

LTX	-	+	-	+	-	+	-	+	-	+	-	+	-	+	-	+
Complexin	-	-	+	+	-	-	-	-	-	-	-	-	-	-	-	-
Syntaxin	-	-	-	-	+	+	-	-	-	-	+	+	+	+	+	+
SNAP25	-	-	-	-	-	-	+	+	-	-	+	+	-	-	+	+
Synaptobrevin	-	-	-	-	-	-	-	-	+	+	-	-	+	+	+	+

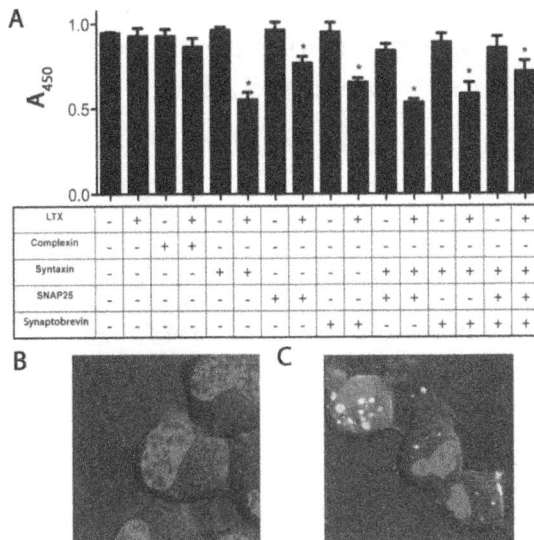

Fig. 2. (a) Cell survival assay of Neuro2A cells treated with SNARE peptide fragments. Neuro2A cells with FITC-syntaxin (45 aa) in the absence (b) or presence (c) of Lipofectamine LTX.

synaptobrevin peptides products, now separated from their transmembrane anchors and still able to form SNARE complexes (determined by GST-pull down experiments), could wreak havoc on the intracellular trafficking machinery. This mechanism was confirmed by the direct cellular penetration of the SNARE peptide fragments potentiated by the very same transfection reagents, Lipofectamine LTX. As can be seen in Figure 2A, SNARE peptides, mimicking the intracellular protease products, cause cytotoxic effects mirroring the loss of viability seen with the proteases. The viability of Neuro2A cells was strongly reduced in the presence of syntaxin as well as syntaxin plus synaptobrevin. The addition of all three SNARE subunits might partially rescue this cytotoxicity mechanism as they could stoichiometrically compete with each other preventing them from binding onto SNARE domains crucial for proper intracellular trafficking (Figure 2A last column). Figure 2B shows Neuro2A cells treated with the 45 aa fluorescent (FITC) syntaxin fragment while Figure 2C shows the same amount of FITC-syntaxin added in the presence of Lipofectamine LTX. A clear elevation of cell penetration can be observed. The freely diffusible syntaxin fragments, shown to cause cytotoxic effects in these tested cell types, were also shown to inhibit neurite outgrowth in differentiated PC12 cells. The treatment of ex vivo cortical cells with the holoenzymes and SNARE peptides showed a parallel cytotoxic effect and an observable Wallerian degeneration. Here we show for the first time that the type C's cytotoxic effect can also be observed in non-neuronal cells. Following our investigation we determine that the free floating, unanchored, syntaxin fragment is the exacerbating cytotoxic factor. Since syntaxin peptide fragments, compounded by synaptobrevin peptide fragments, cause cytotoxicity, we propose that the subsequent formation of erroneous SNARE complexes would have disastrous effects upon cell viability. These results open the door for a dual-acting pharmacological intervention that could simultaneously inhibit secretion while also destroying tumors cells. Our results thus open the door for translational trials of SNARE proteases and SNARE peptides as anti-neoplastic agents. This concomitant inhibition of exocytosis and cytotoxicity could also yield benefits for the targeted treatment of neuroendocrine disorders [3].

Acknowledgments

We would like to thank the supporting staff at MRC LMB for helping with this work. Supported by MRC grant U10578791.

References

1. Zhao, L.C., et al. Neuroreport 21, 14-18 (2010).
2. Kuo, C.L., et al. Toxicon. 55(2-3), 619-629 (2010).
3. Arsenault, J., et al. J. Neurochem. in press, (2013).

Proceedings of the 23rd American Peptide Symposium
Michal Lebl (Editor)
American Peptide Society, 2013

Solid/Solution Phase Peptide Synthesis of Novel Tris-PEGylated Reagents Using Aminocaproic Acid Spacers and Orthogonal Protecting Groups

Arthur M. Felix and Alicia Miller

Ramapo College of New Jersey, Theoretical and Applied Science, Mahwah, NJ, 07430, U.S.A.

Introduction

PEGylation is a process by which linear non-toxic polyethylene glycol (PEG) chains, composed of repeating $(CH_2-CH_2-O)_n$ subunits, are covalently attached to biologically important molecules such as peptides and proteins. The resultant PEGylated conjugates have been reported to possess increased solubility, increased resistance to proteolysis, improved pharmacokinetic and pharmacodynamic properties, and reduced renal clearance [1-4]. Branched PEGylated therapeutic drugs have been reported to exhibit even more protection against degradation [5]. Branched PEGylating reagents usually consist of two units of PEG attached to a trifunctional amino acid. We have been interested in designing and preparing novel tris-PEGylating reagents that will provide greater protection to the peptide or protein substrates to which they are conjugated. It is anticipated that tris-PEGylated peptides and proteins will exhibit even better biological profiles due to an increased resistance to enzymatic degradation [6].

Results and Discussion

Aminocaproic acid was used as a spacer-linker to maximize the distance between the PEGylation sites and produce less sterically hindered tris-PEGylating reagents. As shown in Figure 1 (bottom), the procedure was carried out by a combination of solid phase and solution phase peptide synthesis starting with Fmoc-Aca-Lys(Boc)-Wang-resin. This intermediate was divided in half for the synthesis of both tris-PEGylating reagents (III) and

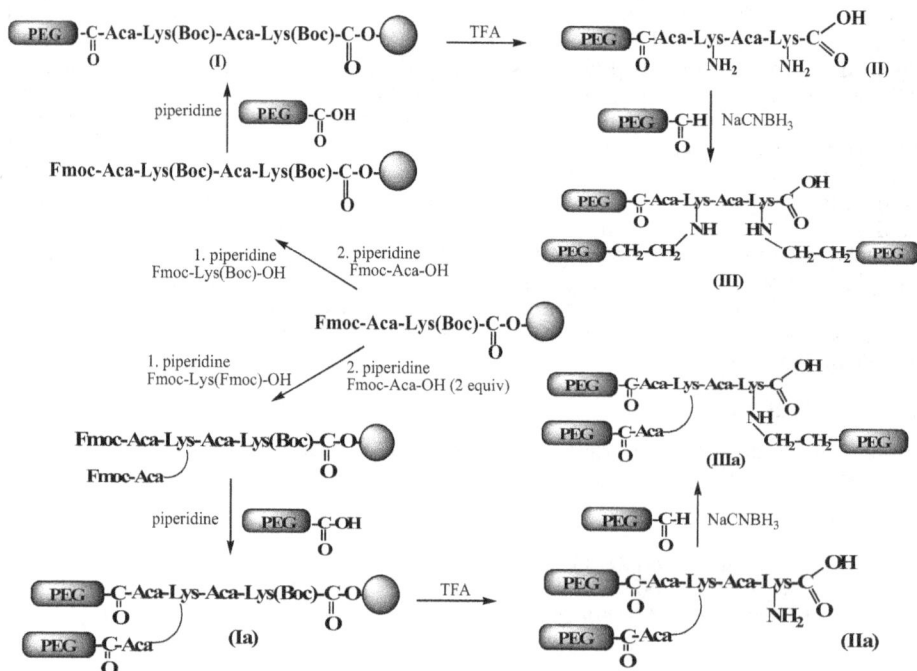

Fig. 1. Procedure for synthesis of tris-PEGylating reagents, III (top) and IIIa (bottom).

(IIIa). One portion of Fmoc-Aca-Lys(Boc)-Wang-resin was subject to two stages of SPPS to give Fmoc-Aca-Lys(Boc)-Aca-Lys(Boc)-Wang-resin. This was followed by deprotection with piperidine and solid phase PEGylation with PEG-COOH (mw 2000) using HBTU to give the mono-PEGylated intermediate, I. Following treatment of I with TFA, the deprotected intermediate, II, was subject to solution phase PEGylation using two equivalents of PEG-CHO (mw 2000) followed by reduction with $NaCNBH_3$ using aqueous trifluoroethanol [7] to yield the tris-PEGylating reagent, III.

The analogous tris-PEGylating reagent, IIIa, was synthesized from another portion of Fmoc-Aca-Lys(Boc)-Wang-resin (Figure 1, top). It was subject to two stages of SPPS to give the branched pentapeptide resin, [Fmoc-Aca]$_2$-Lys-Aca-Lys(Boc)-Wang-resin. This intermediate was deprotected with piperidine to generate two free NH_2-functions that were subject to solid phase PEGylation with PEG-COOH using HBTU to give the di-PEGylated intermediate, Ia. Treatment with TFA deprotected the remaining Boc-group and cleaved the PEGylated peptide from the resin giving intermediate IIa. This was subject to solution phase PEGylation using PEG-CHO followed by reduction with $NaCNBH_3$. The resulting tris-PEGylating reagent, IIIa, is structurally similar to III and may offer some additional steric advantages since the PEG groups are more sterically separated.

The free carboxylic acid components of these two novel PEGylating reagents are available for coupling to biologically active peptides and proteins. It is anticipated that tris-PEGylation of these peptides and proteins will improve their biological profiles.

Acknowledgments

This work was supported by funding from the Ramapo Foundation. We would also like to thank the American Peptide Society and the 23rd American Peptide Symposium organization team for a travel grant issued to Ms. Alicia Miller.

References

1. Harris, J.M. (Ed.) *Poly(ethylene glycol) Chemistry: Biotechnical and Biomedical Applications.* Plenum Press, New York, 1992.
2. Nucci, C.L., Shorr, R., Abuchowski, A. *Adv. Drug Delivery Res.* **6**, 113-151 (1991).
3. Delgado, C., Francis, G.E., Fisher, D. *Critical Reviews in Therapeutic Drug Carrier Systems* **9**, 249-304 (1992).
4. Katre, N.V. *Adv. Drug Delivery Res.* **10**, 91-114 (1993).
5. Kozlowski, A., Charles, S.A., Harris, J.M. *Biodrugs* **15**, 419-429 (2001).
6. Felix, A.M., Bandaranayake, R.M. *J. Peptide Res.* **65**, 71-76 (2005).
7. Felix, A.M. and Veech, S. In Lebl, M. (Ed.) *Breaking Away (Proceedings of the 21st American Peptide Symposium)*, American Peptide Society, San Diego, 2009, 211-212.

Proceedings of the 23rd American Peptide Symposium
Michal Lebl (Editor)
American Peptide Society, 2013

Non-Toxic Delivery and Gene Silencing by siRNA-Peptide Complex

Ayumi Takashina, Jyunichi Obata, Shutaro Fujiaki, and Masayuki Fujii*

Department of Biological & Environmental Chemistry, Kinki Univeristy, 11-6 Kayanomori, Iizuka, Fukuoka, 820-8555; Japan; mfujii@fuk.kindai.ac.jp

Introduction

Recently, small interfering RNA (siRNA), one kind of RNA interference (RNAi) technology represent the most common and, to date, the most effective method to inhibit target gene expression in human cells. It is also a common recognition that non-toxic delivery of siRNA is an urgent problem for the therapeutic application of siRNA. For the efficient gene silencing *in vivo*, prolonged circulation of siRNA with efficient and non-toxic cellular uptake and resistance against enzymatic degradation are indispensably required [1].

Telomerase activity has been regarded as a critical step in cellular immortalization and carcinogenesis and because of this, regulation of telomerase represents an attractive target for anti-tumor specific therapeutics.

Results and Discussion

In this paper, we present the efficient and non-toxic cellular uptake of siRNA using novel amphiphilic peptides and its application to the silencing of hTERT and bcr/abl in human cancer cell lines, Jurkat and K562.

As shown in Figure 1, siRNA-Pfectβ7 complex was efficiently taken up into cells. siRNA-Pfectβ7 complex was completely non-toxic against human cancer cells and the half-life time of siRNA was largely extended up to over 48 h in the complex, whereas naked siRNA and even siRNA in the complex with commercially available lipofection reagents were promptly degraded in a few hours (data not shown).

The silencing effect of siRNA targeting hTERT mRNA were also evaluated in Jurkat

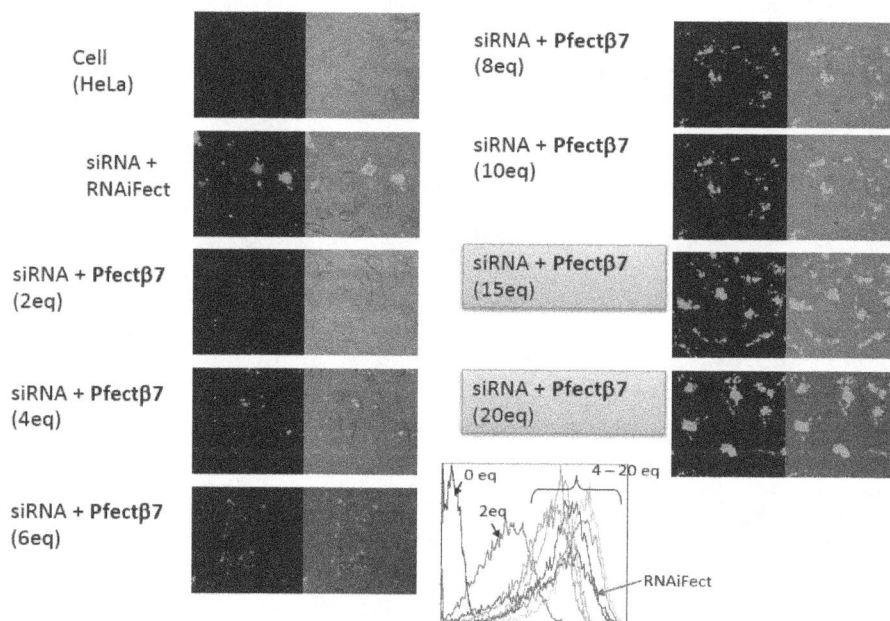

Fig. 1. Cellular uptake of siRNA –Pfectβ7 complex.

and HeLa cells, and the results were summarized in Figure 2. The complex of siRNA and some amphiphilic peptides or its hybrid with an intracellular transport signal peptides could be effectively taken up into cells. The complex also showed a high silencing effect against hTERT and bcr/abl mRNA. Moreover, the combination of siRNA-NES (nuclear export signal) peptide conjugates and the amphiphilic peptides improved silencing effects on the bcr/abl gene in K562 up to 95.2%.

Thus shown above, the complex of siRNA and some amphiphilic peptides or its hybrid with intracellular transport signal peptides could be effectively taken up into cells. The complex also showed a high silencing effect against hTERT and bcr/abl mRNA. The amphiphilic peptides and their hybrids showed almost no cytotoxicity and protected siRNA against intracellular nuclease digestion. We believe that the therapeutic application of these siRNA-amphiphilic peptide complexes is very promising.

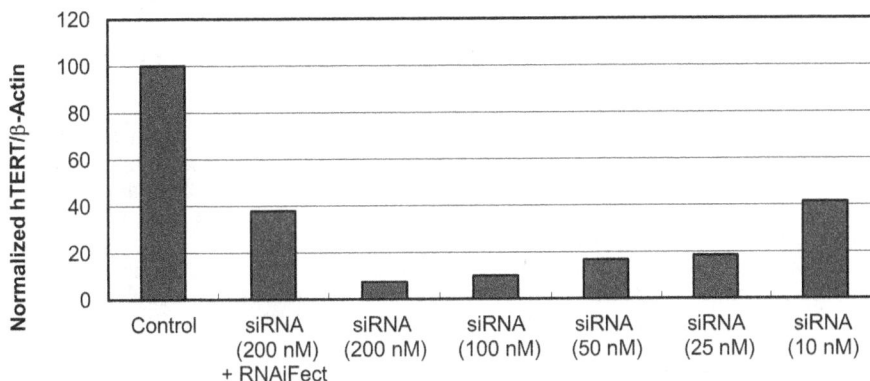

Fig. 2. Silencing of hTERT by siRNA-Pfectβ7 complex.

References

1. Gaynor, J.W., Campbell, B.J., Cosstick, R. *Chem. Soc. Rev.* **39**, 4169-4184 (2010).

Proceedings of the 23rd American Peptide Symposium
Michal Lebl (Editor)
American Peptide Society, 2013

Methotrexate Containing Oligopeptide Conjugates: Synthesis and *in vitro* Cytostatic Effect

Zoltán Bánóczi[1], Márton Flórián[1], Erika Orbán[1], Ildikó Szabó[1], and Ferenc Hudecz[1,2]

[1]*MTA-ELTE Research Group of Peptide Chemistry, Hungarian Academy of Sciences, Budapest; Hungary;* [2]*Department of Organic Chemistry, Eötvös Loránd University (ELTE), Budapest, H-1117, Hungary*

Introduction

Methotrexate (4-amino-10-methylfolic acid, MTX) was the first antimetabolite, used in the treatment of childhood acute lymphoblastic leukemia. After the internalization MTX inhibits key enzymes in the folic acid metabolism, therefore it disturbs the purine and pyrimidine biosynthesis, inhibits the DNA replication and causes cell death.

MTX undergoes polyglutamylation after the cellular uptake and these conjugates inhibit irreversibly the dihydrofolate reductase and the thymidylate synthase [1]. These derivatives containing three or more Glu residues (MTX-Glu$_n$) arrest the efflux of MTX from cells [2] and interestingly cannot be internalized into cells *via* the transport system responsible for MTX influx [3]. One reason responsible for MTX-resistance is the defective antifolate polyglutamylation due to decreased enzyme expression and/or inactivating mutations. Therefore the transport of MTX-Glu$_n$ into cells may result in effective drug which is able to avoid the development of resistance.

Cell penetrating peptides may be useful tools to deliver MTX-Glu$_n$ into cells [4]. In addition, oligo- or polypeptides coupled with MTX or its derivative could improve its biological activity, like antitumour or antileishmania effect [5-7].

Results and Discussion

Methotrexate or its pentaglutamylated derivatives were conjugated with penetratin and octaarginine as cell-penetrating peptides (CPP) with or without spacer (Gly$_3$). These conjugates were synthesized on Rink-amide MBHA resin using Fmoc/^{t}Bu strategy [8]. The final step was the coupling of MTX to the peptidyl-resin, which may result in two isomers dependent on how MTX is attached to the peptide - *via* its α- or γ-carboxylic group (Figure 1). The detection and isolation of isomers was possible only in few cases (see Table 2). In order to study the internalization ability of CPP conjugates with Glu$_5$ or Glu$_5$-Gly$_3$, carboxyfluorescein labeled derivatives were also prepared by solid phase peptide synthesis.

The cellular uptake of these conjugates was studied on human leukemia (HL-60) cells by flow cytometry. The results show that the presence of pentaglutamyl part in the conjugates decreases the internalization dramatically (Table 1). The insertion of the spacer (Gly$_3$) could increase the internalization, but in case of penetratin only at low concentration.

The cytostatic effect of MTX-conjugates was examined on HL-60 cells and also on different breast cancer cells (MCF-7 as MTX sensitive and MDA-MB-231 as MTX resistant cells) (Table 2). MTX was very effective against HL-60 and MCF-7 cells, but not on MDA-MB-231 cells. The conjugates exhibited modest cytostatic effect (IC$_{50}$= 4.96 – 20.04 μM) on HL-60 cells, except of MTX-Glu$_5$-Gly$_3$-Arg$_8$ possessing low activity. The isomer pairs had no very different IC$_{50}$ values. Data also suggest no influence of the nature of CPP studied in the conjugate on the cytostatic effect against HL-60 cells. In contrast, the appearance of the

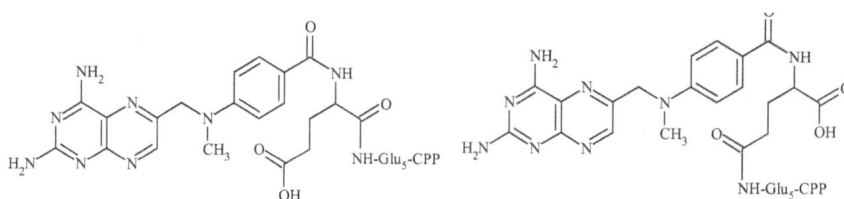

Fig. 1. Schematic structure of MTX(α)-Glu$_5$-CPP (left) , MTX(γ)-Glu$_5$-CPP (right).

Table 1. Internalization of Cf labeled Glu₅ peptide conjugates by HL-60 cells.

Compound	Fmean (sd)		Fluorescent cells % (sd)	
	1 µM	10 µM	1 µM	10 µM
Cf-Arg$_8$	2569(35)	185413(25267)	100 (0)	100 (0)
Cf-Glu$_5$-Arg$_8$	55(2)	493(31)	3 (0)	92 (2)
Cf-Glu$_5$-Gly$_3$-Arg$_9$	335 (21)	2881 (105)	62.8 (5.2)	100 (0)
Cf-PenC(desMet^{12})	4129(744)	22421(863)	100 (0)	100 (0)
Cf-Glu$_5$-Pen(desMet^{12})	172(27)	3450(336)	13 (2)	100 (0)
Cf-Glu$_5$-Gly$_3$-Pen(desMet^{12})	343 (12)	3540 (372)	73.9 (2)	100 (0)

HL-60 cells were treated for 90 min with the solution of compound. After washing and trypsin treatment, the fluorescence intensity of cells was measured by flow cytometry.

spacer in the octaarginine conjugate resulted in marked change. Although the presence of pentaglutamyl part decreased significantly the cellular uptake (Table 1), it had essentially no influence on the cytostatic effect (Table 2).

On MCF-7 cells only one of the isomers, MTX-Pen(desMet^{12}) (2) proved to be slightly cytotoxic. The pentaglutamylated MTX containing penetratin conjugate was more effective even than the free MTX on MCF-7 cells.

Table 2. IC$_{50}$ value of MTX and isomer MTX-conjugates on different cells.

Compound	IC50 (sd) (µM), HL-60	Compound	IC50 (sd) (µM) HL-60	IC50 (sd) (µM) MCF-7	IC50 (sd) (µM) MDA-MB-231
MTX	0.15 (0.007)			0.56 (0.57)	> 100*
MTX-Arg$_8$	15.00*	MTX-Pen(desMet^{12}) (1)	20.04 (1.3)	> 100*	72.6*
		MTX-Pen(desMet^{12}) (2)	8.82 (1.2)	50.4*	15.7*
MTX-Glu$_5$-Arg$_8$	16.62*	MTX-Glu$_5$-Pen(desMet^{12}) (1)	14.5 (2.7)	> 100*	0.09*
		MTX-Glu$_5$-Pen(desMet^{12}) (2)	10.73 (8.3)	n.d.	n.d.
MTX-Glu$_5$-Gly$_3$-Arg$_8$	73.1 (7.8)	MTX-Glu$_5$-Gly$_3$-Pen(desMet^{12}) (1)	13.9 (0.5)	n.d.	n.d.
		MTX-Glu$_5$-Gly$_3$-Pen(desMet^{12}) (2)	4.96 (0.3)	n.d.	n.d.

HL-60, MCF-7 and MDA-MB-231 cells were treated with the solution of conjugates for 3 hrs at 2.56×10^{-4}–100µM concentration range. After 3 days at 37C, MTT-assay was carried out. (2 parallel measurements, * only one measurement, n.d. no data).

In conclusion we found that the influence of pentaglutamylation on cytostatic effect of MTX depends mainly on the cell type, but Gly$_3$ spacer may have an effect too. However, we have identified a conjugate isomer of pentaglutamylated MTX (MTX-Glu$_5$-Pen(desMet^{12}) (1) that highly active on MTX resistant cells. This finding could open a new line of research for the treatment of resistant tumors.

Acknowledgments

This study was supported by grants from Hungarian Research Fund (K104385, PD-83923).

References

1. Schirch, V., et al. *Arch. Biochem. Biophys.* **269**, 371-380 (1989).
2. Zeng, H., et al. *Cancer Res.* **61**, 7225-7232 (2001).
3. Matherly, L.H., et al. *Vitamins and Hormones* **66**, 403-456 (2003).
4. Hudecz, F., et al. *Med. Res. Rev.* **25**, 679-736 (2005).
5. Bai, B.K., et al. *Bioconjugate Chem.* **19**, 2260-2269 (2008).
6. Kóczán, Gy., et al. *Bioconjugate Chem.* **13**, 518-524 (2002).
7. Silva, A., et al. *Trop. Med. Int. Health* **16**, 171-172 (2011).
8. Bánóczi, Z., et al. *Bioconjugate Chem.* **21**, 1948-1955 (2010).

Proceedings of the 23rd American Peptide Symposium
Michal Lebl (Editor)
American Peptide Society, 2013

Development of an Albumin-Bound PSA-Activated Delivery System of a Potent PACE4 Inhibitor for the Treatment of Prostate Cancer

Anna Kwiatkowska, Frédéric Couture, Christine Levesque, Kévin Ly, and Robert Day

Institut de pharmacologie de Sherbrooke (IPS) et département de chirurgie/service d'urologie, Faculté de médecine et des sciences de la santé (FMSS), Université de Sherbrooke, 3001, 12e Ave. Nord, Sherbrooke, J1H 5N4, Canada

Introduction

Recently we have developed a potent and selective inhibitor against the proprotein convertase PACE4, known as the Multi-Leu (ML) peptide inhibitor [1]. The development of this inhibitor was motivated by our previous report showing the overexpression of PACE4 and its validation as a drug target in prostate cancer [2]. Although the ML displays potent antiproliferative effects on prostate cancer cell lines (DU145 and LNCaP), improving its stability and potential tumor-targeting efficiency will be necessary in order to generate a novel ML-based anti-prostate cancer agent. To achieve this goal, we developed a tumor-specific delivery system for the ML-inhibitor. We used a prodrug concept exploiting endogenous albumin as a drug carrier [3]. We present the synthesis and biological evaluation of a series of ML-based prodrugs, consisting of (i) a maleimide group as a thiol-binding moiety, (ii) a prostate specific antigen (PSA) cleavable peptide linker, and (iii) an analogue of the ML-peptide inhibitor (Figure 1).

Results and Discussion

All designed peptides (ML prodrugs and their cleaved forms) were obtained by solid phase peptide synthesis according to standard coupling procedures and Fmoc/But strategy. The identity and purity of the compounds were determined by HPLC and MALDI-TOF mass spectrometry. In order to verify the coupling rate of ML-prodrugs for mouse serum albumin (MSA), HPLC-studies were performed. As an example chromatograms of incubation studies of EMC-RSSYYSL[4-Apaa]-ML with MSA are shown in Figure 2. The peptide was incubated with MSA at molar ratio 1:1. Next, samples were collected after 5 min, 1 h, and were analyzed by HPLC. Chromatograms of MSA and the prodrug are introduced as references. The signal corresponding to the free prodrug (29.6 min) disappeared completely after 1 h incubation with albumin, and a single peak eluting at the retention time of MSA (31.4 min) was observed. The other ML-prodrugs showed similar binding profiles.

The albumin conjugates of the ML prodrugs were incubated with enzymatically active PSA at different time points and the reaction efficiency was monitored by HPLC and MALDI-TOF. As an example chromatograms of incubation studies of the albumin conjugate of EMC-RSSYYSL[4-Apaa]-ML in the presence of human PSA are shown in Figure 3.

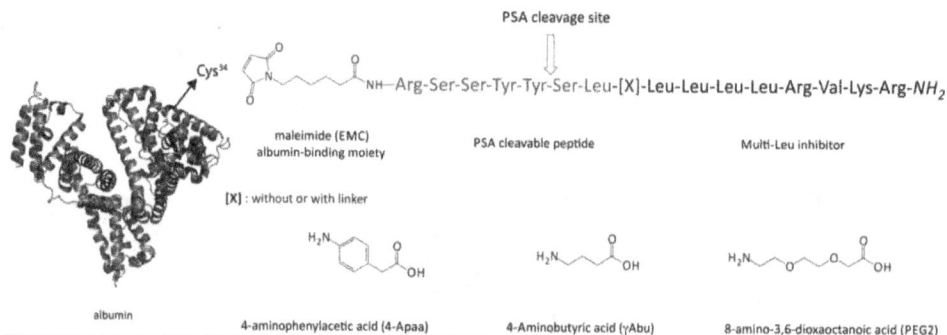

Fig. 1. Structure of the PSA-cleavable ML prodrugs.

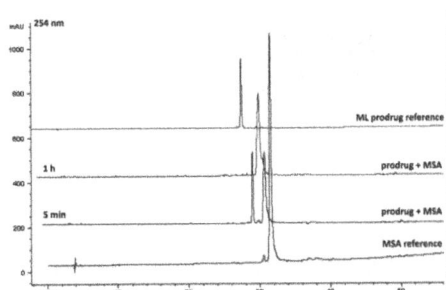

Fig. 2. Chromatograms of incubation studies of the ML-prodrug with mouse serum albumin (MSA).

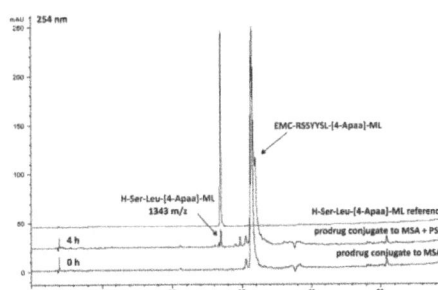

Fig 3. Chromatograms of incubation studies with our leading ML-prodrug in the presence of human PSA.

Table 1. Cleavage efficiency of the ML prodrugs conjugated to MSA.

ML-prodrug	cleavage % after 8h
MSA-EMC-RSSYYSL-ML	0
MSA-EMC-RSSYYSL-[4-Apaa]-ML	4
MSA-EMC-RSSYYSL-[γAbu]-ML	4
MSA-EMC-RSSYYSL-[PEG2]-ML	11

The cleavage properties of MSA-EMC-RSSYYSL[4-Apaa]-ML were evaluated using enzymatically active PSA and analyzed by HPLC. The cleavage product is included as a control. The chromatograms show the enzymatic cleavage of the albumin bound prodrug by PSA. After 4 h a new peak appears and its retention time (26.7 min) and its molecular weight (identified by MALDI-TOF) corresponds to the peptide: H-Ser-Leu-(4-Apaa)-ML.

As shown in Table 1, among all albumin-bound conjugates tested, only the prodrug with the 4-aminophenylacetic acid linker (4-Apaa) was efficiently cleaved by PSA (approximately 11%). Interestingly, no cleavage was observed for the conjugate design by the direct linkage of the ML analogue and a PSA-specific peptide. The other two conjugates (with a PEG2 or an γAbu linker) were cleaved less efficiently (approximately 4%).

Our study provides interesting insights for designing an albumin-bound ML-prodrug:

- The incorporation of a linker between the PSA-specific peptide and the ML-inhibitor is necessary for PSA cleavage.
- Among all the linkers, the aromatic 4-Apaa resulted in a prodrug with the best cleavage profile.

We conclude that our leading prodrug is a suitable candidate for further investigations (plasma stability and *in vivo* studies).

Acknowledgments

This work was awarded by Prostate Cancer Canada and is proudly funded by the Movember Foundation-Grant #2012-951 and the Canadian Cancer Society Grant #701590. AK gratefully acknowledges support from the Bentham Science Publishers Travel Award Grant. FC, CL, KL acknowledge the Fonds de Recherche du Québec-Santé (FRQS) for studentship support.

References

1. Levesque, C., et al. *J. Med. Chem.* **55**, 10501-10511 (2012).
2. D'Anjou, F., et al. *Transl. Oncol.* **4**, 157-172 (2011).
3. Elsadek, B., Kratz, F. *J. Control Release* **157**, 4-28 (2012).

Proceedings of the 23rd American Peptide Symposium
Michal Lebl (Editor)
American Peptide Society, 2013

Identification of FXI Binding Peptides by Solid Phase Peptide Library Screening: Structural and Functional Characterization

Søren Østergaard[1], Szu Wong[2], Susanne Bang[3], Jonas Emsley[2], and Henning Stennicke[1]

[1]Novo Nordisk A/S, Novo Research Park, 2760 Maaloev, Denmark; [2]Centre for Biomolecular Sciences, School of Pharmacy, University of Nottingham, United Kingdom; [3]Present Address: Biogen Idec Manufacturing, Biogen Idec Allé 1, 3400 Hilleroed, Denmark

Introduction

Solid phase peptide libraries also referred as the one-bead-one-compound (OBOC) approach [1] were used in the search for novel peptide ligands that could bind specifically to zymogen FXI and subsequently be used as an affinity purification tool. Array and X-ray contributed to the understanding of the interactions of these peptides at the molecular level.

Results and Discussion

Bead Library screening: In two generic libraries containing only proteinogenic amino acids only one consensus motif was readily identified. This core motif Asp-Phe-Pro was fixed in the design of a sub-library (X_7-DFP-X_4) with the aim to identify peptides with even higher affinity. Other critical residues were also identified, and one peptide from the sub-library, YPRHIYP**DFP**TDTT, was selected both due to its improved affinity and due to the presence of a histidine residue. It was anticipated that the histidine could be used as a pH dependent switch in the binding to FXI.

Array Screening: In order to determine the relative binding of the hits, all the peptide hits from the library screenings were synthesized as arrays on paper sheet using the spot method by Frank [2]. As expected most of the peptides selected from the sub-library also displayed relatively higher affinity compared to peptides from the generic libraries and ELISA measurement of selected peptides indicated up to a 100-fold increase in binding (data not shown). The affinity lead candidate YPRHIYPDFPTDTT was also subjected to substitution analysis using the array technology (Figure 1), and the result shows the importance of the amino acids in position 1 to 7 in the peptide. No attempt was made to analyze binding interactions on the *C*-terminal side of the DPF motif and in addition no particular preference was observed in the sub-library.

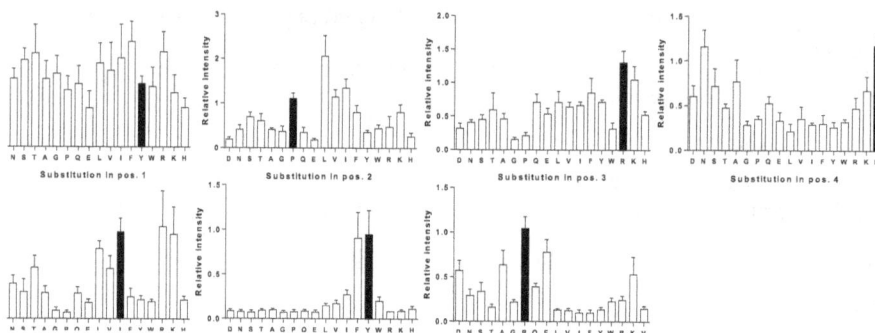

Fig. 1. Array analyses of peptide <u>YPRHIYP</u>DFPTDTT, underlined residues were subjected to replacement analysis. The bars in black indicate the residue present in the peptide. Recombinant FXI labeled with Alexa Fluor 488[TM] was used in the screening (at pH 7.4) and quantification was done using Typhoon scanner 9410 and analyzed by Array-Pro Analyzer.

Affinity Purification: Since the peptide was selected both due to its improved affinity and due to the presence of histidine, it was anticipated that histidine could be used as a pH dependent switch in binding to FXI. The binding of the peptide to FXI was analyzed as a function of pH and indeed a maximum binding was observed at pH above 7.5 whereas binding was lost at pH below 6.0, indicating the histidine was critical for binding. The peptide was synthesized directly on Toyopearl^{TM} AF-amino-650M by standard SPPS and was used for the affinity purification of FXI.

The recombinant FXI, which was prior to peptide affinity chromatography isolated by cation exchange (first purification step after fermentation), was loaded at pH 8.0, and after elution of unbound material, the pH was lowered to pH 5.0 and the FXI was isolated. Purity analysis of the FXI sample after cation exchange (Figure 2A) and subsequently after peptide affinity purification (Figure 2B) on peptide Toyopearl^{TM} affinity resin clearly demonstrates that this peptide is very successful in affinity purifying FXI under very mild conditions, and resulting in a high purity after only two purification steps. The obtained purity is comparable to affinity purification using a mAb column, but the peptide is far more stable and therefore this offers a more robust and cost-effective alternative.

X-ray: Protein X-ray crystallography reveals that peptides from the screen with binding motif Y_DFP bind across the anti-parallel β-sheet and opposite the α-helix within one of FXI's apple domain whilst following a hydrophobic groove along the surface which measures 17Å in length and 10Å in width. Phenylalanine of the DFP motif sits within a deep hydrophobic pocket central to the groove, and directly above FXI's β4 strand. A number of Van de Waals interactions are observable between the phenylalanine residue of the peptide and residues Thr132, Ala134, Phe138, His143, Leu148 and Leu163 that form the hydrophobic pocket.

The peptides also utilize its negatively charged Asp residue from the binding motif to interact with the positively charged surface above the β2 and β6 strands. The structure also reveals significant hydrogen bond network between the aspartate and tyrosine residues to the residues on the surface of FXI's apple domain.

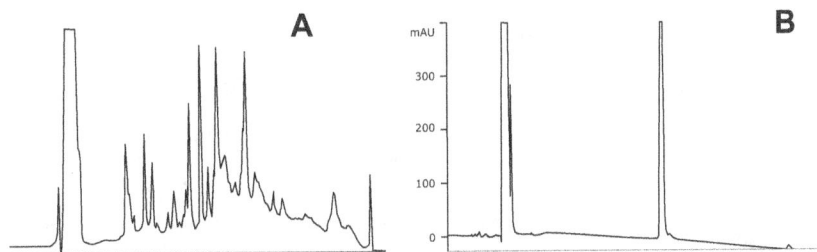

Fig. 2. Purity analysis after cation change of crude recombinant FXI (Fig. 2A) and after affinity purification on peptide-Toyopearl^{TM} (Fig. 2B). HPLC analysis: C4 Phenomenex, buffer A: 0,1% TFA, buffer B 0,07% TFA in AcCN , gradient 25-61% B (18 min.) flow: 1 ml/min.

Acknowledgments

Yvonne B. Madsen and Lisbet L. Hansen are thanked for their excellent technical assistance.

References

1. Lam, K.S. *Nature* **354**, 82-84 (1991).
2. Frank, R. *Tetrahedron* **48**, 9217-9232 (1992).

Proceedings of the 23rd American Peptide Symposium
Michal Lebl (Editor)
American Peptide Society, 2013

On-Demand Assembly of Macromolecules Used for the Design and Application of Targeted Secretion Inhibitors

Jason Arsenault[1], Enrico Ferrari[1,2], Dhevahi Niranjan[1], and Bazbek Davletov[1,3]

[1]MRC Laboratory of Molecular Biology, Cambridge, CB2 0QH, UK; [2]School of Life Sciences, University of Lincoln, Lincoln, LN6 7TS, UK; [3]Department of Biomedical Science, University of Sheffield, Sheffield, S10 2TN, UK

Introduction

Neurological and endocrine pathologies such as acromegalie, Cushing's disease, and neuropathic pain display disregulated exocytosis. Silencing specific cell populations would thus be invaluable to correct these debilitating disorders. To achieve this goal, we re-engineered the Botulinum neurotoxin (BoT), a highly potent pharmaceutical compound capable of inhibiting exocytosis, and fused to it a protein "stapling" domain [1,2]. These peptide motifs, that form an irreversible tetrahelical coiled-coil, are able to link a variety of targeting domains onto the enzyme and thus redirect it towards normally unaffected cells. The conformational diversity of this assembly process greatly supersedes traditional protein expression since multiple targeting domains (homo- and hetero-) can be linked onto one scaffold, larger yields can be produced separately, it permits the combination of solid-phase peptide synthesis with recombinant protein expression, and it can avoid the necessity of an N- to C- translational fusion. With only a few dozen building "blocks" it is possible to construct thousands of different complexes specifically tailored for each purpose as every individual component can be linked onto any other cognate stapling moieties.

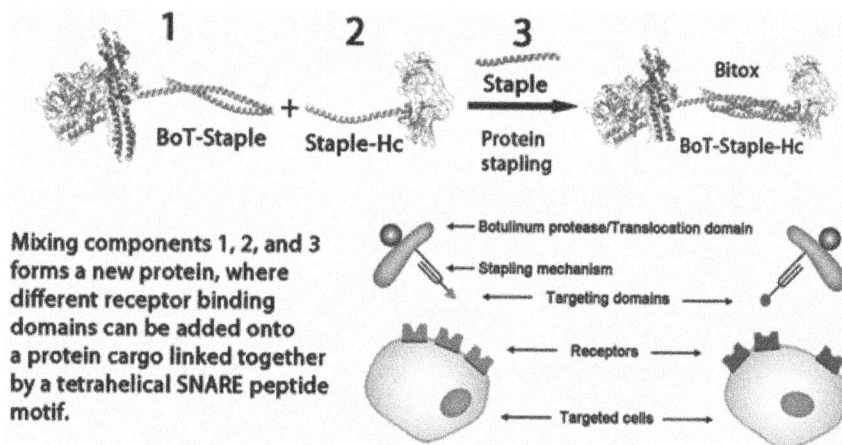

Fig. 1. Schematic representation of SNARE base protein stapling.

We then assayed a range of targeting domains composed of neuropeptides and factors sequences that would exploit each cell's surface receptor expression (see Figure 1).

Results and Discussion

Our results show that Corticotropin Releasing Hormone (CRH), Epidermal growth factor (EGF), Ciliary neurotrophic factor (CNTF), Vasoactive Intestinal peptide (VIP), as well as others, can potentiate the receptor mediated internalization of the BoT enzyme cargo into specific, and pathologically important, cell types [3]. Following this intracellular delivery, those cells displayed significant reductions of hormone and neurotransmitter release.

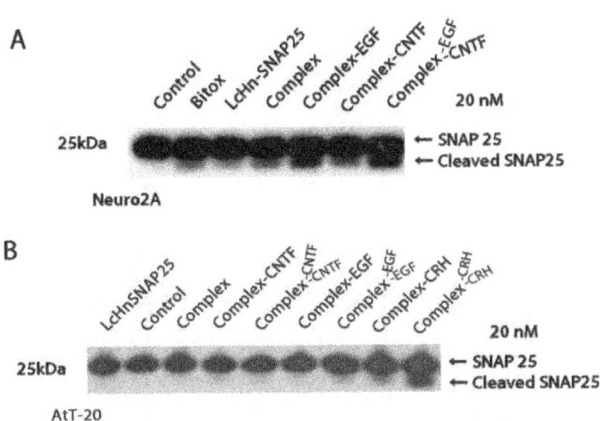

A

Control
Bitox
LcHn-SNAP25
Complex
Complex-EGF
Complex-CNTF
Complex-EGF-CNTF

20 nM

25kDa

← SNAP 25
← Cleaved SNAP25

Neuro2A

B

LcHnSNAP25
Control
Complex
Complex-CNTF
Complex-CNTF-CNTF
Complex-EGF
Complex-EGF-EGF
Complex-CRH
Complex-CRH-CRH

20 nM

25kDa

← SNAP 25
← Cleaved SNAP25

AtT-20

Fig. 2. a) Western immunoblotting of total SNAP25 of neuro2A cells. b) Western immunoblotting of total SNAP25 in AtT-20cells.

Using botulinum neurotoxin's highly sensitive efficacy readout, which is the intracellular cleavage of SNAP25, we have been able to characterize these new cargo amenable targeting domains. This strategy enabled the retargeting of BoNT/A to neuroendocrine cells that are normally unaffected by the native toxin's receptor binding domain. These new toxin complexes can bind to- and be internalized into PC12, Neuro2A, SH-SY5Y, AtT20, and Min6 cells as well as different subpopulations of ex-vivo rat cortical cells. Once internalized by receptor-mediated endocytosis, the enzyme cargo cleaves SNAP25 and thus inhibits exocytosis. Furthermore, the facility at which one can recombine the different targeting domains onto the enzymatic cargo permits us to explore novel conformations that protein recombination does not permit. As shown in Figure 2, using single EGF and CNTF targeting domains, domain on Neuro2A cells permits protease entry. However, when they are both included as a hetero bifunctional targeting domains onto a single complex, the efficacy of internalization is increased (Figure 2A). Such approaches could undoubtedly heighten the cell selectivity to the cargo delivery. While testing the corticotroph AtT-20 cells, which are amenable to CRH targeting, we saw that the homodimer targeting domains greatly enhanced the cytosolic entry compared to the monomer domain (Figure 2B). Such approach enables receptor dimer targeting while also elevating the probability of statistical rebinding. The functional targeting domains such as EGF, CNTF, CRH, and VIP, identified in this study, can be complexed with other stapled domains of diverse utilities (e.g. fluorophores, chelators, reactive groups). Combinations of stapled domains enable a "swiss-army knife" configuration to the pharmaceutical molecule. Such as ameliorated efficacy through heightened avidity, a strategy that remains intransigent for single strand protein expression. This targeted secretion inhibitor assembly platform has numerous biotechnological and medical implications [1,3]. Current endeavors are aimed towards testing our compounds in animal models, and subsequently, for clinical trials. Future work will be directed towards monitoring *in vivo* activity of our compounds to treat endocrinopathies and extend this approach to other pharmaceutically useful cargos.

Acknowledgments

We would like to thank the supporting staff at MRC LMB for helping with this work. Supported by MRC grant U10578791.

References

1. Darios, F., et al. *Proc Nat. Acad. Sci.* USA **107**, 18197-18201 (2010).
2. Ferrari, E., et al. *Bioconjug. Chem.* **23**, 479-484 (2012).
3. Arsenault, J., et al. *J. Neurochem.* in press, (2013).

Proceedings of the 23rd American Peptide Symposium
Michal Lebl (Editor)
American Peptide Society, 2013

Solid-Phase Synthesis of *C*-Terminal Peptide Libraries for Studying the Specificity of Enzymatic Protein Farnesyltransferase

Yen-Chih Wang and Mark D. Distefano

Department of Chemistry, University of Minnesota, Minneapolis, MN, 55455, U.S.A.

Introduction

Protein prenylation is a common post-translational modification of specific protein-derived cysteine residues in eukaryotic cells. Previous experiments with tetrapeptide sequences identified the so-called *C*-terminal CaaX box motif as the key structural element recognized by protein farnesyltransferase (PFTase) and geranylgeranyltransferase (GGTase) [1]. To study the substrate specificity of these enzymes, the primary strategy employed to date has involved the synthesis, purification and assaying of individual peptides [2]. As an improvement, here the construction of peptides containing free *C*-termini by SPOT synthesis is described. The specificity of protein farnesyltransferase was probed by screening the peptide arrays with the target enzyme.

Results and Discussion

We applied a previously developed method [3] to synthesize peptides containing free *C*-termini. In order to analyze peptides with free *C*-termini on solid phase, a linker should be installed so that the peptides can be cleaved from the beads and analyzed by MS. A photocleavable linker was chosen because it is orthogonal to other functional groups and the cleavage conditions require no other reagents except light. The photocleavable linker we chose has the advantage that the byproducts of photolysis remain on solid phase so that they do not interfere with the subsequent MS analysis (Figure 1).

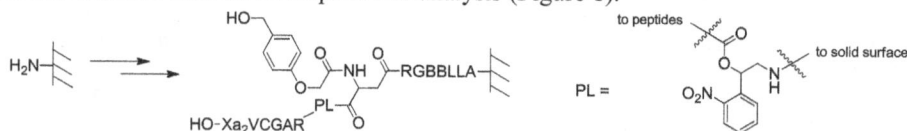

Fig. 1 Synthesis of inverted peptides.

Because *Candida albicans* is an opportunistic pathogen associated with a variety of diseases and the protein prenylation pathway is a potential new target for antifungal treatment, we are interested in comparing the substrate specificity of fungal and mammalian PFTases. The whole library was subjected to enzymatic prenylation by two different PFTases with an alkyne analogue of farnesyl diphosphate followed by click reaction with biotin-azide and binding to streptavidin-alkaline phosphatase. The membrane was then stained with BCIP (Figure 2a). The results showed that rPFTase can accept more peptide substrates than CaPFTase but there are a few peptides that showed higher activity for CaPFTase. Key individual peptides are being synthesized to determine the origin of the different reactivities for the two PFTases. The ultimate goal is to derive lead compounds for selective inhibitors of *C. albicans* PFTases as potential antifungal treatments.

Because the crystal structures of PFTase complexed with both substrates showed that the peptide and isoprenoid are in contact [1], we want to know if the peptide specificity changes when we change the structure of isoprenoid. Thus a CVa_2X library was subjected to rPFTase-catalyzed prenylation with alkyne analogues of farnesyl diphosphate with different lengths and visualized by the above method (Figure 2b). The results showed that rPFTase can prenylate more peptides with C15-alkyne than with C5-alkyne. This is consistent with the function of PFTase since FPP (15 carbons) is the enzyme's natural substrate. Our previous report [4] showed that C5-alkyne is a very slow substrate when the peptide substrate was CVIA which is also consistent with the screening results. Other peptides are being synthesized to compare the activity between C5-alkyne and FPP. Computational modeling is also being performed to explain how the isoprenoid structure influences the peptide specificity. This screening methodology should be useful for identifying many possible substrates for prenylation. This methodology should be applicable to a wide range of studies of enzyme-catalyzed *C*-terminal modification and subsequent processing.

(a)

	A	R	N	D	C	Q	E	G	H	I	L	K	M	F	P	S	T	W	Y	V
E	2 4	2 3	3 2	2 1	2 1	2 1	2 1	1 2	2 2	2 2	1 2	2 2	2 2	2 2	2 2	2 2	2 2	3 2	3 2	3 2
Q	47 45	3 4	26 29	3 2	46 12	24 33	2 2	6 3	45 25		59 47	3 6	7 12	66		9 35	60 34	65	3 4	41 26
D	2 2	3 2	3 2	3 2	3 2	3 2	3 2	3 2	3 2	2 2	2 2	3 2	3 2	1 2	2 2	2 2	2 2			
N	3 5	3 6	4 7	1 4	7 5	2 8	2 3	3 4	2 5	59 37	3 4	2 4	2 3	3 3	2 5	3 4	26 6	3 4	4 4	43 17
R	2 3	2 4	3 3	2 2	3 3	2 3	2 2	2 4	4 2	3 2	2 5	2 6	4 7	1 6	2 6	2 5	2 10	2 7	3 6	2 17
K	3 10	2 5	3 8	3 5	4 9	4 10	4 4	7 8	5 7	7 28	5 9	3 3	4 11	4 5	3 9	3 21	3 6	2 6	3 33	
H	3 4	3 6	5 7	2 5	6 6	3 7	2 5	3 5	2 6	57 30	3 6	2 5	2 4	2 3	2 5	21 7	58 10	4 5	4 5	45 25
A	4 5	3 4	3 3	2 2	17 4	3 5	1 2	1 3	7 10		6 6	3 6	4 6	29 29	2 7	3 7	42 11	2 4	4 5	
V	4 16	3 4	2 8	2 5	5 12	2 9	2 5	3 6	5 37	35	5 17	7 11	6 7	6 27	4 17	4 13	22 44	2 4	2 5	47
I	3 7	3 5	4 7	3 5	3 7	2 9	2 5	2 6	3 7	7 25	3 5	4 7	3 6	3 7	3 8	2 10	3 13	2 7	2 6	9 40
L	3 10	4 4	4 5	4 4	4 10	2 5	2 5	2 4	2 7	5 38	4 4	5 5	4 6	3 7	2 7	3 10	3 15	2 6	2 4	18 52
F	9 3	8 4	6 4	4 6	34 7	5 6	5 4	4 6	3 9	19 13	4 5	4 6	4 5	3 4	37 7	23 7	59 9	2 3	2 3	39 14
Y	3 3	2 5	3 6	2 5	7 5	2 8	2 5	2 4	3 6	23 7	2 5	2 8	3 4	4 6	4 7	4 7	21 8	2 6	2 4	19 7
W	4 5	3 5	2 5	3 4	3 5	2 5	2 4	2 4	3 5	3 2	2 3	3 6	4 6	3 6	3 6	2 6	2 6	2 6	2 5	2 5
G	3 5	2 6	2 5	3 4	6 4	3 4	2 3	2 4	2 5	48 9	3 5	3 6	3 6	4 5	3 6	3 6	4 5	2 5	6 12	7
C	4 5	3 7	3 8	3 7	13 10	3 9	3 5	3 6	4 9	40 27	12 6	3 7	4 5	12 5	9 5	13 5	48 10	3 5	2 4	44 25
M	16 24	2 6	5 20	2 4	26 17	4 4	3 10	9 24		10 25	4 8	3 9	8	3 4	6 17	57 39	3 10	13 28		
S	52 14	3 5	12 5	2 2	6 57	12 2	7 6	15 15		32 20	3 13	5	25 12	29 10	43 12	55	4 7	4 7		
T	4 32	4 5	3 13	3 5	13 26	3 15	2 7	3 11	5 22		4 37	3 10	2 6	13 12	2 11	4 18	45 57	2 3	2 5	

(b)

	A	R	N	D	C	Q	E	G	H	I	L	K	M	F	P	S	T	W	Y	V
E	7 2	6 2	5 3	4 2	4 2	5 2	4 2	5 1	4 2	5 2	5 1	5 2	5 2	5 2	5 2	5 2	5 2	5 3	5 3	5 3
Q	6 47	7 3	5 26	3 3	20 46	18 24	3 2	5 6	4 43		59 8	3 14	7 60	66 8	9 8	60 13	25 3	8 41		
D	6 2	7 3	8 3	8 3	8 3	7 3	7 3	9 3	10 3	8 3	7 3	9 3	7 5	7 5	7 5	1				
N	9 3	9 3	7 4	4 1	9 7	9 2	8 2	10 3	7 2	12 59	10 3	9 2	8 2	10 3	8 2	7 3	8 26	11 3	9 4	8 43
R	8 2	6 2	10 3	7 2	8 3	7 2	6 2	11 2	8 2	7 2	10 2	7 2	9 4	8 1	12 2	10 2	11 2	10 2	11 3	11 2
K	11 3	9 2	11 3	9 3	10 4	9 4	11 4	14 7	9 5	9 7	7 5	7 3	9 9							
H	6 3	7 3	6 5	7 2	8 6	7 3	8 2	7 3	5 2	9 57	9 3	10 2	9 2	8 2	7 2	8 21	8 58	10 4	10 4	9 45
A	7 4	7 3	6 3	5 2	8 17	7 3	5 1	7 1	5 7	9	9 6	8 3	7 4	12 29	9 2	7 7	8 42	7 2	8 4	10
V	10 4	11 3	9 2	8 2	10 5	8 2	9 2	11 3	9 5	9 35	8 5	12 7	8 6	7 6	7 4	7 4	7 22	7 2	7 2	6 47
I	9 3	10 3	7 4	6 3	9 3	8 2	7 2	8 2	7 3	7 7	8 3	10 4	10 3	9 3	8 3	9 2	9 3	11 2	11 2	8 9
L	10 3	9 4	12 4	8 4	7 4	7 2	6 2	6 2	5 2	5 5	11 4	10 5	7 4	7 3	8 2	9 3	7 3	9 2	7 2	7 18
F	13	14 3	14 4	12 4	14 34	11 5	13 3	15 4	15 3	15 19	15 4	9 4	13 4	19 3	8 37	9 23	8 59	7 2	6 2	5 39
Y	6 3	5 2	4 3	4 2	5 7	5 2	4 2	7 2	6 3	6 23	7 2	6 2	6 3	7 4	8 3	5 4	7 21	9 2	9 2	6 19
W	14 4	12 3	10 3	10 2	9 3	6 2	5 3	4 2	10 2	6 2	5 3	4 2	10 2	12 2	10 2	8 2				
G	12 3	13 2	13 2	13 2	11 3	12 6	11 3	13 2	13 2	12 2	8 48	8 3	9 3	8 3	10 4	7 3	7 3	6 4	6 2	5 2	4 12
C	9 4	8 3	6 3	5 3	10 13	9 3	7 3	8 3	8 4	38 40	20 12	11 3	11 4	18 12	10 9	9 13	11 48	11 3	9 2	15 44
M	7 16	9 2	8 5	8 2	10 26	9 4	5 3	8 4	9 2	24	19 10	10 4	11 3	15 24	9 4	10 57	12 3	10 13	24	
S	10 52	12 3	12 12	11 2	19	19 57	10 2	11 7	7 15		52 8	3 8	5 33	25 9	29 8	43 10		12 4	6 4	
T	7 4	11 4	7 3	7 3	10 13	9 3	8 2	9 3	7 5	15	8 4	8 3	7 2	8 13	6 2	7 4	7 45	13 2	7 2	7

C5-alkyne =

C15-alkyne =

Fig. 2. Evaluation of the extent of farnesylation of a RAGCVa₂X library of peptides. Rows represent the a_2 position and columns represent the X position. (a) For each sequence the left numbers represents R. norvegicus PFTase results while the right one represents the C. albicans PFTase results. (b) For each sequence, the left figure indicates screening using OPP-C5-alkyne, the right ones using OPP-C15-alkyne. Color intensity was quantified by Image J software. For comparison, the intensity was normalized relative to that observed with CVIS position. The library was synthesized and screened at least two times and the average color intensities were color coded. The intensities below 33% are shown in white, the intensities between 34 to 66% are shown in yellow and the intensities above 66% are shown in red.

Acknowledgments

We thank the center for mass spectrometry and proteomics from University of Minnesota for MS and MS/MS analysis. We also thank Dr. Lorena S. Beese from Duke University Medical Center for providing *C. albicans* PFTase. This research was supported by the National Institutes of Health (GM084152, MDD).

References

1. Reid, S.T., Terry, K.L., Casey, P.J., Beese, L.S. *J. Mol. Bio.* **343**, 417-433 (2004).
2. Hougland, J.S., Hicks, K.A., Hartman, H.L., et al. *J. Mol. Biol.* **395**, 176-190 (2010).
3. Wang, Y.-C., Distafano, M.D. *Chem. Commun.* **48**, 8228-8230 (2012).
4. Wollack, J.W., Silverman, J.M., Petzold, C.J., Mougous, J.D., Distefano, M.D. *ChemBioChem* **10**, 2934-2943 (2009).

Proceedings of the 23rd American Peptide Symposium
Michal Lebl (Editor)
American Peptide Society, 2013

Peptide-Boronic Acid Libraries for Saccharide Recognition

Wioleta Kowalczyk, Julie Sanchez, Philippe Kraaz, Laurence Meagher, David Haylock, Oliver E. Hutt, and Peter J. Duggan

CSIRO Materials Science & Engineering, Bag 10, Clayton South, VIC, 3169, Australia

Introduction

Cells are covered by a dense and complex array of carbohydrates. These glycans mediate and/or modulate many cellular interactions. Proteins that bind sugars, i.e. lectins, play a key role in the control of various normal and pathological processes. Lectins are highly specific for a particular sugar sequence expressed on the cell surface [1]. Boronic acids enhance sugar binding *via* their ability to form covalent yet reversible bonds to the 1,2- and 1,3-diols present on many saccharides [2,3]. Compounds modified with boronic acids are the most extensively explored amongst the artificial receptors for the carbohydrates – synthetic lectins [4-6]. As benzoboroxoles (Figure 1) in particular are well known for the covalent binding to *cis*-diols present in saccharides at physiological pH [7], 5-carboxybenzo-boroxole (5CBB) was used to incorporate benzoboroxole functionality into peptides. Such a combination should lead to libraries of peptide boronic acids that show variable carbohydrate binding strength and selectivity. Synthetic lectins may be used in cell sorting or cell immobilization on artificial surfaces.

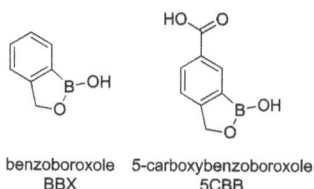

benzoboroxole   5-carboxybenzoboroxole
BBX   5CBB

Fig. 1. Benzoboroxoles.

Results and Discussion

A 54-member peptide-boronic acid library was prepared using combinatorial approach and examined using a dye displacement assay to select binders with favorable properties. The peptide library was prepared using Mimotopes TranSort technology, which allows the identity of the peptide on each lantern to be tracked using an attached radio-frequency tag. The general structure of the peptide library is shown on Figure 2. All peptides were produced by solid phase peptide synthesis on Fmoc-Rink Amide-PA-*SynPhase* Lanterns (8μmol) using Fmoc/t-Bu chemistry [8]. The N^ε-amino group of one lysine residue was protected with the Mtt group. After Mtt removal [9] 5CBB was introduced. Crude peptides were characterized by analytical HPLC (purity ~70%) and MALDI-TOF MS.

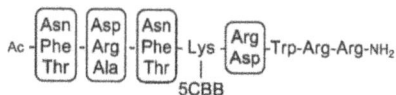

Fig. 2. Peptide library structure.

A displacement assay with Alizarin Red S (ARS) was used for the determination of binding constants of a saccharide-boronic acid interaction. This approach was initially developed by Springsteen and Wang [10,11] to measure the interaction of sugars with boronic acids by monitoring fluorescence changes of ARS. The determination of binding constant is complicated because not only the binding equilibrium between host and guest has to be considered, but also the equilibrium between the dye (reporter) and the host has to be known and evaluated. The determination of binding constants in solution in this way is still superior to a similar approach on the solid-phase, a method we used previously and which involved many complicating factors [12].

We have adapted Wang's ARS method to a 384 well plate format using a FlexStation 3 fluorescent plate reader. This allows us to determine binding constants with milligram quantities of peptide in a high throughput manner. In the first step binding affinity between peptide boronic acid and ARS were measured. All synthesized peptides (compounds **1-54**) and positive controls: BBX and Lys(5CBB) (Figure 3) were tested. Data analysis was performed using GraphPad Prism 6.01 for Windows (fit: binding saturation - one site). Representative data for some peptides are shown in Figure 4. Association constants (K_a) with ARS are shown in the Figure 5.

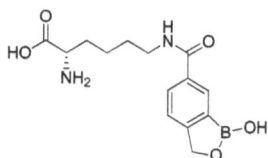

Fig. 3. Lys(5CBB).

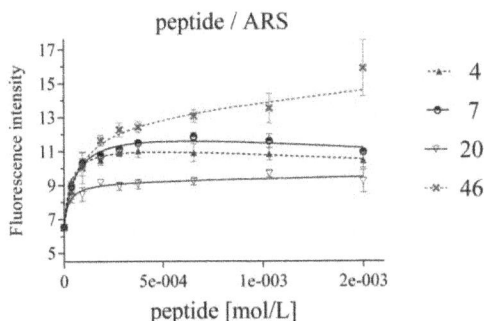

Fig. 4. Titration of peptide into solution of ARS (Em. λ=560 nm, Exc. λ=490 nm).

Preliminary data show significant differences of binding affinity between ARS and peptide library members. The structure of the peptide is definitely influencing ARS binding – up to 180x binding constant cf. Lys(5CBB). This is a promising result in the search for selective and strong binding partners. Comparing two groups of peptides compounds **1-27** (with Arg^5 modification) and **28-54** (Asp^5), the first group give higher binding affinity. This is consistent with previous findings [12] that the presence of arginine residue improves binding between peptide boronic acids and diols.

The second step of the dye displacement assay, which will allow us to determine the binding selectivity towards various saccharides, is currently being performed. Techniques such as ITC will be used for peptides with interesting binding properties.

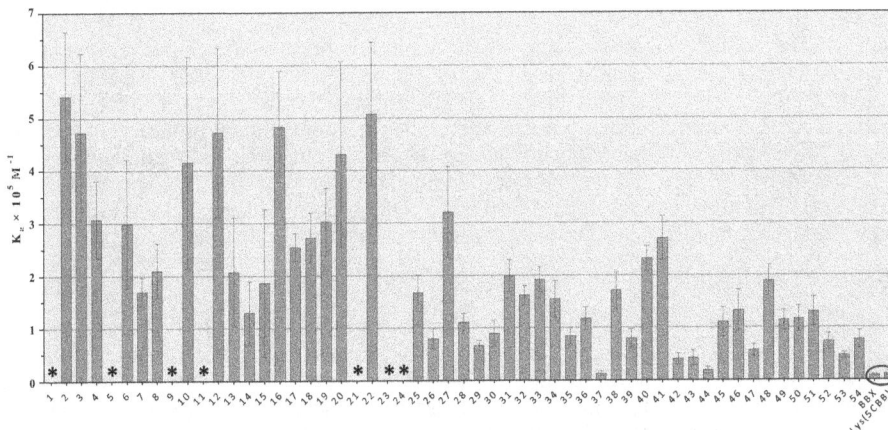

Fig. 5. Association constants (K_a) with ARS at pH 7.4, 0.1 M phosphate buffer; all experiments were carried out in triplicates; * very weak binding or data fit unsuccessful; red circle – positive controls BBX and Lys(5CBB).

Acknowledgments

We would like to acknowledge the CSIRO Advanced Materials Transformational Capability Platform – AMTCP for funding this project.

References

1. Varki, A. *Nature* **446**, 1023-1023 (2007).
2. Lorand, J.P., Edwards, J.O. *J. Org. Chem.* **24**, 769-774 (1959).
3. Springsteen, G., Wang, B. *Tetrahedron* **58**, 5291-5300 (2002).
4. Pal, A., Berube, M., Hall, D.G. *Angew. Chem.* **49**, 1492-1495 (2010).
5. Duggan, P.J., Offermann, D.A. *Tetrahedron* **65**, 109-114 (2009).
6. Bicker, K.L., et al. *Chem. Sci.* **3**, 1147-1156 (2012).
7. Bérubé, M., Dowlut, M., Hall, D.G. *J. Org. Chem.* **73**, 6471-6479 (2008).
8. http://www.mimotopes.com/files/editor_upload/File/CombinatorialChemistry/Mim-SAN002.pdf.
9. Li, D., Elbert, D.L. *J. Peptide Res.* **60**, 300-303 (2002).
10. Springsteen, G., Wang, B. *Tetrahedron* **58**, 5291-5300 (2002).
11. Springsteen, G., Wang, B. *Chem. Commun.* 1608-1609 (2001).
12. Duggan, P.J., Offermann, D.A. *Aus. J. Chem.* **60**, 829-834 (2007).

Proceedings of the 23rd American Peptide Symposium
Michal Lebl (Editor)
American Peptide Society, 2013

Development of an Effective Dual Ring-Opening/Cleavage Approach for the Synthesis and Decoding of One-Bead One-Compound Cyclic Peptide Libraries

Xinxia Liang[1,2], Anick Girard[1,2], and Eric Biron[1,2]

[1]Faculty of Pharmacy, Université Laval, Québec (QC), G1V 0A6, Canada; [2]Laboratory of Medicinal Chemistry, CHU de Québec Research Center, Québec (QC), G1V 4G2, Canada

Introduction

Cyclic peptides are useful tools in chemical biology and medicinal chemistry. Their great therapeutic potential has prompted their use in combinatorial chemistry. The one-bead-one-compound (OBOC) approach, in which each bead carries many copies of a unique compound, has become a powerful tool in drug discovery [1]. Such libraries have been successfully used to discover ligands for a wide variety of macromolecular targets. However, the use of the OBOC technology with cyclic peptides has been limited by difficulties in sequencing hit compounds after the screening. Lacking a free *N*-terminal amine, Edman degradation sequencing cannot be used and complicated fragmentation patterns are obtained by MS/MS. In this regard, the Pei group used a one-bead-two-compound approach on topologically segregated bilayer beads in which the cyclic peptide is exposed on the surface and its linear counterpart is found inside as a tag for sequencing [2]. More recently, a ring-opening strategy on cyclic peptoids has been developed to reduce the need for encoding in OBOC libraries [3,4]. In this strategy, a cleavable residue is introduced in the cycle backbone to allow a linearization of the molecule under specific conditions and sequencing of the linear variant by MS/MS. Based on this ring-opening strategy, our objective was to develop a simple and efficient approach to prepare OBOC cyclic peptide libraries that would allow fast sequence determination after simultaneous macrocycle opening and release from the resin.

Results and Discussion

Our strategy was based on the incorporation of a cleavable residue within the cycle and as a linker (Figure 1) [4]. Amongst the different linkers and cleavable residues readily available, we were particularly interested in methionine (Met). Being an amino acid, Met can be used in standard solid phase peptide synthesis and is stable in acidic, basic or reductive conditions used to remove protecting groups. Moreover, the reaction conditions used to cleave Met are very selective and compatible with free amino acid side chains. Met has been widely used as a linker in OBOC peptide libraries and can be selectively cleaved upon treatment with CNBr to yield a *C*-terminal homoserine lactone. In the proposed approach, the cleavable residue can be introduced at two different positions in the macrocycle to allow its transformation into a linear peptide. Depending on the Met position, the pattern of the linearized peptides will be completely different. Met can be introduced as the first amino acid before peptide synthesis or the last amino acid during peptide elongation and before cyclization.

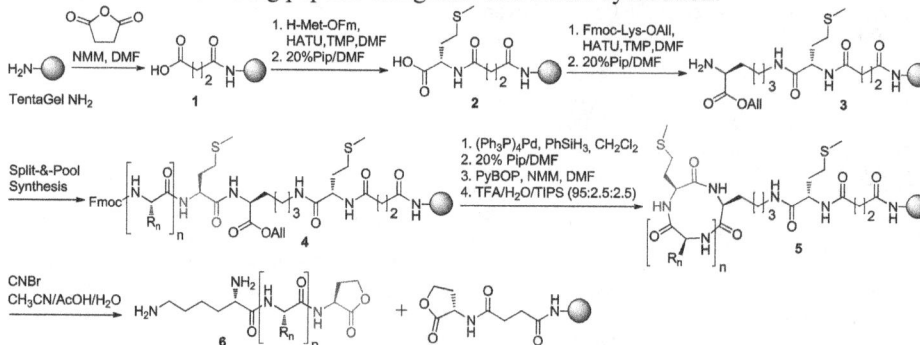

Fig. 1. Synthesis of OBOC cyclic peptide libraries on a reverse Met handle.

Unfortunately, both approaches have generated linear peptides bearing two homoserine lactone residues, yielding complicated MS/MS spectra very difficult to sequence. Also, when Met was coupled as the last residue, the cyclization efficiency was moderate. This behaviour was also observed by Simpson and Kodadek with cyclic peptoids [4]. To overcome these problems, the Met in the cycle was introduced as the first amino acid and the Met linker was inverted to eliminate one *C*-terminal homoserine lactone (Figure 1). The reverse Met handle has been used by Kappel and Barany for the synthesis of lysine-containing cyclic peptides [5]. To do so, H-Met-OFm was coupled to the carboxylic acid derived resin **1** and following 9-fluorenylmethyl ester cleavage, Fmoc-Lys-OAll was anchored to resin **2** *via* its side chain. After standard Fmoc solid phase synthesis on resin **3**, the supported fully protected linear peptide **4** was treated with Pd(PPh$_3$)$_4$ to cleave the *C*-terminal allyl ester and piperidine to remove the *N*-terminal Fmoc group. Afterwards, head-to-tail cyclization was performed and the side chains deprotected to yield the anchored cyclic peptides **5**. The introduction of a Met residue directly after the side chain anchored Lys in the cyclic peptide allowed a simultaneous ring-opening and cleavage from the resin upon treatment with a CNBr solution to yield a linear peptide **6** bearing a single *C*-terminal homoserine lactone and an *N*-terminal lysine (Figure 1). The invariable residues at the *C*- and *N*-termini can be used as starting points in the MS/MS spectra analysis, significantly helping the sequencing process. Moreover, with its positive charge, the lysine residue facilitates ionization of the peptide during MALDI MS analyses. The generated peptides were analyzed by MALDI MS and showed clear MS/MS fragmentation patterns, allowing a rapid and accurate sequence determination (Figure 2).

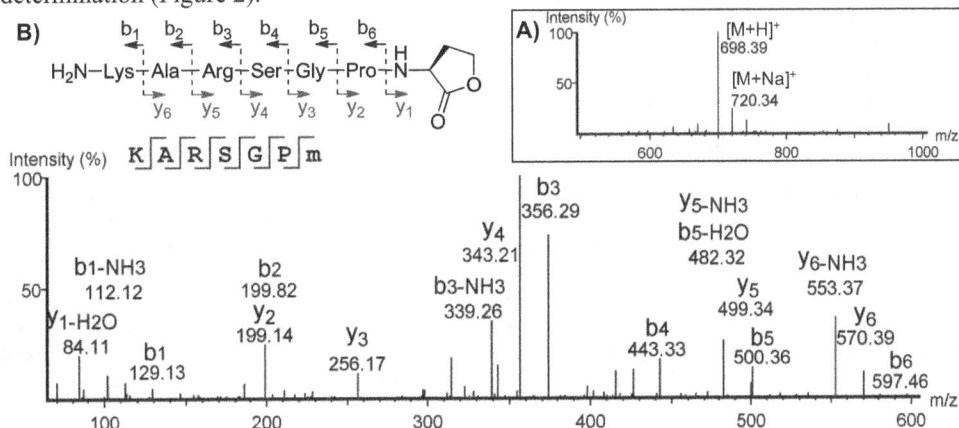

Fig. 2. MALDI-TOF/TOF MS analysis of peptide H$_2$N-KAYKPFNh* released from a single bead; (a) MS spectra (b) MS/MS of the molecular ion (698.39) (h* = homoserine lactone).

The developed alternative tandem ring-opening/cleavage approach is compatible with commonly used amino acids and can be used on a single bead to release linear peptides that can be clearly sequenced by tandem mass spectrometry. The described procedure for the preparation of the Met handle and unencoded cyclic peptide libraries is simple and affordable for any laboratories involved in peptide synthesis or combinatorial chemistry.

Acknowledgments

We thank Isabelle Kelly of the Genomics Center of Quebec at CHU de Québec Research Center for MALDI-TOF MS analysis. Xinxia Liang thanks the China Scholarship Council for Ph.D. scholarship. This work was supported by the NSERC of Canada and FQR-S of Quebec.

References

1. Lam, K.S., Krchnak, V., Lebl, M. *Chem. Rev.* **97**, 411 (1997).
2. Joo, S.H., Xiao, Q., Ling, Y., Gopishetty, B., Pei, D. *J. Am. Chem. Soc.* **128**, 13000 (2006).
3. Lee, J.H., Meyer, A.M., Lim, H.S. *Chem. Comm.* **46**, 8615 (2010).
4. Simpson, L.S., Kodadek, T.A. *Tetrahedron. Lett.* **53**, 2341 (2012).
5. Kappel, J.C., Barany, G. *Lett. Peptide Sci.* **10**, 119 (2003).

Proceedings of the 23rd American Peptide Symposium
Michal Lebl (Editor)
American Peptide Society, 2013

Discovery of Peptidomimetic Death Ligands Against Ovarian Cancer Through OB2C Combinatorial Library Approach

Ruiwu Liu, Tsung-Chieh Shih, Xiaojun Deng, Lara Anwar, David Olivos, Mary Saunders, and Kit S. Lam

Department of Biochemistry and Molecular Medicine, University of California Davis, Sacramento, CA, 95817, U.S.A.

Introduction

Ovarian cancer is the second most common gynecologic cancer and the deadliest in terms of absolute number. There is a great need for the discovery of new, effective, and less toxic therapies for ovarian cancer. We have recently developed a powerful "one-bead two-compound" (OB2C) combinatorial library method to discover synthetic death ligands against cancer cells in an ultra-high throughput fashion [1]. In the OB2C library, each bead displays on its surface a cell capturing molecule and a random library compound. The chemical coding tag (bar code) resides in the interior of each bead (Figure 1, top). When live cells are incubated with such novel OB2C libraries, every bead will be coated with a monolayer of cancer cells. The cell membranes of the captured cells facing the bead surface are exposed to the library compounds displayed on each bead. Cells undergoing apoptosis will be readily detected by a caspase 3-specific fluorescent protease substrate (Figure 1, bottom). We applied this approach to the discovery of novel death ligands against ovarian cancer cell line SKOV3 using LXY30 as a cell capturing ligand. LXY30 is a highly potent and specific peptide ligand of α3β1 integrin which is highly expressed in SKOV3 cells [2].

Results and Discussion

A benzimidazole-based peptidomimetic OB2C library OB2C-S3 (Figure 1, top), with three diversities, was synthesized on topologically segregated TentaGel resin beads [3]. Each of the three diversity points R_1 (42 primary amines), R_2 (42 aldehydes), and X_3 (42 amino acids) were encoded with amino acid X_1, X_2 and X_3, respectively. To screen for death ligands against SKOV3 cells, the beads were incubated with SKOV3 cells for 24 hours. Apoptosis of the bead-bound cells was detected with immunocytochemistry (ICC) using anti-cleaved caspase 3 antibody-horse radish peroxidase conjugate and 3,3'-diaminobenzidine as a substrate. Caspase-3 was chosen because it is a pivotal executioner of apoptosis and its activation via cleavage into fragments is an excellent indication of early apoptosis [4]. Four positive beads (A1-A4) were identified from a total of about 40,000 library beads screened. Their structures were decoded using Edman microsequencing [3] (Figure 2).

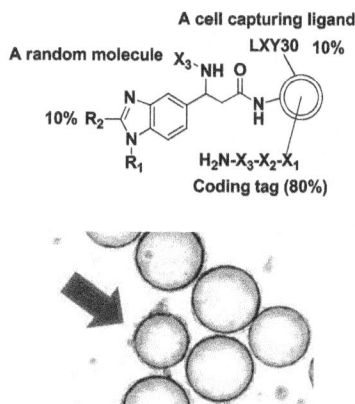

To validate the pro-apoptotic activity of these death ligands, we first resynthesized them on beads, and then repeated the staining of the bead-bound SKOV3 cells as described above. A1 and A2 were able to induce cell death, but A3 and A4 only had very weak activity (data not shown). We also validated the killing effect by propidium iodide (PI), which stains late apoptotic cells [5,6]. After 48 hours of binding to A1 and A2 beads, SKOV3 cells were stained with PI. The number of PI positive cells was found to increase markedly (Figure 3). Compounds A1 and A2 were also synthesized in soluble form for further characterization. MTT assay was used to determine whether these compounds could kill cancer cells. The result showed that both A1 and A2 alone as a monomeric form did not have any significant effect on cell death at concentration as high

Fig. 1. Top: OB2C-S3 library bead. Bottom: Screening library for death ligands against SKOV3 cells. Red arrow points to a positive bead.

as 200 µM. We also constructed tetrameric compounds by mixing neutravidin with biotinylated compounds. Notably, we found that A2 tetramer killed SKOV3 cells with the IC_{50} value of 41.4 µM, but A1 tetramer was inactive even after 72 hours of incubation. Formation of apoptotic body was also observed after 24 hours incubation with tetrameric A2 (data not shown). The pro-apoptotic effects of A2 tetramer is cell type specific, as it has no effect on prostate cancer (PC3) and lung cancer (A549) cell lines as high as 250 µM.

In summary, four death ligands against ovarian cancer SKOV3 cells have been identified from library OB2C-S3. Their pro-apoptotic functions have been evaluated on bead as well as in soluble form. Ligand A2 is the most potent and specific compound to induce cell death against SKOV3 cells. These death ligands, once fully optimized, by themselves or when chemically linked to a cancer cell surface capturing ligand, can potentially be developed into a novel, effective, but less toxic therapy against ovarian cancer.

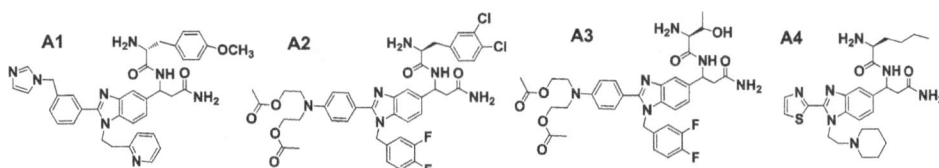

Fig. 2. Structures of death ligands.

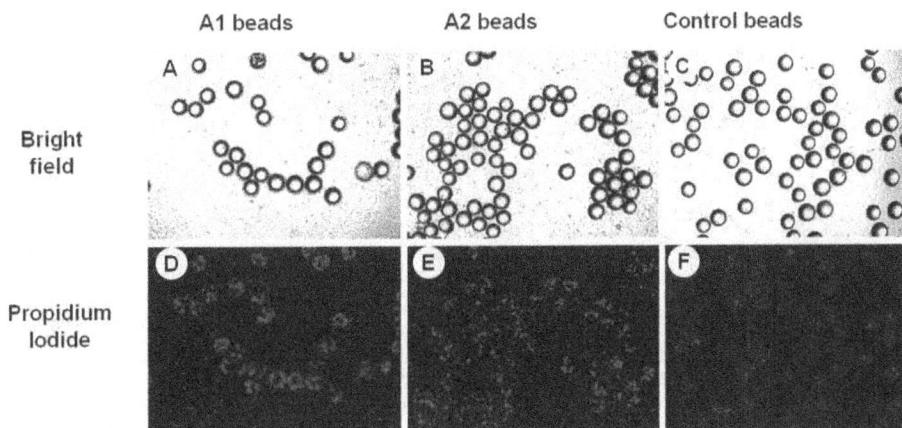

Fig. 3. Propidium Iodide (PI) staining for dead cells on beads displaying pro-apoptotic ligands. SKOV3 cells were incubated with A1, A2, and blank control TentaGel beads for 48 hours. Significantly more dead cells were found on both A1 and A2-beads.

Acknowledgments

This work was supported by NIH R33CA160132 and institutional fund from UC Davis.

References

1. Kumaresan, P.R., Wang, Y., Saunders, M., Maeda, Y., Liu, R., Wang, X., Lam, K.S. *ACS Comb. Sci.* **13**, 259-264 (2011).
2. Xiao, W., Liu, R., Bononi, F.C., Lac, D., Liu, Y., Sanchez, E., Mazloom, A., Lin, J., Lam, K.S. Manuscript in preparation.
3. Liu, R., Marik, J., Lam, K.S. *J. Am. Chem. Soc.* **124**, 7678-7680 (2002).
4. Nicholson, D.W., Thornberry, N.A. *Trends Biochem. Sci.* **22**:299-306 (1997).
5. Hug, H., Los, M., Hirt, W., Debatin, K.M. *Biochemistry* **38**, 13906-13911 (1999).
6. Darzynkiewicz, Z., Li, X., Gong, J. *Methods Cell Biol.* **41**, 15-38 (1994).

Proceedings of the 23rd American Peptide Symposium
Michal Lebl (Editor)
American Peptide Society, 2013

Development of Peptide Inhibitors Disrupting PCSK9-LDLR Protein-Protein Interactions

Kévin Ly[1], Anna Kwiatkowska[1], Sophie Routhier[1], Roxane Desjardins[1], Monika Lewandowska[2], Adam Prahl[2], Josée Hamelin[3], Nabil G. Seidah[3], Yves Dory[1], and Robert Day[1]

[1]Institut de Pharmacologie de Sherbrooke et Département de chirurgie/Service d'urologie, Université de Sherbrooke, Sherbrooke, J1H5N4, Canada; [2]Faculty of Chemistry, University of Gdańsk, Gdańsk, 80-952, Poland; [3]Institut de recherches cliniques de Montréal, Montréal, H2W2R7, Canada

Introduction

Cardiovascular disease (CVD) is the leading cause of global mortality. Hypercholesterolemia, characterized by increased plasma low-density lipoprotein (LDL) cholesterol, is a major determinant of CVD risk. Proprotein convertase subtilisin/kexin 9 (PCSK9) plays a critical role in cholesterol homeostasis by regulating LDL receptor (LDLR) protein levels. PCSK9 binds to the EGF-A domain of the LDLR and promotes its internalization and degradation in endosomal/lysosomal compartments [1]. Inhibition of PCSK9 action on LDLR has emerged as a novel therapeutic target for hypercholesterolemia and the prevention of CVD [1]. The present study attempts to evaluate if small peptides can be used to interfere with the protein-protein interactions of PCSK9 with EGF-A domain. Initially we used a strategy with positional scanning of synthetic peptide combinatorial libraries (PS-SPCL) from a L-hexa and L-decapeptide library to analyze the potential of small peptides to inhibit PCSK9 (Figure 1) in a cell-based assay measuring fluorescent LDL incorporation. Our results from the SP-SPCL approach showed the potential of small peptides to prevent reduction in LDL uptake from PCSK9 effect. Also, based on the 40 aa sequence of EGF-A domain and molecular modeling from the co-crystal of PCSK9:EGF-A [2], we synthesized a cyclic peptide of 13 aa that prevents PCSK9 effect on LDL uptake.

Results and Discussion

Using a cell-based assay with HepG2 cells and exogenous purified recombinant hPCSK9 R218S+D374Y (hPCSK9 RSDY), we screened the L-hexa and L-decapeptide combinatorial library. LDL uptake in HepG2 cells is greatly reduced by the addition of exogenous PCSK9 in the media resulting from PCSK9-LDLR interaction. In presence of PCSK9 and L-hexapeptides mixture from the library, we did not observe significant increases in LDL uptake compared to the positive control as seen for the position 4 (Figure 1). The screening

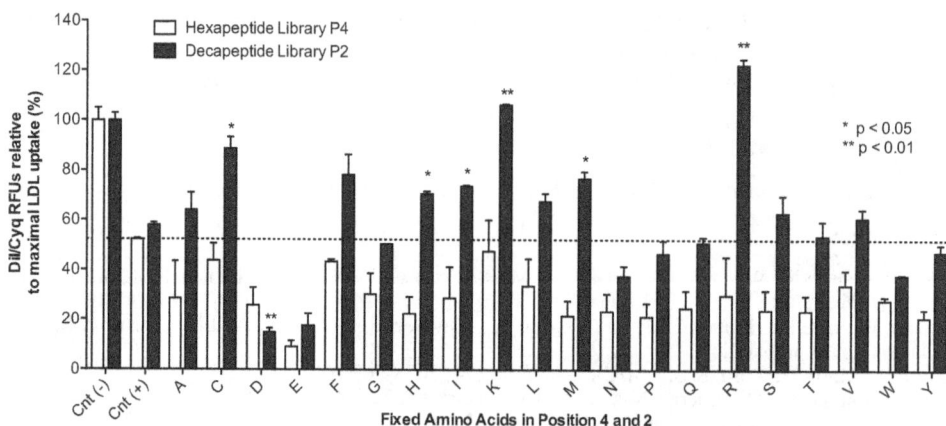

Fig. 1. Screening of L-hexa and L-decapeptide library by PS-SPCL in a cell-based assay with HepG2 cells.

from each position of L-hexapeptide library revealed that short peptides of 6 aa could not block PCSK9 action upon LDLR degradation in HepG2 cells. Instead, we see a greater decrease when both components are added in the cell-based assay. However, the screening of L-decapeptide library led to the identification of multiple amino acids for each position that were able to increase LDL uptake in presence of PCSK9 (Figure 1). The amino acids Cys, Ile, Lys and Arg in position P2 seems to be important to block PCSK9 effect and subsequent dose-response experiment were done to confirm the effect of those peptide mixtures. The PS-SPCL functional approach suggests the potential for small peptides of 10 aa to interfere with the PCSK9-LDLR interaction.

According to literature and also based on our results (Figure 2), we know that the EGF-A polypeptide of 40 aa can functionally block PCSK9-LDLR interactions [2-4]. The EGF-A was prepared by solid-phase peptide synthesis and Cys groups were orthogonally protected to direct disulfide formation in the proper configuration. EGF-A polypeptide has an IC50 of 1.1 μM and completes LDLR for binding to PCSK9. However, the EGF-A polypeptide has a low potency to block the PCSK9 effect and the development of smaller peptides with higher potency would be a better alternative. Additionally, we showed (Figure 3) that cyclic peptide of 13 aa can block PCSK9-LDLR interaction and increase LDL uptake in HepG2 cells in very high concentration compare to its linear homologous sequence which showed no significant changes in LDL uptake. From this design, it will be possible to develop further short sequences of peptides based on the interaction between the PCSK9 and EGF-A domain, and the library screening to improve the potency of our small peptides.

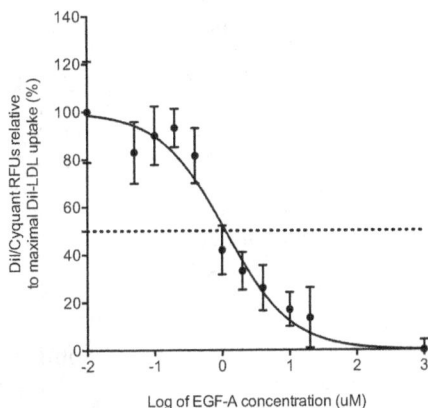

Fig. 2. Inhibition of hPCSK9 RSDY by polypeptide EGF-A.

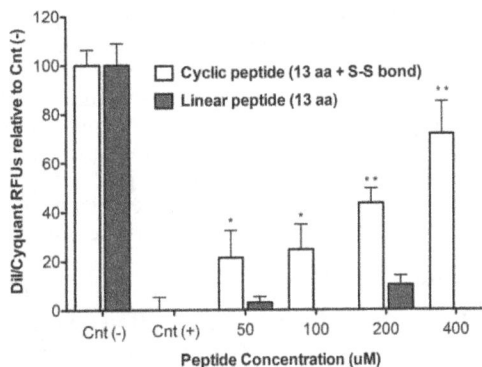

Fig. 3. Inhibition of hPCSK9 RSDY by linear and cyclic peptides.

Acknowledgments

We gratefully acknowledge Dr. Jon Appel and Dr. Richard A. Houghten for the hexa/decapeptide librairies (Torrey Pines Institute for Molecular Studies (TPIMS), San Diego, California, USA). We acknowledge FRSQ for postgraduate scholarships and Fondation Leducq (Transatlantic Networks of Excellence) for financial support.

References

1. Seidah, N.G., Prat, A. *Nat. Rev. Drug Discov.* **11**, 367-383 (2012).
2. Kwon, H.J., Lagace, T.A., McNutt, M.C., et al. *Proc. Nat. Acad. Sci. USA* **105**, 1820-1825 (2008).
3. Poirier, S., Mayer, G. Poupon, V., et al. *J. Biol. Chem.* **284**, 28856-28864 (2009).
4. Shan, L., Pang, L., Zhang, R., et al. *Biochem. Biophys. Res. Commun.* **375**, 69-73 (2008).

Proceedings of the 23rd American Peptide Symposium
Michal Lebl (Editor)
American Peptide Society, 2013

Discovery of Novel Selective Melanotropin 3 Ligands: A New Candidate for the Study of Inflammatory Diseases

Minying Cai[1], Florian Rechenmacher[2], Jennifer Bao[1], Morgan R. Zingsheim[1], Florian Opperer[2], Johannes G. Beck[2], Lucas Doedens[2], Horst Kessler[2], and Victor J. Hruby[1]

[1]Department of Chemistry and Biochemistry, The University of Arizona, Tucson, AZ, 85721, U.S.A.;
[2]Center of Integrated Protein Science Institute for Advanced Study at the Technische Universitat Munchen, Lichtenbergster. 4, 85747, Garching, Germany

Introduction

The melanocortin system involves numerous physiological functions and is associated with many diseases such as skin cancer, obesity and diabetes, sexual dysfunction, neuropathic pain, inflammatory diseases etc. Our primary interests have been focusing on the human melanocortin 3 receptor (hMC3R) due to its direct involvement in the regulation of feeding behavior, energy homeostasis, as well as inflammatory diseases. The hMC3 receptor has been reported as an inhibitory autoreceptor on POMC neurons based on the observed stimulation of food intake by peripheral administration of an hMC3R-selective agonist, and hMC3R agonist-induced inhibition of spontaneous action of POMC neurons. Nevertheless, the full scope of physiological functions of this receptor is still poorly understood. There has been a resurgence of interest in peptide pharmaceuticals recently as they have an advantage of potency, selectivity and less toxicity compared with small-molecule therapeutics. The main drawback of peptides is lack of stability in biological media. N-Methylation of peptides have been one of the approaches to improve in vivo stability of the peptides [1]. Several new modalities in constraining peptides have been developed over recent years and this work highlights some of the new developments in our lab [2,3]. The newer methylation strategies have rendered, in some cases, oral activity, cell permeability, improved potency at the target receptor, selectivity against receptor subtypes and improved stability to enzymes. Further understanding rules governing cell permeability, oral absorption and enhancing stability of peptides can help peptides to enter the clinic for many unmet medical needs.

Results and Discussion

During the last decade, great efforts have been made to develop selective melanotropins. Nevertheless, the potential of potent and selective peptides as drug candidates is challenged by their poor pharmacokinetic properties. Many peptides have a short half-life in vivo and a lack of oral availability. Inspired by the excellent pharmacokinetic profile of cyclosporine, a natural, multiply N-methylated cyclic peptide, we visualized multiple N-methylation as a promising way to rationally improve key pharmacokinetic characteristics of melanotropins. The incorporation of this strategy to multiple N-methylated analogs of melanotan II (MTII) has successfully reached this goal by increasing receptor selectivity and peptide stability [4]. Here, we extended our efforts toward modulating the properties of peptides by multiple N-methylation of SHU9119 (Ac-Nle[4]-c[Asp[5], DNal(2')[7], Lys[10]]α-MSH(4-10)-NH$_2$) [5], which is a selective agonist of the hMC1R and hMC5R and a selective antagonist of the hMC3R and hMC4R. Figure 1 illustrates the strategy of systematic N-methylation of SHU9119. Mono- and multiple N-methylations of SHU9119 peptides were synthesized and investigated to elucidate their remarkable conformational modulation ability by imparting steric constraints in the peptide backbone and to improve the pharmacokinetic profile of the peptides for use as drug leads. Our results indicate that the systematic N-methylations of SHU9119 led to the most selective antagonists of the hMC1R, hMC3R and hMC5R. Table 1 show that the triple-N-methylations of SHU9119 at the position of His[6], D-Nal(2')[7] and Lys[10] leads to the most selective antagonist of the hMC3R. NMR studies demonstrate that triple-N-methylations strongly constrained the β-turn pharmacophore, except for the position of Trp[9], which is critical for the hMC3R selectivity (data not shown here). This can explain the reason why triple-N-methylations-SHU9119 leads to hMC3R selectivity. Further investigations of the triple-N-methylation of SHU9119, a selective hMC3R antagonist, have

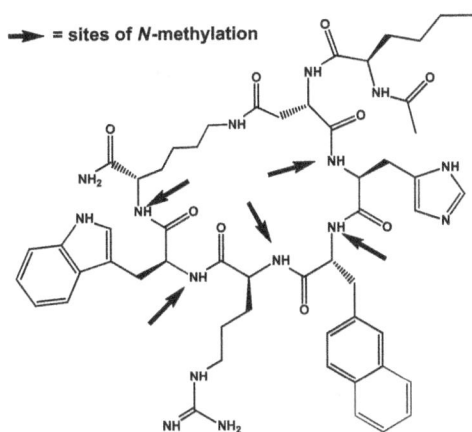

= sites of *N*-methylation

Fig. 1. Multiple N-methylation of SHU-9119.

been performed to study the inflammatory functions (data now shown), which can be used as a powerful inhibitor for the treatment of inflammatory diseases.

The improvement of oral bioavailability by multiple *N*-methylation is a significant advance toward the development of peptide-based therapeutics, which has been hampered over the years due to poor pharmacokinetic properties. Multiple *N*-methylation of MTII and SHU9119 result in enhancement in the activity and selectivity of receptor subtypes using either library or designed approaches and helps in understanding the finer details of the bioactive conformation. Thus, with these diverse properties, we foresee a bright future for peptide chemistry by multiple *N*-methylation toward their development as therapeutic prototypes.

Table 1. Binding assay of N-methylated SHU9119 analogues at hMCRs.

	hMC1R		*hMC3R*		*hMC4R*		*hMC5R*	
	IC_{50} nM	Binding Efficiency	IC_{50} nM	Binding Efficiency	IC_{50} nM	Binding Efficiency	IC_{50} nM	Binding Efficiency
SHU9119	1±0.1	100	3.7±0.5	97	2.4±0.2	98	6.9±1.2	98
*N-(3Me)*SHU9119	820±100	68	13.7±2	38	8700±840	59	390±41	70

IC_{50} =concentration of peptide at 50% specific binding (N=4). NB = 0% of ^{125}I-NDP-α-MSH displacement observed at 10 μM. Percent Binding Efficiency = maximal % of ^{125}I-NDP-α-MSH displacement observed at 10 μM.

Acknowledgments

This work is supported by in part by a grant from the US Public Health Service and NIDDK.

References

1. Kessler, H. *Angew. Chem. Int. Ed.* **21**, 512-523 (1992).
2. Hruby, V.J. *Nature Reviews Drug Discovery* **1**, 847-858 (2002).
3. Li, P., Roller, P.P. *Curr. Topics Med. Chem.* **2**, 325-341 (2002).
4. Doedens, L., Opperer, F., Cai, M., Beck, J.G., Dedek, M., Palmer, E., Hruby, V.J., Kessler, H. *J. Amer. Chem. Soc.* **132**, 8115-8128 (2010).
5. Hruby, V.J., Lu, D., Sharma, S.D., Castrucci, A.L., Kesterson, R.A., Al-Obeidi, F.A., Hadley, M.E., Cone, R.D. *J. Med. Chem.* **38**, 3454-3461(1995).

Proceedings of the 23rd American Peptide Symposium
Michal Lebl (Editor)
American Peptide Society, 2013

Furan Oxidation Cross-Linking: A New Approach for the Study and Targeting of Peptide/Protein and Nucleic Acid Interactions

Lieselot L.G. Carrette[1], Takashi Morii[2], and Annemieke Madder[1]

[1]Organic and Biomimetic Research Group, Department of Organic Chemistry, UGent, Krijgslaan 281-S4, 9000, Gent, Belgium; [2]Laboratory for Biofunctional Science, Institute of Advanced Energy, Kyoto University, Uji, Kyoto, 611-0011, Japan

Introduction

A novel mildly inducible cross-linking methodology for DNA interstrand cross-linking (ICL) was developed by our group, inspired by the metabolic activation of furan moieties (Figure 1). Nucleosides, modified in various ways with furan and easily accessible due to the stability and wide commercial availability of furan derivatives, are incorporated into oligonucleotides (ODNs). After hybridization into a duplex they are oxidized to a very reactive ketobutenal, resulting in selective and high yielding cross-link formation with the opposite base [1].

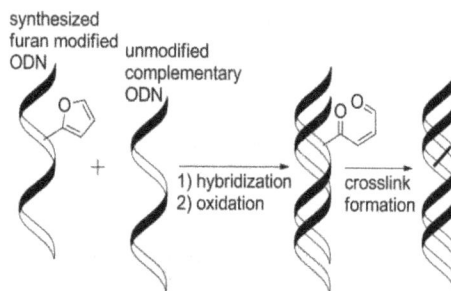

Fig. 1. The furan oxidation cross-linking method for ICL.

The strategy has now been investigated for broader application, as the method further holds potential for cross-linking of DNA binding proteins to their target.

Results and Discussion

Understanding the remarkable selectivity and affinity of nucleic acid binding proteins for their targets (DNA and RNA) in the complex cellular environment, which is essential in the regulation and execution of biological processes, is a central goal in their study. Formation of a covalent cross-link facilitates identification of the interaction partners and can be used to map the interaction interface. The obtained information can than in return be used for the design of novel drugs that alternate or compete in such interactions or improved antisense drugs that can escape interactions with proteins to fulfill their function [2].

Attempts have been undertaken to expand the furan oxidation cross-link methodology to protein - DNA interactions. For this purpose two approaches can be followed. Similarly as the ICL approach, the DNA can be modified with a furan containing nucleoside. However in this case, competitive ICL can occur and the cross-linking will proceed through a different

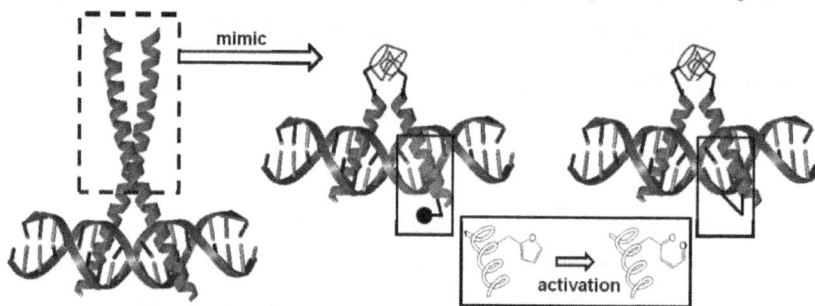

Fig. 2. Schematic representation of the approach. Furylalanine was used during the synthesis of a miniature transcription factor. Upon binding of the DNA the furan was oxidized for cross-link formation.

reaction mechanism targeting amino acids. Although the alternative approach starts by using different building blocks, furan modified amino acids to modify the protein; it proceeds through a similar cross-linking mechanism targeting the exocyclic amines of the bases of the target nucleic acid. Therefore this second approach was followed. From initial studies, the prerequisite of a good major groove binding synthetically accessible peptide could be concluded, to ensure proximity between the reactive enal and the attacking nucleophile [3]. For this reason we decided to work with the well studied non-covalent miniature transcription factor of GCN4, described by Morii, et al. based on a cyclodextrin - adamantane inclusion complex (Figure 2) [4].

The furan was introduced in the DNA binding peptide through commercially available furylalanine, replacing a lysine (K231) or alanine (A239) residue positioned in the major groove [5]. Binding of these modified protein mimics was ensured by the generation of heterodimers and verified by EMSA experiments. Optimization of the oxidation conditions was carried out by reaction with hydrazine and monitoring on HPLC. However, furan oxidation in the presence or just before the addition of DNA did not result in observable cross-link formation.

Though unfavorable positioning and/or linking cannot be completely excluded, the results strongly suggest that the involvement of the targeted exocyclic amines in Watson-Crick base pairing, strongly reduces their reactivity towards cross-linking. This rationale is further supported by the reported use of aldehydes for the detection of single stranded regions of DNA [6] and their use to elucidate RNA folding pathways [7]. Figure 3 depicts the difference between ICL and the targeting of double stranded DNA with a furan modified peptide. While the reactive moiety on a peptide has to intrude in and interfere with the highly organized duplex structure for protein – DNA cross-linking, the nucleophile of the base opposite to the furan modified ODN is readily available for ICL.

To obtain protein-DNA cross-linking we are currently exploring the reverse approach, where the furan moiety is incorporated into DNA, for cross-linking to a non-modified protein.

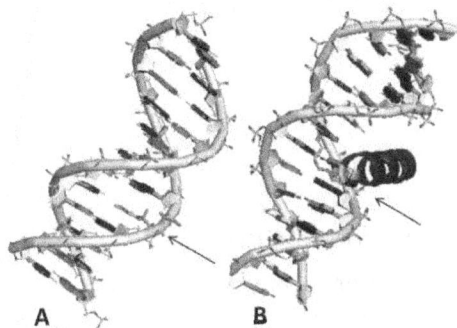

Fig. 3. ICL vs. cross-linking from a modified peptide to double stranded DNA (modified residues indicated by arrow). In A the oxidized furan moiety faces an unpaired C base, while in B the oxidized furan moiety faces a Watson-Crick base paired C.

Acknowledgments

Lieselot L.G. Carrette, as a recipient of an AAPPTec Travel Award for attending the 23rd APS, is further indebted to the Fund for Scientific Research Flanders for an aspirant position, the collaboration of the Fund for Scientific Research Flanders with the Japanese Society for the Promotion of Science [project VS.01S.10N] and the European Cooperation in Science and Technology [action TD0905].

References

1. Op de Beeck, M., Madder, A. *J. Am. Chem. Soc.* **134**, 10737-10740 (2012); Carrette, L.L.G., Gyssels, E., Madder, A. *Curr. Prot. Nucl. Acid Chem.* (2013 *accepted*) and references therein.
2. Verzele, D., Carrette, L.L.G., Madder, A. *Drug Discov. Today: Tech.* **7**, e155-e123 (2010).
3. Deceuninck, A., Madder, A. (unpublished results).
4. Ueno, M., et al. *J. Am. Chem. Soc.* **115**, 1257-12575 (1993).
5. Keller, W., König, P., Richmond, T.J. *J. Mol. Biol.* **254**, 657-667 (1995).
6. Murray, V. *Prog. Nucleic Acid Res. Mol. Biol.* **63**, 367-415 (1999).
7. Mathews, D.H., Turner, D.H. *Curr. Prot. Nucl. Acid Chem.* 11.9.1-11.9.4 (2002).

Proceedings of the 23rd American Peptide Symposium
Michal Lebl (Editor)
American Peptide Society, 2013

Differences in HPLC Retention Times of Peptides with the Same Amino Acid Composition – Problems of Prediction Algorithms

Zuzana Flegelová and Michal Lebl

Spyder Institut Praha, Nad Safinou II 365, 252 42 Jesenice u Prahy, Czech Republic

Introduction

It is convenient to predict retention time of the newly synthesized peptide to be able to choose the appropriate chromatographic conditions for its analysis. In the field of proteomics it is actually one of the tools to identify appropriate peptide in the complex mixture. However, it is quite well known that the amino acid composition is just one parameter defining the retention characteristics of a peptide. All prediction algorithms taking into account only amino acid composition (e.g. [1]) obviously fail to predict differences in peptides of the same length composed of the same amino acid set. Factors which clearly affect the retention of peptides in RP-HPLC include also peptide chain length and other sequence-dependent effects [2-9]. Krokhin, et al. [2-5] improved the prediction by including the effect of the proximity of the first three amino terminal residues. This prediction is now available from the web site http://hs2.proteome.ca/SSRCalc/SSRCalcX.html. Petritis, et al. [8,9] added to their vast analysis of hundreds of thousands peptides by artificial neural network also information about sequence and improved significantly their prediction algorithm.

Results and Discussion

During testing the concept of our new peptide synthesizer, we prepared arrays of simple peptides. In one experiment we prepared 24 mutants of Leucine-enkephalin, YGGFL. We

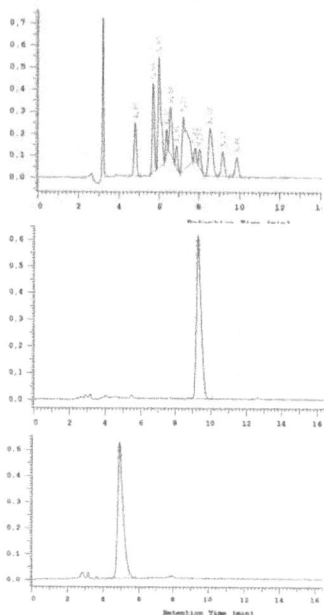

Fig. 1. RPHPLC traces of pentapeptides with the same amino acid composition. Top: mix of all 24 peptides; middle: YGGFL; bottom LYGGF.

observed quite significant differences of their retention times on RPHPLC (Table 1). Figure 1 shows that we can separate most of these mutants using slow gradient (22-32%ACN/0.05%TFA/20 min) on XBridge C18 5um, 240 mm column. Based on the predictions using just amino acid coefficients, obviously, they should all be eluted at the same time. Using prediction algorithms from the papers of Krokhin, et al. [2,5] and Tripet, et al. [6] we calculated expected retention characteristics and correlated them with our observed values. We obtained the following R2 values: 0.74 according to [5], 0.59 [2], 0.09 [6]. Obviously, the correlations are not extremely good or not existent. Comparing the retention of longer sequences containing the same composition, we can see that the difference in retention times is much smaller (Table 2).

Our speculation that distribution of hydrophobic residues in positions 1,3,5 would improve affinity to the hydrophobic phase, especially in sequence YGFGL, would result in the largest retention time was not confirmed (even though it is the sequence with 1,3,5 distribution of hydrophobic residues with highest retention time). The prediction by Krokhin, et al. [5] (the best for our peptide set) correctly identified GGYLF as the slowest peptide and confirmed the significant influence of the free amino terminus on the retention coefficients of hydrophobic residues, especially on Leucine residue. Another speculation about the effect of clustering or separation of aromatic residues was not

Table 1. Observed retention times and predicted retention orders.

Peptide	Rt	Predicted		
		[5]	[6]	[2]
FYLGG	4.35	-15.5	47.3	20.2
FLYGG	4.83	-28.2	47.3	19.1
LYGGF	4.88	-16.9	56.9	18.3
LFYGG	5.41	-30.8	44.6	19.4
YFLGG	5.73	-3.2	50.2	21.5
LFGGY	5.84	-30.4	52.8	19.1
LYFGG	6.05	-20.3	44.6	20.1
FGGYL	6.72	-9.8	56.5	20.5
LGYGF	6.99	-15.6	56.9	19.2
LGFGY	7.01	-18.7	52.8	20.6
FGLGY	7.04	-13.9	55.5	20.6
FGYGL	7.12	-10.2	56.5	20.4
GFYLG	7.2	3.6	45.6	22.4
YLFGG	7.49	-5.4	50.2	21.3
GFLYG	7.79	-0.1	45.6	22.7
GYFLG	8.19	14.1	45.6	23.5
GLYFG	8.29	0.8	45.6	22.3
GLFYG	8.51	-2.3	45.6	22.8
GGFLY	8.96	15.7	53.8	23.6
YGLGF	9.2	12.0	62.5	21.4
YGGFL	9.25	16.0	59.4	22.4
YGFGL	9.31	12.6	59.4	22.6
GYLFG	9.92	13.5	45.6	23.2
GGYLF	11.25	18.8	57.9	22.1

Table 2. Observed and predicted properties of decapeptides.

Peptide	Rt	Predicted [2]
FLYGGFLYGG	4.84	5.03
LYGGFLYGGF	4.86	5.21
GGFLYGGFLY	4.91	5.67
GFLYGGFLYG	4.94	5.49
YGGFLYGGFL	4.96	5.70
YGFGLGYGFL	5.07	5.64
YGGGGFLYFL	5.13	5.55

confirmed either. The largest difference from prediction to reality was observed for sequences YFLGG and LFGGY.

We tried to modify Krokhin's formula for the calculation of hydrophobicity

$$\sum R_c + 0.42R^1_{cNt} + 0.22R^2_{cNt} + 0.05R^3_{cNt}$$

(where R_c is retention coefficient of given amino acid; R^1_{cNt} is retention coefficient of given amino acid if it is in N-terminal position; R^2_{cNt} is retention coefficient of given amino acid if it is in second to N-terminal position; and R^3_{cNt} is retention coefficient of given amino acid if it is third from N-terminus) and we obtained better correlation with different coefficients for contribution of the first three amino terminal amino acids

$$\sum R_c + 0.69R^1_{cNt} + 0.38R^2_{cNt} + 0.16R^3_{cNt}$$

pointing to the more significant influence of the proximity to the charged peptide terminus in shorter peptides. However, due to the limited set of studied peptides and amino acid residues, we cannot generalize these findings. It would be interesting to see prediction of retention of our peptides by Petritis et al. [8] method.

In conclusion, obviously, the prediction algorithms are still inadequate for short peptides. We observed that the distance of an amino acid residue from amino terminus influences its retention characteristics in short peptides even more significantly than previously reported. On the other hand, clustering of aromatic residues does not seem to have strong effect on retention times.

References

1. Meek, J.L, Rosetti, Z.L. *J. Chomatogr.* **211**, 15-28 (1981).
2. Krokhin, O.V., Spicer, V. *Analytical Chemistry* **81**(22), 9522-9530 (2009).
3. Krokhin, O.V., Ying, S., Cortens, J.P., Ghosh, D., Spicer, V., Ens, W., et al. *Analytical Chemistry* **78**(17), 6265-6269 (2006).
4. Spicer, V., Yamchuk, A., Cortens, J., Sousa, S., Ens, W., Standing, K.G., Wilkins, J.A., et al. *Analytical Chemistry* **79**(22), 8762-8768 (2007).
5. Krokhin, O.V., Craig, R., Spicer, V., Ens, W., Standing, K.G., Beavis, R.C., Wilkins, J.A. *Molecular Cellular Proteomics MCP* **3**(9), 908-919 (2004).
6. Tripet, B., Renuka Jayadev, M., Blow, D., Nguyen, C., Hodges, R., Cios, K. *International Journal of Bioinformatics Research and Applications* **3**(4), 431-445 (2007).
7. Tripet, B., Cepeniene, D., Kovacs, J.M., Mant, C.T., Krokhin, O.V., Hodges, R.S. *J. Chromatogr. A.* **1141**(2), 212-225 (2007).
8. Petritis, K., et al. *Analytical Chemistry* **78**(14), 5026-5039 (2006).
9. Petritis, K., et al. *Analytical Chemistry* **75**(5), 1039-1048 (2003).

Proceedings of the 23rd American Peptide Symposium
Michal Lebl (Editor)
American Peptide Society, 2013

Left-Handed Helical Preference in an Achiral Peptide Chain is Induced by an L-Amino Acid in an N-Terminal Type II β-Turn

Matteo De Poli, Marta De Zotti, James Raftery, Juan A. Aguilar, Gareth A. Morris, and Jonathan Clayden

School of Chemistry, University of Manchester, Oxford Road, Manchester, M13 9PL, UK

Introduction

Oligomers of the achiral quaternary α-amino acid Aib (2-aminoisobutyric acid) adopt helical conformations known as 3_{10} helices, in which each monomer is hydrogen-bonded to the monomer three positions further along the chain [1]. Without any stereochemical bias, these Aib homopeptides display no preference for left- (*M*) or right-handed (*P*) helicity and undergo fast conformational inversion (k ~ 10^{-3} s). Such dynamic equilibrium may be biased in favour of either the right- or left-handed screw sense by capping the peptide *N*-terminus with a chiral residue. Our group has published a simple NMR method to report on the degree of the helical ratio obtained (*h.r.*) [2].

In the case of peptides 1 and 2 (Figure 1), the screw-sense induced in the helix depends not only on the configuration of the chiral *N*-terminal residue but also on whether it was *tertiary* (as in proteinogenic amino acid L-Val) or *quaternary* (L-α-methylvaline [L-(αMe)Val]) [3]. L-Val induces left-handed helicity in an oligo-Aib helix, while L-(αMe)Val induces right-handed helicity. This behaviour is in contrast with the right-handed screw sense preference for natural peptide helices made of C^{α}-trisubstituted L-amino acids.

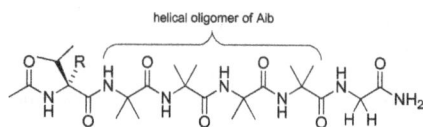

helical oligomer of Aib

peptide	R	screw-sense	h.r.
1	H (L-Val)	*M*	76:24
2	Me (L-(αMe)Val)	*P*	75:25

Fig. 1. Helicity ratios and screw-sense observed for peptide Cbz-Val-Aib$_4$-GlyNH$_2$ (1) and Cbz-(αMe)Val-Aib$_4$-GlyNH$_2$ (2).

Results and Discussion

X-ray conformational analysis of the structurally related peptides 3 and 4 (Figure 2) gave the first experimental evidence for the correlation between peptide helix screw sense and different types of turns at their *N*-termini. The values of the torsion angles (φ, ψ) in the two peptides are of similar magnitude but of opposite sign, reflecting the opposite screw sense observed in the two structures. Sign inversion takes place at ψ of Val^1 of peptide 3 which corresponds to position *i* + 2 of the first helix β-turn [4]. This reflects the opposite sense of twist imposed by different types of turns: left-handed for type II *vs.* right-handed for type III.

To detect the proposed *N*-terminal type-II β-turn as the origin of *M* helicity in peptide 1, we used two different NMR methods. We started by estimating interproton distances using NOE spectroscopy in MeOH-d_3 solution and comparing them with those found both in calculated structures [3] and in the *N*-terminal segment of the X-ray crystal structure of the related peptide 3. The strong cross-peak (Figure 3) detected between Val^1 C$^{\alpha}$H and Aib^2 NH immediately suggests the presence of a type II β-turn at the peptide *N*-terminus, this being the only possible conformation where the such protons are in close spatial proximity [4]. All other expected, diagnostic, C$^{\alpha(\beta)}$H(*i*)–NH(*i*+3), C$^{\alpha(\beta)}$H(*i*)–NH(*i*+2) and sequential NH–NH cross peaks

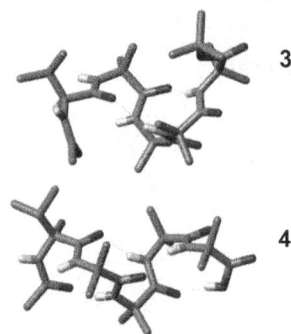

Fig. 2. X-ray crystal structures of Ac-Val-Aib$_4$-O^{t}Bu (3) and Ac-(αMe)Val-Aib$_4$-OH (4).

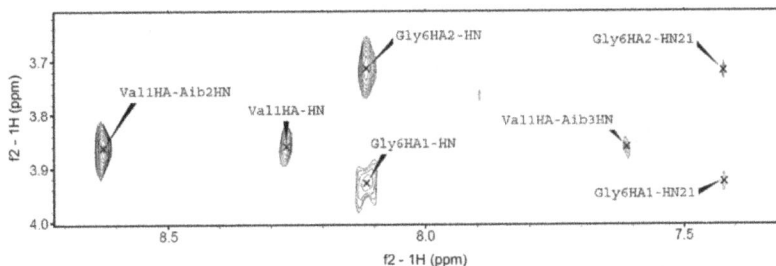

*Fig. 3. C$^\alpha$H-NH region of the NOESY spectrum of peptide **1** in CD$_3$OH.*

are visible and confirm the presence of a mainly 3$_{10}$-helical structure throughout the rest of the sequence.

We then used experimental coupling constants to calculate dihedral angles at peptide 1 N-terminus using the Karplus equation. $^1J_{C\alpha H\alpha}$ and $^3J_{NH\alpha}$ of Val^1 were measured using non-decoupled HMQC and phase-sensitive DQF-COSY. The $^3J_{NH\alpha}$ coupling constant is related to the torsion angle ϕ by equation (1):

$$^3J_{NH\alpha} = 6.4 \, cos^2 \, \theta - 1.4 \, cos \, \theta + 1.9 \text{ (where } \theta = |\phi - 60°|) \text{ [5]} \tag{1}$$

For J = 5.4 Hz, and taking into account the allowed (ϕ, ψ) torsion angle regions [4] of the Ramachandran plot for L-Val, the dihedral angle ϕ was calculated to be −70°. This value was used to determine ψ dihedral angles by means of equation (2):

$$^1J_{C\alpha H\alpha} = 140.3 + 1.4 \, sin \, (\psi + 138°) - 4.1 \, cos^2 \, (\psi + 138°) + 2.0 \, cos \, 2 \, (\phi + 30°) \text{ [6]} \tag{2}$$

By analogy, the calculated, allowed ψ torsion angles are ψ = +106° and ψ = +160°, both positive values. This fact explains why the N-terminal β-turn gives rise to a *left-handed screw-sense* in peptide **1**. The angles (ϕ, ψ = -69.0°, +166.4°) observed at Val^1 in the X-ray

Table 1. Comparison between observed and characteristic coupling constants in different protein secondary structures.

Constant	Found	α-Helical	Random coil	Poly-(Pro)II
$^3J_{HN\alpha}$	5.4	4.8	7.5	6.5
$^1J_{C\alpha H\alpha}$	141.1	146.3	141.5	142.6

structure of peptide **3** fit nicely with one of the two possible pairs of (ϕ, ψ) values obtained by combining equations (1) and (2), specifically (ϕ, ψ) = (-70°, +160°). This findings prove that the N-terminal segment of peptide **1** is folded into a type II β-turn.

As regards peptide **2**, the analysis was complicated by extensive overlapping in both the NH–NH and in the $^\beta$CH$_3$ - NH proton correlation regions. To overcome this problem a range of temperatures was screened. The variation of the NMR spectrum with temperature was consistent with the presence of highly populated 3$_{10}$-helical structures, where the chemical shift of the two N-terminal NHs show a higher temperature dependence than all other amide protons. Furthermore, L-(αMe)Val is well known to fold preferentially into right-handed 3$_{10}$ helices, as seen in the X-ray crystal structure of peptide **4**.

In summary, we conclude that the *M* helicity found in these Aib-rich peptides is the result of an N-terminal type II β-turn conformation occurring in both solution and in the solid state and is not the result of crystal packing forces.

Acknowledgments

Matteo De Poli is recipient of an APS Travel Grant Award. We thank ERC for funding.

References

1. Hummel, R.P., Toniolo, C., Jung, G. *Angew. Chem. Int. Ed. Engl.* **26**, 1150 (1987).
2. Clayden, J., Castellanos, A., Solà, J., Morris, G.A. *Angew. Chem. Int. Ed.* **48**, 5962 (2009).
3. Brown, R.A., Marcelli, T., De Poli, M., Solà, J., Clayden, J. *Angew. Chem. Int. Ed.* **51**, 1395 (2012).
4. Venkatachalam, C.M. *Biopolymers* **6**, 1425 (1968).
5. Pardi, A., Billeter, M., Wüthrich, K. *J. Mol. Biol.* **180**, 741 (1984).
6. Vuister, G.W., Delaglio, F., Bax, A. *J. Am. Chem. Soc.* **114**, 9674 (1992).

Proceedings of the 23rd American Peptide Symposium
Michal Lebl (Editor)
American Peptide Society, 2013

Thioamides as CD and NMR Reporters of Helical Screw-Sense

Matteo De Poli and Jonathan Clayden

School of Chemistry, University of Manchester, Oxford Road, Manchester, M13 9PL, UK

Introduction

In the absence of a stereochemical influence, oligomers of the achiral, quarternary amino acid Aib (α-aminoisobutyric acid) exist as a rapidly inverting racemic mixture of enantiomeric left- (*P*) and right-handed (*M*) 3_{10} helices [1] in solution. Incorporation of an *N*-terminal chiral amino acid can bias this equilibrium and favour one of the two screw-sense conformations, which can be "read" spectroscopically [2]. Such helical preference can be induced in Aib peptides by covalently attaching an amino acid to their *N*-terminus as a chiral controller. Our group developed a simple NMR method to quantify the degree of screw-sense preference (*h.r.*), which relies on a pair of diastereotopic nuclei located within the helix [3]. When the rate of helix inversion is slow on the timescale of their chemical shift separation, the two reporter nuclei will give rise to *anisochronous* signals $\Delta\delta_{slow}$. At fast exchange (for example at higher temperature) a new averaged pair of anisochronous signals arises with peak separation $\Delta\delta_{fast}$. In the absence of a chiral influence, $\Delta\delta_{fast}$ must be 0. The '*helicity excess*' can be quantified by dividing $\Delta\delta_{fast}$ by $\Delta\delta_{slow}$ (Figure 1). VT-NMR analysis allowed the measurement of the h.e. of a small library of peptides using a ^{13}C-labeled Aib residue (A, B = ^{13}CH$_3$, Figure 1) [3]. We then extended this method by comparing the observed $\Delta\delta_{fast}$ values of the geminal protons in a *C*-terminal glycinamide (A, B = ^{1}H, measured in parts per billion) with those previously obtained with the ^{13}C-labeled Aib probe [4].

$$\text{h.e.} = \frac{\Delta\delta}{\Delta\delta} = \frac{([\ ]-[\ ])}{([\ ]+[\ ])} = \frac{(\ -1)}{(\ +1)}$$

Fig. 1. NMR method to derive the helicity excess.

Results and Discussion

Although NMR allows us to quantify the level of helical screw-sense control, it fails to provide information about the absolute handedness ((*P*) or (*M*)) adopted by each helix. CD spectroscopy is generally used to investigate the conformation of helical structures although it does not offer a residue-specific level of detail. Replacement of amide bonds with thioamide bonds (ψ[CSNH]) represents a non-intrusive, isosteric modification of the peptide backbone which displays several interesting chemical and physical properties [5].

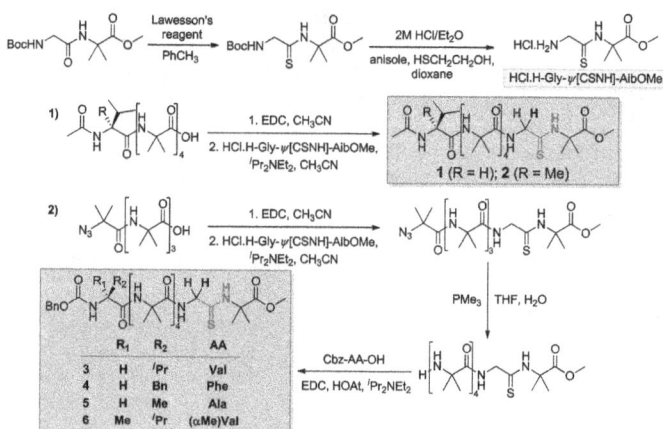

Fig. 2. Synthesis of the endothiopeptides.

We chose to use the red-shifted absorption of a thionated glycinamide as an NMR- and CD-active helicity reporter capable to sense the local handedness adopted by the helix. We explored the utility of this method by synthesizing a set of Aib oligomers (Figure 2) bearing different N-terminal chiral controllers and a Gly-ψ[CS-NH]-Aib-OMe helicity reporter at their C-terminus.

A correlation is observed (Figure 3) between the measured ppb values of the thioglycinamide probe and the relative values obtained using the non-thionated GlyNH$_2$, confirming the accuracy of Gly-ψ[CSNH] as a reporter for the helical screw-sense preference. The more bulky, branched amino acids give greater levels of control.

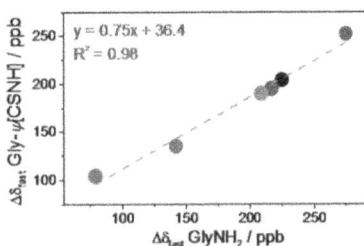

Fig. 3. Comparison between ppb separation observed for the thionated and non-thionated glycinamide probes in peptides 1-6.

Figure 4 shows the CD spectra of endothiopeptides **1-6**. The distinct Cotton effect at about 268 nm is assigned to the π-π* transition of the thioamide chromophore and is clearly separated from the rest of the peptide bond absorptions. The n-π* band is hardly visible (~340 nm). As hoped, the sign of the ψ[CSNH] π-π* absorption band centered at 268 nm reflects the local orientation of the helicity as experienced by the thioglycinamide probe. Endothiopeptides capped with C$^\alpha$-tetrasubstituted amino acids show positive Cotton effect, while the opposite trend is observed with C$^\alpha$-tertiary (proteinogenic) amino acids, despite all sharing the same (L) configuration. This is in agreement with the reported behaviour of peptides -Xxx-Aib$_n$-, which adopt (M) helicity if Xxx is a *tertiary* L-amino acid and (P) helicity if Xxx is a *quaternary* L-amino acid [6].

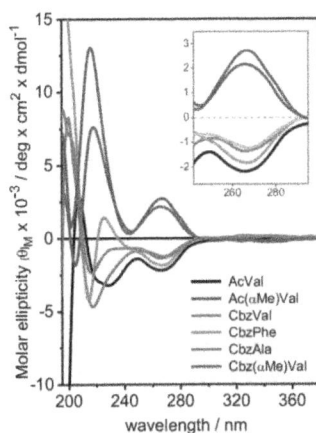

Fig. 4. CD spectra of endo-thiopeptides 1-6.

Comparison of the absolute molar ellipticities ([θ]$_M$) at 268 nm (ψ[CSNH] π-π* absorption) parallels the extrapolated h.e. values obtained previously using different NMR probes (GlyNH$_2$ or ^{13}C-labeled Aib[3]).

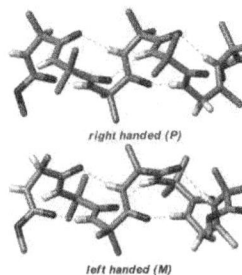

Fig. 5. X-ray structure of peptide 5.

X-ray conformational analysis of peptide **5** (Figure 5; phenyl ring removed for clarity) shows that this peptide folds into both M and P 3$_{10}$-helices in the solid state, initiated by different β-turns [6]. This confirms the poor level of screw-sense control exhibited by Ala.

In conclusion, we have demonstrated the ability of a new thioglycinamide probe to report on the orientation and the magnitude of the M or P helicity adopted by Aib oligomers capped by chiral amino acids at their N-teminus. The versatility of the probe is shown by being responsive towards both NMR and CD spectroscopy.

Acknowledgments

Matteo De Poli is recipient of an APS Travel Grant Award. We thank ERC for funding.

References

1. Hummel, R.P., Toniolo, C., Jung, G. *Angew. Chem. Int. Ed. Engl.* **26**, 1150 (1987).
2. Solà, J., Helliwell, M., Clayden, J. *J. Am. Chem. Soc.* **132**, 4548 (2010).
3. Clayden, J., Castellanos, A., Solà, J., Morris, G. A. *Angew. Chem. Int. Ed.* **48**, 5962 (2009).
4. Brown, R.A., Marcelli, T., De Poli, M., Solà, J., Clayden, J. *Angew. Chem. Int. Ed.* **51**, 1395 (2012).
5. Alemán, C. *J. Phys. Chem. A* **105**, 6717 (2001); Zhao, J., et al. *Chem. Eur. J.* **10**, 6093 (2004); Reiner, A., Wildemann, D., Fischer, G., Kiefhaber, T. *J. Am. Chem. Soc.* **130**, 8079 (2008).
6. De Poli, M., et al. *J. Org. Chem.* **78**, 22486 (2013).

Proceedings of the 23rd American Peptide Symposium
Michal Lebl (Editor)
American Peptide Society, 2013

Amyloid Peptide Self-Assembly in Protic Ionic Liquids

Natalie J. Debeljuh[1], Colin J. Barrow[1], and Nolene Byrne[2]

[1]*School of Life and Environmental Sciences, Deakin University, Geelong, Victoria, 3216, Australia;*
[2]*Institute for Frontier Materials, Deakin University, Geelong, Victoria, 3216, Australia*

Introduction

Amyloid proteins are renowned for their ability to self-assemble into highly ordered fibrils which are associated with diseases such as Parkinson disease, type II diabetes and Alzheimer's disease [1]. These fibrils are the result of a complex protein folding pathway that is still intensely debated among scientists worldwide. Currently, the neurotoxic species in the amyloid fibril assembling process is thought to be some form of soluble oligomer, most likely those that form from the early stages of Aβ aggregation [2,3]. Recent trends have been directed at finding ways to control amyloid self-assembly to allow for early state oligomer characterization [4]. New designer solvents such as protic ionic liquids (pILs) have received increased attention in protein refolding and renaturing studies [5,6].

Because the formation of amyloid fibrils proceeds *via* the formation of intramolecular hydrogen bonds, a selection of pILs with distinctly different hydrogen bonding abilities were investigated to better control the fibrilization process of Aβ$_{16-22}$. These pILs consist of the common cation, triethylammonium (Tea), and various anions including dihydrogen phosphate (H_2PO_4), hydrogen sulfate (HSO_4), trifluoro acetate (Tfac), lactate (La), triflate (Tf) and mesylate (Ms).

Results and Discussion

The kinetics of Aβ$_{16-22}$ fibrilization was investigated in these pILs at 90 wt% concentration using the Thioflavin T (ThT) fluorescence assay (Figure 1). Since ThT is sensitive to beta sheet structure, an increase in fluorescence observed is representative of a higher organized amyloid state that is rich in beta sheet structure. As illustrated, both TeaH$_2$PO$_4$ and TeaHSO$_4$ show very fast fibrilization with a maximum ThT intensity observed after 1 min. TeaTfac, TeaLa and TeaTf show lag phases with fibrilization proceeding to a measured maximum

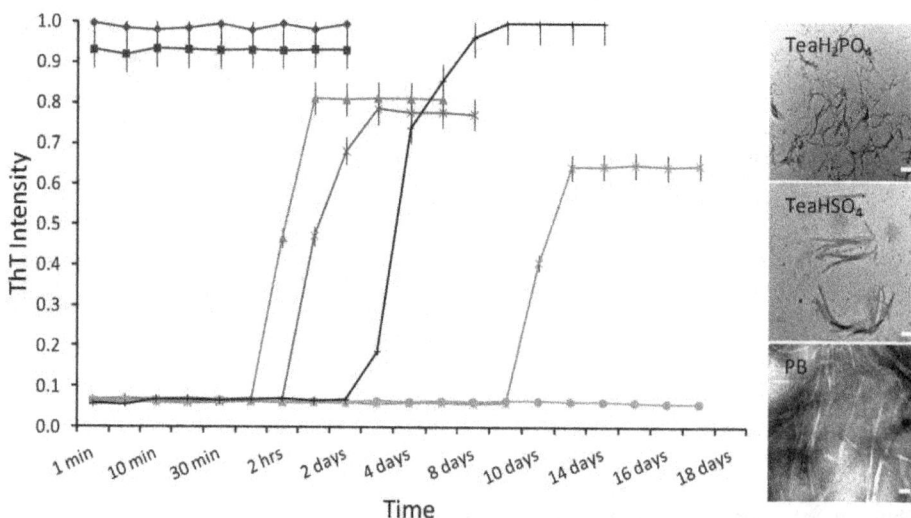

Fig. 1. *ThT intensity as a function of time for Aβ$_{16-22}$ in 90 wt% TeaH$_2$PO$_4$ (dark blue diamonds), TeaHSO$_4$ (red squares), TeaTfac (green triangles), TeaLa (purple crosses), phosphate buffer pH 7 (black line), TeaTf (pale blue stars) and TeaMs (orange circles) Data points were measured within 5% error. TEM images of Aβ$_{16-22}$ fibrils in TeaH$_2$PO$_4$, TeaHSO$_4$ and phosphate buffer after self-assembly. Scale bar 200 nm.*

after 2 hours, several days and weeks respectively. Surprisingly, no increase in ThT intensity was observed for TeaMs even after several months. In the case for TeaTfac, TeaLa and TeaTf, the maximum ThT signal measured is less than 1, suggesting the presence of aggregated states or different amyloid structures. Despite the large variation in fibril kinetics for $TeaH_2PO_4$ and TeaMs, the peptide remains in an extended 3(10)-helical conformation according to CD analysis. This implies that the initial stages of self-assembly are very similar and that an alternative explanation, such as specific hydrogen bonding interactions, could be causing these trends. The pK_a of each pIL is different and does not follow the trend observed in the fibrilization kinetics. Since the kosmotropic anions $H_2PO_4^-$ and HSO_4^- allow fibrilization to occur within seconds, and the chaotropic mesylate anion suppresses fibrilization, this competitive hydrogen bonding between the anions of the pILs and water molecules are driving the self-assembly of amyloid fibrils.

We further examined the hydrogen bonding environment of these two pILs by conducting a simple proton chemical shift analysis as a function of concentration (Figure 2). In the case of $TeaH_2PO_4$, as the pIL content is increased, a large downfield chemical shift is observed for water. This downfield shift implies that water is strongly interacting with the pIL. Since the $H_2PO_4^-$ anion lacks any detectable protons, a ^{31}P NMR titration was carried out and confirmed that the anion was interacting strongly with water due to a large change in chemical shift of the phosphorus atom (data not shown). Interestingly, TeaMs as a function of concentration shows virtually no chemical shift in the water peak, and the proton designated to the mesylate anion shows a slight upfield chemical shift as the pIL concentration is increased. Due to the kosmotropic nature of the $H_2PO_4^-$ anion, water is becoming less shielded and less electron dense because of its strong hydrogen bonding interaction with the anion. In terms of amyloid fibrilization, this confirms that the competitive hydrogen bonding between the anion and water are driving the self-assembly of amyloid fibrilization.

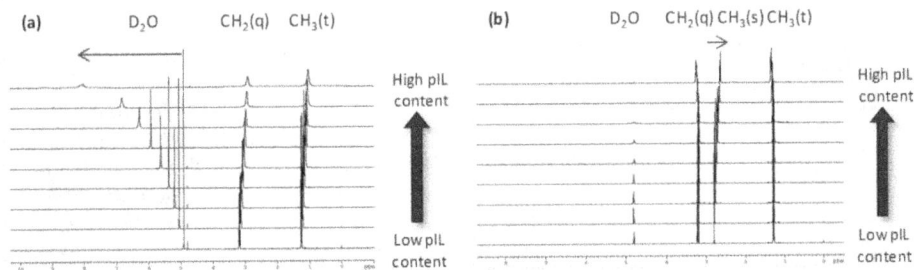

Fig. 2. ^1H NMR chemical shift of (a) $TeaH_2PO_4$ and (b) TeaMs as a function of water. Spectra were individually calibrated with TMS using an internal capillary.

Acknowledgments

The authors would like to thank the Centre for Biotechnology, Chemistry and Systems Biology for financial assistance. Dr. Nolene Byrne would like to acknowledge the Australian Research Council for an APD.

References

1. Dobson, C.M. Trends Biochem. Sci. 24, 329-332 (1999).
2. Kirkitadze, M.D., Bitan. G., Teplow. D.B. J. Neurosci. Res. 69, 567-577 (2002).
3. Haass, C., Selkoe. D.J. Nat. Rev. Mol. Cell. Biol. 8, 101-112 (2007).
4. Broersen, K., Rousseau, F., Schymkowitz, J. Alzheimer's Research & Therapy 14, 12 (2010).
5. Summers, C.A., Flowers, R.A. Protein Sci. 9, 2001-2008 (2000).
6. Lang, C., Patil, G., Rudolph, R. Protein Sci. 14, 2693-2701 (2005).

Proceedings of the 23rd American Peptide Symposium
Michal Lebl (Editor)
American Peptide Society, 2013

Structure of the Human AT1 Receptor Bound to Angiotensin II: Structural Basis of Previous SAR Findings

Jérome Cabana*, Dany Fillion*, Gaétan Guillemette, Richard Leduc, Pierre Lavigne, and Emanuel Escher

Department of Pharmacology, Faculty of Medicine & Health Sciences, Université de Sherbrooke, Sherbrooke, Québec, Canada, J1H 5N4; *contributed equally to this work.

Introduction

Structure-activity relationship studies have been a mainstay activity for MedChem approaches on GPCR. On the Angiotensin II (AngII) system a great wealth of such information has been accumulated, predominantly on the AT1 receptor. The present contribution on a first experimentally determined structure of the AT1 receptor (Figure 1) intends therefore to validate previously found SAR observations by evidencing the molecular interactions of the ligand-receptor interactions responsible for the respective SAR. In particular it was observed on position 4 (Tyr) that an electronegativity correlation existed that necessitated simultaneous H-bridge donor and acceptor functions in the para-position of the aromatic nucleus. A second feature was that enhanced hydrophobicity in position 8 increased duration of action and antagonistic behaviour resulting in peptides that had duration of action *in vivo* comparable to ACEi and ARB compounds.

Hypothesis: If the receptor structure is correct then the respective molecular interactions responsible for these properties should be present and identifiable in MD simulations.

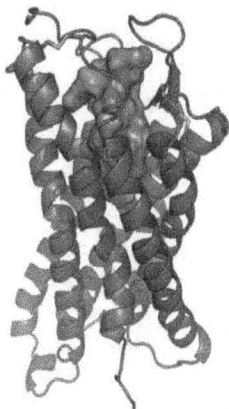

Fig. 1. Homology model of the AT1 receptor based on the chemokine CXCR4 receptor. The model was generated using the LOMETS web server [1] by vectoring in distance restrictions from 52 experimentally determined ligand-receptor contact points. These contacts were obtained from photolabeling experiments on a combined Met-walk of the receptor AT1 and a photolabel walk in the angiotensin ligand (Methionine proximity assay, MPA) [2]. The AngII molecule (pdb 1N9V) was manually positioned based on MPA results. The strongest MPA-reported contacts for each ^{125}I-AngII analogs were used as molecular distance restraints for a short (1ns) MD simulation to further refine the model of the complex. The OPLS/AA force field with the GBSA implicit solvent model was used and the backbone atoms of the transmembrane domains were fixed during this step. AT1 elements are depicted in blue, AngII elements in orange.

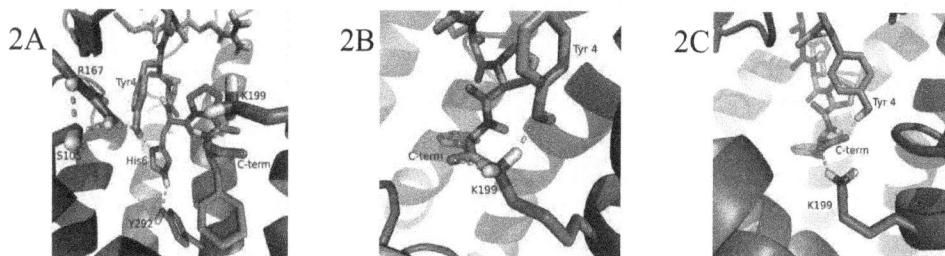

Fig. 2A, 2B, 2C. Ph–OH acts as acceptor (Fig. 2A and 2B) and donor (Fig. 2A and 2C). Fig. 2 shows the AT1R-Ang-II complex through the 1ns MD simulation, as described in Fig. 1. This model was used as the starting point for more extensive MD simulations of the complex without any restraints in the system. The GROMACS software was used to perform ten 40 ns simulations using the wild type receptor and five using the constitutively active N111G AT1). Figures represent snapshots of the 40 ns simulations.

Results

Position 4, Tyr: SAR has always shown the essential nature of this residue for binding and activation. In particular, the affinity of analogues modified in pos. 4 was inversely proportional to electronegativity of this aromatic residue and optimal in presence of a hydrogen bridge donor and acceptor function [3]. Molecular Dynamics simulations showed continuous interaction of Tyr4 with two aromatic residues of the 2nd extracellular loop, F182 and Y184 (not shown). Intense hydrogen-bridge donor-acceptor interactions with position His6 of the ligand and R167 of the receptor intercede with interactions between the C-terminal carboxylate of AngII and K199 where the phenolate acts as acceptor and donor (Figure 2).

Position 8, Phe: Residue 8 was always considered essential for activity initiation since substitution of this residue by aliphatic residues produced antagonistic behaviour (blood pressure reduction, smooth muscle contraction, IP3 production). With increasing bulk and hydrophobicity progressive increase in duration and reduced efficacy was observed. Extreme substitutions, e.g. Br_5Phe and pyrenyl Ala (PyA), were neutral antagonists with long duration of action, $[Sar^1, Br_5Phe^8]$AngII had several hours antihypertensive action in a hypertensive dog model and non-measurable off-rate in binding studies [4,5]. To investigate this particular property $[Sar^1, PyA^8]$AngII was compared to AngII in MD simulations. The Phe8-residue was moving in and out of the hydrophobic receptor core formed by F77, V108, L112, W253, I288, A291 and Y292 to structures shown in Figure 3A and 3B whereas the PyA residue did never dissociate from this hydrophobic core (Figure 3C and 3D).

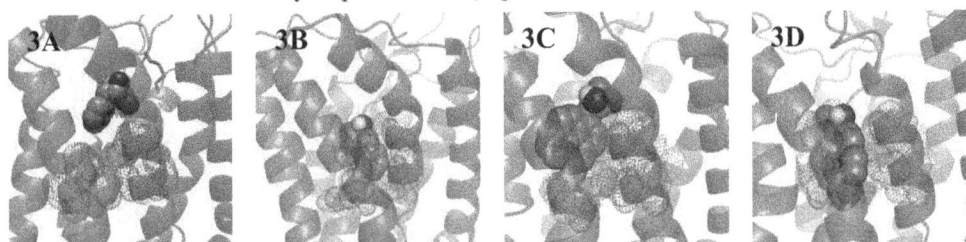

Fig. 3A-3D. The hydrophobic core element is indicated in grey dotted surface outline. MD calculations were carried out as indicated for Fig. 2. Figs. 3A and 3B are snapshots of AngII (Phe⁸) during MD simulations, Figs. 3C and 3D are snapshots from [Sar¹,PyA⁸]AngII MD simulations. AT1 transmembrane elements are, from left to right and in front, TMD5, TMD6, TMD7 and TMD1.

Discussion and Conclusion

These two examples of particular SAR features from earlier studies are quite well explained by the present Molecular Dynamics experiments on the generated AT1 structure, lending credibility and confidence to the structure. In order to refine it further, several other SAR features of different parts of the AngII molecule and, eventually, of non-peptide ligands and their SAR, will be explored by this approach. Exact molecular structures are needed to perform meaningful docking studies for the discovery of synthetic ligands for therapeutic purposes; but highly exact molecular structures, including their structural mobility and energy barriers in-between, are of particular importance today: Structural changes induced by biased agonists, the mode of action of biased ligands and their therapeutic potential renewed interest into "old" drug targets to address unmet medical needs and rekindled the hope to develop drugs with more efficacy and less unwanted side-effects.

Acknowledgements

These research activities were supported by CIHR; JC holds a studentship from FRQNT and EE is a J.C.Edwards professor of cardiovascular research.

References

1. Wu, S., et al. *Nucl. Acids Res.* **35**, 3375 (2007).
2. Fillion, D., et al. *J. Biol. Chem.* **288**, 8187 (2013).
3. Guillemette, G., et al. *J. Med. Chem.* **27**, 315 (1984).
4. Bossé, R., et al. *J. Cardiovasc. Pharmacol.* **16**, Suppl 4: S50 (1990).
5. Holck, M., et al. *Biochem. Biophys. Res. Comm.* **160**, 1350 (1989).

Proceedings of the 23rd American Peptide Symposium
Michal Lebl (Editor)
American Peptide Society, 2013

Design, Synthesis, and Conformational Analysis of Melanotropin Analogues of MTII and SHU9119

Paolo Grieco[1], Ali M. Yousif[1], Alfonso Carotenuto[1], Antonio Limatola[1], Diego Brancaccio[1], Ettore Novellino[1], and Victor J. Hruby[2]

[1]Department of Pharmacy, University of Naples, Naples, 80131, Italy; [2]Department of Chemistry and Biochemistry, University of Arizona, Tucson, AZ, 85721-0041, U.S.A.

Introduction

The melanocortin receptors are involved in many physiological functions, including pigmentation, sexual function, feeding behavior, and energy homeostasis, making them potential targets to treat obesity, sexual dysfunction, etc. [1]. Previously, we have demonstrated that replacing His[6] by Pro[6] in the well-known agonist MTII or antagonist SHU-9119 resulted in potent and/or selective ligands at MCRs [2,3]. Further replacement of a DPhe or DNal at position 9 brought active compounds [3,4]. To further investigate the effect of a D-residue at position 9, we considered the derivatives PG987 and PG988 bearing a DTrp[9] and derivatives PG960 and PG961 bearing a DNal[9]-Gly[10] sequence (Figure 1).

Results and Discussion

All peptides were prepared on acid-labile 2Cl-Trt resin by solid-phase synthesis of linear peptide sequences, using the Fmoc protection strategy, followed by cyclization and side-chain deprotection in solution. Cyclization was carried out after removal of the Allyl/Alloc protection according to strategy reported by Grieco, et al. [5]. Peptide PG987 showed a very interesting activity/selectivity profile. It is about 200-fold more selective for the hMC4R vs. the hMC3R (IC$_{50}$ = 2.23 and 464 nM, respectively). More interestingly, it shows full antagonist activity at the hMC4R regardless of the presence of a DPhe residue at position 7.

Code	Sequence
MTII	*Ac-Nle-[Asp-His-DPhe-Arg-Trp-Lys]-NH₂*
SHU9119	*Ac-Nle-[Asp-His-DNal-Arg-Trp-Lys]-NH₂*
PG987	*Ac-Nle-[Asp-Pro-DPhe-Arg-DTrp-Lys]-NH₂*
PG988	*Ac-Nle-[Asp-Pro-DNal-Arg-DTrp-Lys]-NH₂*
PG960	*Ac-Nle-[Asp-Pro-DPhe-Arg-DNal-Gly-Lys]-NH*
PG961	*Ac-Nle-[Asp-Pro-DNal-Arg-DNal-Gly-Lys]-NH₂*

Fig. 1. Sequence of novel analogues of MTII and SHU-9119.

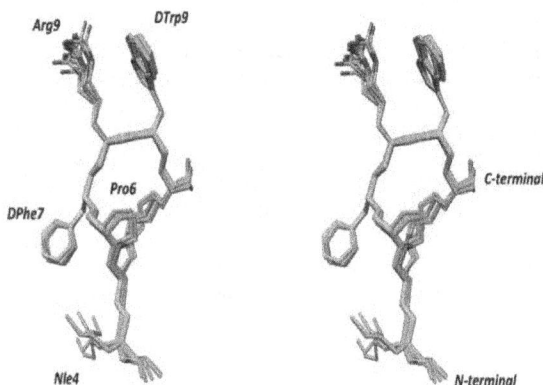

Fig. 2. Stereoviews of the 10 lowest energy conformers of PG987. Structures were superimposed using the backbone heavy atoms.

Table 1. Binding and intracellular cAMP iccumulation of the α-melanotropin analogues at human melanocortin receptors.

Code	hMC3R			hMC4R			hMC5R		
	IC_{50} (nM)	EC_{50} (nM)	%Act. at $10\mu M$	IC_{50} (nM)	EC_{50} (nM)	%Act. at $10\mu M$	IC_{50} (nM)	EC_{50} (nM)	%Act. at $10\mu M$
MTII	1.25	1.85	100	1.1	2.9	100	7.5	3.3	100
SHU9119	2.3	-	0	0.6	-	0	0.9	1.2	97
PG987	464	1671	14.1	2.23	-	0	0.01	57.6	80.4
PG988	88	112	100	93.4	262.8	38	160.2	116.6	95
PG960	2.52	0.125	88	8.47	14	86	1.68	0.5	102
PG961	2.21	-	0	0.09	0.7	38	0.01	0.16	107

In contrast, peptide PG988, which derives from SHU-9119, henceforth supposed to be an antagonist, showed full agonist activity at hMC3R and partial agonist activity at hMC4R. The addition of a Gly^{10} residue restores the expected agonist activity in PG960 and antagonist activity at hMC3R (partial agonist at hMC4R) in PG961. Conformational analysis of PG987 was performed in DPC micelle solution. NMR parameters indicated a folded conformation with several medium range NOEs. Among those: $d_{\alpha N(i, i+2)}$ between Asp^5 and $DPhe^7$, Pro^6 and Arg^8, and Arg^8 and Lys^{10}; and $d_{NN(i, i+2)}$ between Arg^8 and Lys^{10}. Furthermore, many interchain NOEs and the up-field shifts of Arg^8 signals indicate that its side chain is close to $DTrp^9$. NOE derived constrained were used as input data for subsequent structure calculation by simulated annealing MDs. Ten lowest energy conformers of PG987 are shown in Figure 2. Two β-turns can be identified, centered on Pro^6-$DPhe^7$ and Arg^8-$DTrp^9$. As shown, Nle^4 and $DPhe^6$ as well as Arg^8 and Trp^9 side chains are spatially close. The last disposition is very different from that observed in MTII [6] where the two side chains are far apart. Since Trp^9 has been involved in receptor activation through its interaction with His^{264} [6], conformational results can explain the activity switch from agonist to antagonist observed for PG987. Conformational analysis on the cognate peptide PG988 and docking studies using MCR models are currently in progress.

In conclusion, the results demonstrated that configuration at residue 9 can modulate activity and selectivity of MTII and SHU9119 analogues. In particular, compound PG987 showed a very interesting potency and selectivity as MC4R antagonist.

Acknowledgments

Supported by grant from the Italian Ministry of Education (MIUR) (PRIN n° 2009EL5WBP).

References

1. Cone, R. *The Melanocortin Receptors* Humana Press, Totowa, NJ, 2000.
2. Grieco, P., et al. *Biochem. Biophys. Res. Commun.* **292**, 1075-1080 (2002).
3. Grieco, P., et al. *J. Pept. Res.* **62**, 199-206 (2003).
4. Grieco, P., et al. *Peptides* **27**, 472-481 (2006).
5. Grieco, P., et al. *J. Pept. Res.* **57**, 250-256 (2001).
6. Grieco, P., et al. *Eur. J. Med. Chem.* **46**, 3721-3733 (2011).

Proceedings of the 23rd American Peptide Symposium
Michal Lebl (Editor)
American Peptide Society, 2013

Structure Activity Relationship of Dynorphin A(2-13) Analogs at the Bradykinin-2 Receptor

Sara M. Hall[1], Yeon Sun Lee[1], Cyf Nadine Ramos Colón[1], David R. Rankin[2], Frank Porreca2, Josephine Lai[2], and Victor J Hruby[1]

Departments of [1]Biochemistry and Chemistry, [2]Pharmacology, University of Arizona, Tucson, AZ, 85721, U.S.A.

Introduction

Neuropathic pain affects 100 million Americans and imposes a significant public health problem [1]. This type of pain results from the dysfunction of the central nervous system (CNS) or the peripheral nervous system (PNS) that can occur in the presence or absence of an initial injury [1]. Treatment for this disease is difficult with conventional methods, partly because the mechanisms of this disease are not well known. Thus, there is a pressing need to investigate the mechanisms underlying neuropathic pain with the goal of developing better drugs to treat more targets.

One target for neuropathic pain treatment may be the blockade of Dynorphin A (Dyn A). Dyn A has been found to be upregulated in the dorsal horn of the spinal cord under a number of pathologies that result in chronic pain states [2]. The peptide was first discovered as an endogenous opioid peptide having inhibitory effects in the spinal cord. Its *N*-terminal tyrosine is essential for its high affinity, agonist action at the opioid receptors (μ, δ, κ). Aminopeptidase removal of this tyrosine is sufficient to inactivate the opioid action of Dyn A as the *des*-tyrosyl fragments of Dyn A do not bind opioid receptors ($K_i > 10,000$ nM) [2]. However, these fragments remain biologically active both *in vitro* and *in vivo*. It has been shown previously that Dyn A(2-13) activates bradykinin-2 receptors (B2Rs) in a neuronal cell line to induce calcium influx [3]. This neuronal excitatory effect was proposed to underlie the pronociceptive actions of spinal Dyn A, and may be a critical mediator of experimental models of inflammatory pain [4] and neuropathic pain [3,5]. Therefore, the development of B2R antagonists can be used to block the pronociceptive actions of Dyn A.

Results and Discussion

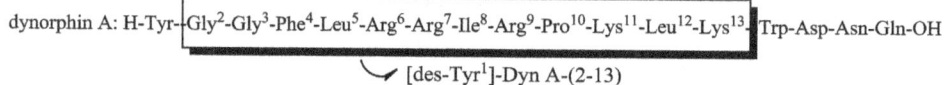

dynorphin A: H-Tyr-$|$Gly^2-Gly^3-Phe^4-Leu^5-Arg^6-Arg^7-Ile^8-Arg^9-Pro^{10}-Lys^{11}-Leu^{12}-Lys^{13}$|$Trp-Asp-Asn-Gln-OH

[des-Tyr^1]-Dyn A-(2-13)

Fig. 1. Structure of dynorphin A. Box shows non-opioid fragment, Dyn A(2-13).

Structure Activity Relationship (SAR) of Dyn A Analogs at the B2R

To explore novel therapeutics for neuropathic pain, a SAR study was performed with Dyn A analogs (Figure 1) at the B2R. It was previously found that LYS1044 ([des-Arg^7] Dyn A(4-11)) ($IC_{50} = 98$ nM) is the shortest pharmacophore for the B2R, and this peptide has shown promising *in vivo* data in a chronic pain animal model as an antagonist. Although this pharmacophore appears to be efficacious, it is composed of natural amino acids and therefore may be susceptible to degradation by peptidases. In an effort to improve the peptide's stability, Dyn A analogs were designed and synthesized that were further tested for binding at the rat brain B2R. Based on the SAR results, we found that *N*-terminal acetylation as well as replacement of the non-natural Nle in place of Leu/Ile retained affinity at the B2R. *C*-terminal amidation as well as inverso modifications (inverse in chirality) resulted in loss of affinity at the B2R. Although a complete reverse of chirality resulted in loss of affinity ($K_i > 10,000$ nM), a D amino acid scan found that a single substitution of DPhe^4, Leu^5, or DArg^6 resulted in a retention of affinity ($K_i \sim 100$ nM). Substitution of Pro^{10} by proline analogs (Tic, DTic, Oic, Thi) resulted in a loss of affinity.

From these peptides and others, we conclude that a basic (Lys, Arg, or Orn) C-terminus is important for high affinity at the B2R, and binding of Dyn A analogs to the B2R is through electrostatic interactions. The replacement of non-natural amino acids for Leu/Ile and some D amino acid substitutions were okay and may improve the stability of these compounds. Also N-terminal acetylation was well tolerated and could also improve stability. With the more potent analogs, *in vitro* plasma stability tests, functional assays, and studies in animal pain models will be carried out.

pH Effect of Dyn A analogs at the Bradykinin 2 Receptor

In the process of screening new compounds, we discovered that Dyn A's interaction with the B2R is surprisingly sensitive to pH within a narrow physiological range (Figure 2). The pH values between 6.8 to 7.4 are the optimal range for binding of Dyn A analogs at the B2R (IC_{50}= 22 nM and 38 nM respectively). The affinity of Dyn A(2-13) for the B2R decreased significantly at the higher pH of 8.0 and 8.5 (IC_{50}= 200 nM and 1700 nM respectively). The pH effect on Dyn A binding to the BRs thus unmasks a selective, high affinity binding site at the B2R for Dyn A. The pharmacological characteristics of Dyn A at the B2R suggest that at physiological concentrations, Dyn A likely acts as an opioid due to its high affinity and efficacy at μ (K_i = 5.3 nM) and κ (K_i = 1 nM) receptors. However, conditions which produce elevated levels of Dyn A and its proteolytic fragments may favor their activity at the B2Rs, particularly if accompanied by a shift in tissue pH, such as under conditions of tissue injury.

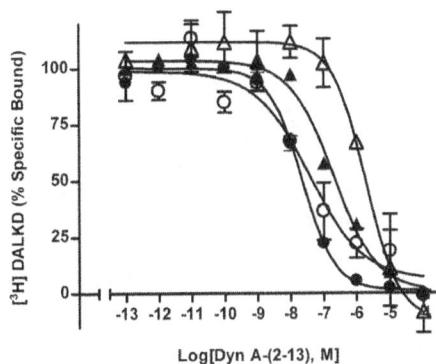

Fig. 2. Competitive binding of Dyn A (2-13) against [³H]DALKD in crude membranes prepared from adult rat brain. The IC_{50} values for Dyn A (2-13) are: 22 nM at pH 6.8 (•), 38 nM at pH 7.4 (○), 200 nM at pH 8.0 (▲), and 1700 nM at pH 8.5 (Δ).

Acknowledgments

We thank Ann Ngyuene, Alyssa Peake, Alice Cai, Robert Kupp, and Lindsay Lebaron for their help in synthesis and bioassay of ligands. Supported by grants from the U.S. Public Health Services, NIH, and NIDA (P01DA006248). We also thank APS as Sara Hall is a recipient of an APS Travel Award.

References

1. Institute of Medicine Report from the Committee on Advancing Pain Research, Care, and Education: *Relieving Pain in America, A Blueprint for Transforming Prevention, Care, Education and Research.* The National Academies Press, 2011.
2. Chavkin, C., Goldstein, A. *Nature* **291**, 591 (1981).
3. Lai, J., Luo, M.C., Chen, Q.M., Ma, S.W., Gardell, L.R., Ossipov, M.H., Porreca, F. *Nature Neuroscience* **9**, 1534-1540 (2006).
4. Luo, M.C., Chen, Q.M, Ossipov, M.H., Rankin, D.R., Porreca, F., Lai, J., *J. Pain* **9**, 1096-1105 (2008).
5. Lai, J., Luo, M.C., Chen, Q.M., Porreca, F. *Neuroscience Lett.* **437**, 175-179 (2008).

Proceedings of the 23rd American Peptide Symposium
Michal Lebl (Editor)
American Peptide Society, 2013

Running Interference on Protein Aggregation: Effects of Non-Specific Peptide Inhibitors of Amylin Aggregation

Hector Figueroa and Deborah L. Heyl

Department of Chemistry, Eastern Michigan University, Ypsilanti, MI, 48197, U.S.A.

Introduction

Protein aggregation is a widespread but poorly understood phenomenon occurring in ailments ranging from Alzheimer's and Mad Cow Disease to Type-2 Diabetes (T2D). In T2D, rational design methods to prevent aggregation of the polypeptide amylin have had moderate success [1]; however, we still do not know whether or not *any* peptide molecule in solution with amylin will alter its behavior. Can non-specific interactions between peptides change the rate of amyloid fibril formation or impact amylin-induced membrane damage? In this study, we present data on a set of 5 heptapeptide sequences with no known relation to amylin and their effect on aggregation kinetics and lipid membrane damage.

Results and Discussion

Table 1. NS peptide compounds.

Peptide	Sequence	MW (g/mol)
NSx5	YAFDVVG	768.87
NSx6	YFSPSFY	909.01
NSx7	YFKPKFY	991.20
NSx8	YAYPYAY	909.01
NSx9	YFEPEFY	993.08

Table 1 shows the non-specific (NS) peptide compounds synthesized and assayed for their effect on amylin aggregation and membrane damage. NSx5 is a modified version of deltorphin-I, which has no known physiological association with the amylin peptide. The remaining compounds show variation in the third position to probe this small region of sequence space. Peptides were synthesized via established Fmoc (fluorenylmethyloxycarbonyl) solid-phase techniques, purified by reverse-phase high-performance liquid chromatography (RP-HPLC), and the identity of each was confirmed with electrospray ionization mass spectrometry (ESI-MS). Synthetic techniques used here produce a *C*-terminal amide group.

Figure 1 shows the effect of each compound on the rate of amylin aggregation. Peptides were combined in various ratios with 10 µM amylin (range: 2-100 µM NS compound) in the presence of uniformly sized model liposomes composed of a 7:3 ratio of the lipids 1,2-dioleoyl-sn-glycero-3-phosphocholine (DOPC) and 1,2-dioleoyl-sn-glycero-3-(phospho-L-serine) (DOPS), respectively. Thioflavin T dye was added to the reaction mixture to monitor fibril formation over the course of at least 4 hours (Ex: 440/30 nm, Em: 485/20 nm). A sigmoidal increase in fluorescence was taken as confirmation of fiber formation. Figure 1 reports normalized $t_{1/2}$ values, allowing direct comparison of assay results performed on different days. NSx6 and NSx8 slow the rate of fiber formation (positive values), whereas all other compounds generally accelerate aggregation. The effect of NSx6 is strongly

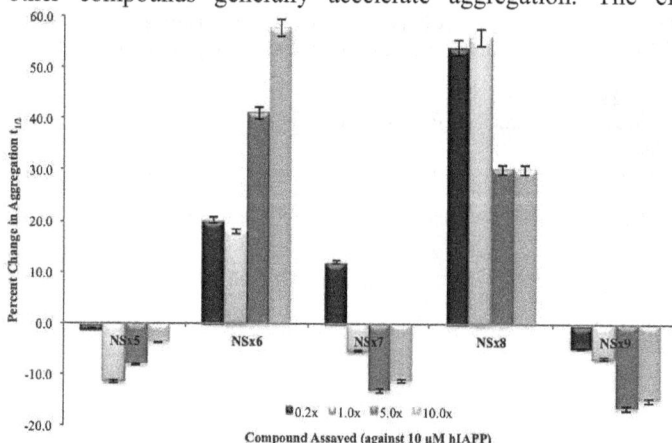

Fig. 1. Normalized $t_{1/2}$ values of amylin aggregation in the presence of varied amount of inhibitor. NSx6 and -8 slow aggregation (positive values); all others generally increase the aggregation rate (negative values). Only the effect of NSx6 is concentration dependent, indicating a specific interaction.

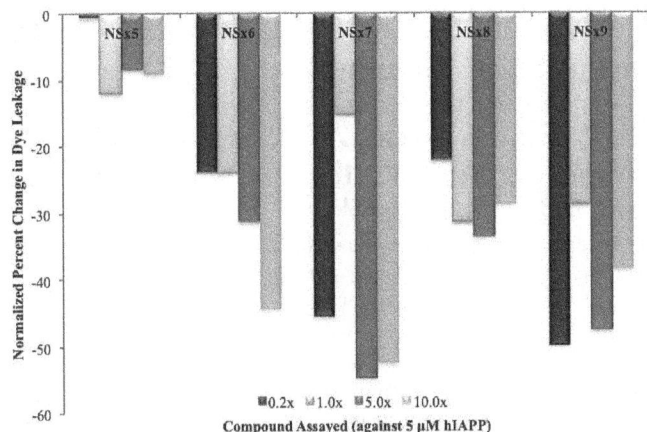

Fig. 2. Normalized change in amylin- induced membrane damage in the presence of varied amount of inhibitors. All compounds decrease the extent of damage (negative values); no concentration-dependence is evident.

concentration-dependent (data not shown, $R^2 = 0.97$), likely indicating a specific interaction. All other compounds lack concentration-dependence, which we suggest is an indicator of non-specific associations with amylin.

Figure 2 shows the average effect of each NS compound on amylin-induced membrane damage. Amylin was assayed at 5 μM against various ratios of each NS compound (range: 1-50 μM NS peptide). Identical model liposomes as in the previous assay were used, except that thioflavin T dye was absent and a carboxyfluorescein (Cbf) dye was encapsulated within the liposomes during preparation. An increase in fluorescence over a control solution of buffered vesicles indicated membrane damage. Detergent-treated vesicles served as a maximal control; assays were run for 3 hours in triplicate. Pore formation by an on-pathway intermediate in amylin aggregation has been proposed as a mechanism of membrane disruption [2]. Thus, changes in the rate of amylin aggregation can increase or decrease the extent of membrane damage depending on the mechanism of inhibition. Negative values in Figure 2 indicate that all compounds decreased the amount of membrane damage at all concentrations; no effect is concentration-dependent.

This experiment demonstrates that even 'random' sequences can have a significant effect on the behavior of an aggregating peptide. Additionally, NSx6 and NSx8 provide two novel lead compounds pursuable as potential aggregation inhibitors. The dye leakage results shown here represent average end-point readings after 3 hr. Sigmoidal behavior in dye leakage assays has recently been reported [3], linking the aggregation process with membrane damage directly. Our results in Figure 2 may still be in the lag phase, and so performing this assay again for a greater time period would be desirable. The results thus obtained may better correlate the change in aggregation kinetics with changes in membrane disruption.

Acknowledgments

The authors thank Dr. Ruth Ann Armitage for conducting the ESI-MS experiments to confirm peptide identity. This work was funded partly through the following academic awards: EMU Undergraduate Research Stimulus Program Award (2013); EMU Honors Undergraduate Fellowship Award (2013); EMU Honors Senior Thesis/ Symposium Award (2013).

References

1. Scrocchi, L., et al. *J. Mol. Biol.* **318**, 697-706 (2002); Scrocchi, L., et al. *Lett. Peptide Sci.* **10**, 545-551 (2003); Yan, L., et al. *Proc. Nat. Acad. Sci. USA* **103**, 2046-2051 (2006); Figueroa, H., et al. *J. Chem. Info. Model.* **52**, 1298-1307 (2012).
2. Khemtemourian, L., et al. *Exp. Diabetes Res.* DOI: 10.1155/2008/421287 (2008); Smith, P., et al. *J. Am. Chem. Soc.* **131**, 4470-4478 (2008); Heyl, D., et al. *Int. J. Peptide Res. Ther.* **16**, 43-54 (2010); Knight, J., et al. *Biochem.* **45**, 9496-9508 (2006); Last, N., et al. *Proc. Nat. Acad. Sci. USA* **108**, 9460-9465 (2011).
3. Engel, M., et al. *Proc. Nat. Acad. Sci. USA* **105**, 6033-6038 (2008).

Proceedings of the 23rd American Peptide Symposium
Michal Lebl (Editor)
American Peptide Society, 2013

Platform Technology to Develop Synthetic Peptide Vaccines to Prevent Viral Infections

Ziqing Jiang[1], Lajos Gera[1], Wendy Hartsock[1], Zhe Yan[1], Brook Hirsch[1], Colin T. Mant[1], Zhaohui Qian[2], Kathryn V. Holmes[2], J Paul Kirwan[1], and Robert S. Hodges[1]

[1]Department of Biochemistry and Molecular Genetics, [2]Department of Microbiology, University of Colorado, Anschutz Medical Campus, School of Medicine, Aurora, CO, 80045, U.S.A.

Introduction

Influenza A viruses spread rapidly, causing widespread seasonal epidemics of respiratory disease worldwide, which result in at least half a million deaths annually [1]. Increasing immunity to the virus in the population selects for viruses with mutations in the receptor binding domain of hemagglutinin (HA) and/or neuraminidase (NA) genes [2]. Current influenza vaccines primarily elicit antibodies to the receptor-binding region in the HA1 domain of HA. This region is hyper-variable and highly mutatable, leading to new forms of the virus that can evade neutralizing antibodies. The stem region of HA plays a very important role in membrane fusion and virus entry. Studies have shown that a few rare neutralizing human monoclonal antibodies can recognize these highly conserved epitopes and neutralize both homotypic and heterotypic influenza strains. Our strategy is to develop a "Universal Influenza Synthetic Peptide Vaccine" based on these highly conserved stem regions of HA. Our platform technology consists of a two-stranded α-helical coiled-coil peptide template of 29 residues per strand, functionalized with a protein carrier. An α-helical sequence from the HA protein is inserted into the template to display the exposed helical surface of the native protein. The templated B-cell epitopes are designed for maximum stability by creating a hydrophobic core consisting of Ile/Leu residues and an interchain disulfide bridge (Figure 1). Positively charged residues are used in the solubility enhancer region to increase overall peptide solubility if required. These stabilized helical epitopes in the two-stranded α-helical coiled-coil generate protective antibodies *in vivo*. This platform technology has general applicability not only for vaccines to any class-1 viral fusion protein but also for generating antibodies (polyclonal or monoclonal) to any α-helix in any protein.

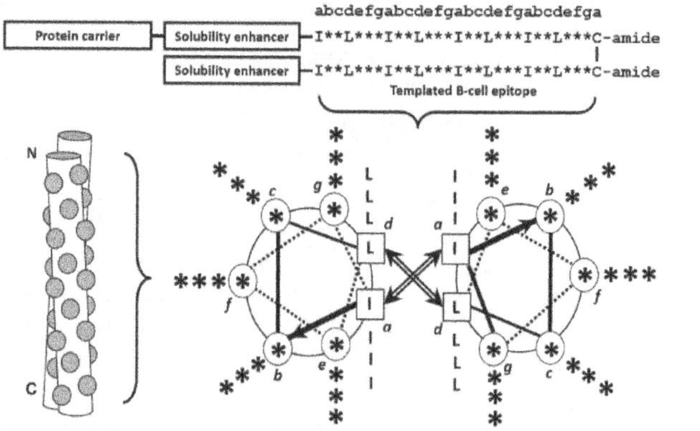

Fig. 1. Cross-sectional view of a two-stranded coiled-coil immunogen displaying two stabilized surface exposed helices to generate antibodies that bind to the native protein.

Results and Discussion

1. Identify conserved α-helical segments in the stem region of HA from different subtypes using our StableCoil algorithm to predict stable amphipathic α-helices (Figure 2).
2. A minimum of five out of seven residues in each heptad are unique to the α-helical sequence of interest. The two-stranded template is used for immunization to generate polyclonal antibodies, which are specific not only to the sequence of interest but also for its α-helical conformation (Figure 3).
3. Elicit rabbit anti-peptide antibodies that cross-react with native HA protein (Figure 3).

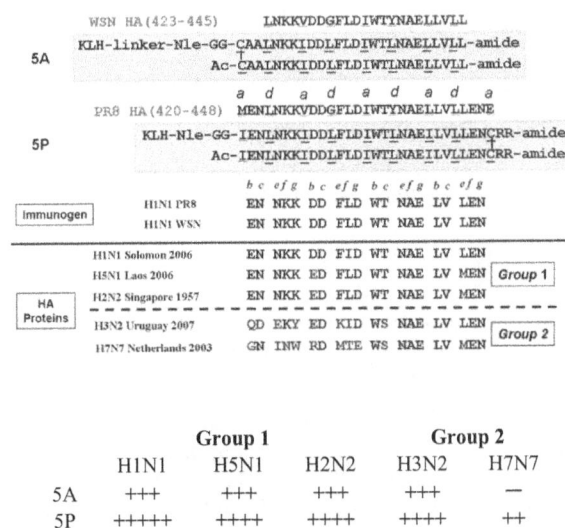

		Epitope 5	Epitope 6	
	defg	abcdefgabcdefgabcdefgabcdefg	abcdefgabcdefgabcdefgabcdefg	abcdef
H1N1 PR8	LEKR	MENLNKKVDDGFLDIWTYNAELLVLLEN	ERTLDFHDSNVKNLYEKVKSQLKNNAKE	IGNGCF
H1N1 WSN	LEKR	MENLNKKVDDGFLDIWTYNAELLVLLEN	GRTLDFHDLNVKNLYEKVKSQLKNNAKE	IGNGCF
H1N1 A/Solomon Islands/3/2006	LERR	MENLNKKVDDGFLDIWTYNAELLVLLEN	ERTLDFHDSNVKNLYEKVKSQLKNNAKE	IGNGCF
H1N1 A/California/07/2009	LEKR	IENLNKKVDDGFLDIWTYNAELLVLLEN	ERTLDYHDSNVKNLYEKVRSQLKNNAKE	IGNGCF
H5N1 A/duck/Laos/3295/2006	LERR	IENLNKKMEDGFLDVWTYNAELLVLMEN	ERTLDFHDSNVKNLYDKVRLQLRDNAKE	LGNGCF
H2N2 A/Singapore/1/1957	LERR	LENLNKKMEDGFLDVWTYNAELLVLMEN	ERTLDFHDSNVKNLYDKVRMQLRDNVKE	LGNGCF
H3N2 A/Uruguay/716/2007	VEGR	IQDLEKYVEDTKIDLWSYNAELLVALEN	QHTIDLTDSEMNKLFEKTKKQLRENAED	MGNGCF
H7N7 A/Netherlands/219/2003	VERQ	IGNVINWTRDSMTEVWSYNAELLVAMEN	QHTIDLADSEMNKLYERVKRQLRENAEE	DGTGCF

Fig. 2. Selected helical epitopes from the HA stem region. Black residues are identical across strains; blue residues (highlighted) are conserved; red (underlined) residues are non-conserved.

Antibodies to 5A bind to four of the five HA proteins tested (H1, H2, H3 and H5) (Figure 3, right panel). Antibodies to 5P bind to all five HA proteins tested and thus are cross-reactive to both Group 1 and Group 2 HA proteins (Figure 3, right panel). We improved the cross-reactivity to HA proteins by subtle changes in immunogen design. The cross-reactivity of antibodies generated to 5P can be explained by the sequence of the 11 exposed surface residues of the helix in the *C*-terminal of the five HA proteins. The antibodies to 6A or 6P show no cross-reactivity to Group 2 HA proteins because the sequences are dramatically different between the HA proteins in this region (data not shown). Thus to date these results show that immunogen 5P is the best immunogen to create a "Universal" synthetic peptide vaccine where antibodies bind to all HA proteins from Group 1 and Group 2.

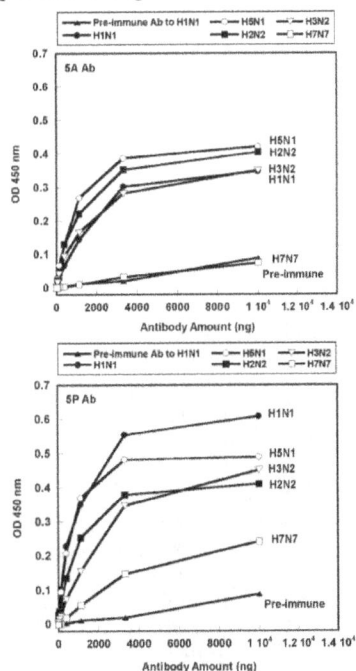

	Group 1			**Group 2**	
	H1N1	H5N1	H2N2	H3N2	H7N7
5A	+++	+++	+++	+++	—
5P	+++++	++++	++++	++++	++

Fig. 3. ELISA results of 5A and 5P peptide immunogens and summary of cross-reactive anti-peptide antibodies.

Conclusions

1. The template provides highly structured peptide immunogens and tolerates a variety of sequences.
2. The templated peptide immunogen technology is a robust approach to elicit antibodies (cross-reactive with HA protein subtypes; protective against influenza challenge in mice).
3. The sequence inserted into the templated peptide immunogen can be optimized to improve cross-reactivity with heterologous HA proteins.

References

1. http://www.who.int/topics/influenza/en/.
2. Compans, R.W. and Orenstein, W.A. (Eds.), *Vaccines for Pandemic Influenza.* Springer, Berlin, 2009.

Proceedings of the 23rd American Peptide Symposium
Michal Lebl (Editor)
American Peptide Society, 2013

[Des-Arg7]-Dynorphin A Analogs for Bradykinin-2 Receptor

Yeon Sun Lee[1], Sara M. Hall[1], Cyf Ramos-Colon[1], Dhana Muthu[1], David Rankin[2], Frank Porreca[2], Josephine Lai[2], and Victor J. Hruby[1]

[1]Department of Chemistry and Biochemisty, University of Arizona, Tucson, AZ, 85721, U.S.A;
[2]Department of Pharmacology, University of Arizona, Tucson, AZ, 85721, U.S.A.

Introduction

Under chronic pain and inflammation, dynorphin A (Dyn A) (Figure 1) is up-regulated in the spinal cord and up-regulated Dyn A interacts with bradykinin-2 receptor (B2R) resulting in hyperalgesia [1,2]. Since there is no evidence for *de novo* synthesis of the endogenous ligand, bradykinin (BK) or its precursors in the spinal cord, it supports the fact that activation of spinal B2R promotes pain, and that they are activated by upregulated Dyn A. This is a non-opioid effect that cannot be blocked by opioid antagonists. On the basis of this fact, we have been developing Dyn A-based ligands to block the hyperalgesic effect of B2R in the spinal cord. Our systematic structure-activity relationships (SAR) study identified a good pharmacophore of Dyn A for the central B2R: amphipathic non-opioid fragment, Dyn A-(4-11), and also validated that the truncation of Arg7 residue in all Dyn A fragments does not affect their respective binding affinity at the receptor (Figure 1, Table 1).

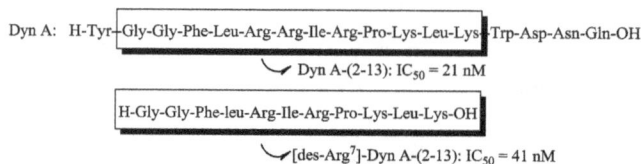

Dyn A: H-Tyr-Gly-Gly-Phe-Leu-Arg-Arg-Ile-Arg-Pro-Lys-Leu-Lys-Trp-Asp-Asn-Gln-OH

Dyn A-(2-13): IC_{50} = 21 nM

H-Gly-Gly-Phe-leu-Arg-Ile-Arg-Pro-Lys-Leu-Lys-OH

[des-Arg7]-Dyn A-(2-13): IC_{50} = 41 nM

Fig. 1. Structure of Dyn A analogs and their binding affinities at rat brain B2R.

Results and Discussion

In our earlier study, it was confirmed that the B2R recognition predominantly depends on the basicity of the *C*-terminal amino acid and the modification of *C*-terminal acid to amide decreased the binding affinity dramatically [3].Therefore our SAR study on [Des-Arg7]-Dyn A analogs was limited to the analogs with a basic amino acid residue at the *C*-terminal.

Dyn A analogs were synthesized by solid phase synthesis using standard Fmoc-chemistry on the Wang resin. Crude peptides could be obtained by the cleavage using 92% TFA cocktail solution containing 4% TIS, 2% water, and 2% anisole in high purity (70-90%) and could be purified by preparative RP-HPLC in a short time (<15 min) thanks to their hydrophilic characters (aLOGPs < 0). The purified analogs were validated by analytical HPLC and Mass spectroscopy and were tested for their binding affinities by competition assay using [3H]DALKD in rat brain membranes where non-specific binding is defined by 10 μM kallidin (Table 1).

As shown in Table 1, [Des-Arg7]-Dyn A analogs **1-7** exhibited very similar pattern of binding affinities with Dyn A analogs **8-14** at the rat brain B2Rs. This confirms that removal of the Arg residue at position 7 did not affect binding affinities at all. This is very remarkable in the sense that truncating one amino acid residue in the middle of a bioactive sequence, in general, can cause significant topographical changes for a different biological profile.

For comparison of the topographical structures of Dyn A and [Des-Arg7]-Dyn A, ^{1}H 2D-NMR spectroscopy was performed in membrane-like SDS micelles for lead ligands **4** and **11**. The NMR study showed that the two ligands **4** and **11** have the same distorted type 1 β-turn structure at the *C*-terminal, which is considered a key region for the binding (Figure 2). This explains how two ligands bind to the receptor in the same range even with a dissimilarity of structure at the *N*-terminal. The *N*-terminal part does not play an important role in the receptor interactions and thus does not affect ligand binding. Therefore one Arg residue at position 3 in ligand **4** seems to be sufficient for the receptor recognition similar to the two Arg residues in ligand **11**.

Table 1. Binding affinities of [Des-Arg^7]-Dyn A analogs at B2R in rat brain membrane.

no	[Des-Arg^7]-Dyn A Analogs	B2R^a				no^b	B2R^a	
		pH 6.8		pH 7.4			pH 7.4	
		log [IC$_{50}$]^c	IC$_{50}$	log [IC$_{50}$]^c	IC$_{50}$		log [IC$_{50}$]^c	IC$_{50}$
1	GFLRIRPKLK	-7.58±0.06	26	-7.39±0.09	41	8	-6.50 ± 0.07	320
2	FLRIRPKLK	-7.20±0.08	63	-6.95±0.22	110	9	-6.41 ± 0.13	390
3	LRIRPKLK	-7.52±0.09	30	-7.17±0.10	68	10	-6.33 ± 0.16	470
4	FLRIRPK	-7.16±0.09	69	-6.71±0.11	190	11	-6.86 ± 0.06	140
5	LRIRPK	-6.71±0.24	190	-6.74±0.09	180	12	-6.55 ± 0.06	280
6	GFLRIR	-6.56±0.10	280	-6.21±0.09	620	13	-6.11 ± 0.09	780
7	FLRIR	-6.12±0.07	760	-5.67±0.15	2100	14	-5.67 ± 0.15	1300

^{a}Competition assays against [^{3}H]DALKD were carried out at pH 6.8 or 7.4 using rat brain membrane; ^{b}Respective Dyn A analogs; ^{c}Logarithmic values determined from the nonlinear regression analysis of data collected from at least two independent experiments.

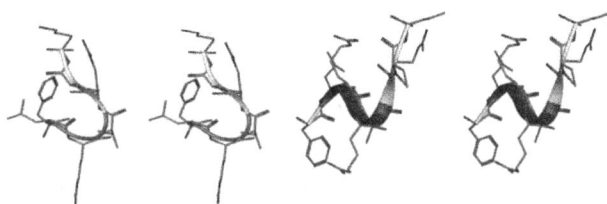

Fig. 2. Stereoviews of [Des-Arg^7]-Dyn A-(4-11)(4, left) and Dyn A-(4-11) (11, right).

Furthermore, it was observed that in relatively longer length of analogs such as ligand **1**, truncation of the Arg residue increased their binding affinities (log[IC$_{50}$] = -7.39±0.09, IC$_{50}$ = 41 nM at pH 7.4, cf. **8**) Interestingly, binding affinities of [Des-Arg^7]-Dyn A analogs depend on pH and increased affinities were obtained at lower pH, 6.8 (up to 2.8 fold). The same trend was also observed in the Dyn A analogs (not shown here). This result validates the optimal binding conditions for the Dyn A analogs to be at pH 6.8, which is consistent with the binding conditions previously established for the B2R.

In summary, we have discovered a lead ligand **4** and further SAR study on [Des-Arg^7]-Dyn A showed that the removal of Arg^7 residue does not affect the receptor recognition thanks to the conserved type I β–turn structure at the C-terminal. To validate the therapeutic potential of the [Des-Arg^7]-Dyn A analogs in chronic pain states, we are currently investigating ligand **4** *in vivo*.

Acknowledgments

Supported by grant P01DA006284 (U.S.Public Health Services, NIH, and NIDA).

References

1. Lai, J., Ossipov, M., Vanderah, T.W., Malan, T.P., Porreca, F. *Mol. Interventions* **1**, 160-167 (2001).
2. Lai, J., Luo, M.C., Chen, Q., Ma, S., Gardell, L.R., Ossipov, M., Porreca, F. *Nat. Neurosci.* **9**, 1534-1540 (2006).
3. Lee, Y.S., Rankin, D., Paisely, B., Muthu, D., Ortiz, J.J., Porreca, F., Lai, J., Hruby, V.J, In Lebl M. (Ed.) *Peptides: Building Bridges. (Proceedings of the 22nd American Peptide Symposium)*, Prompt Scientific Publishing, San Diego, 2011, p 348.

Proceedings of the 23rd American Peptide Symposium
Michal Lebl (Editor)
American Peptide Society, 2013

TNFα-Based Peptides as Bioprobes for Exosites of ADAM Proteases

Mare Cudic, Nina Bionda, Marc Giulianotti, Laura Maida, Richard A. Houghten, Gregg B. Fields, and Dmitriy Minond

Torrey Pines Institute for Molecular Studies, 11350 SW Village Parkway, Port Saint Lucie, FL, 34987, U.S.A.

Introduction

ADAM proteases are implicated in multiple diseases, but no drugs based on ADAM inhibition exist. Most of the ADAM inhibitors developed to date feature Zn-binding moieties that target the active site Zn, which leads to a lack of selectivity and off-target toxicity.

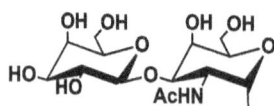

(EDANS)-E-^{72}PLAQAVRSSS^{81}-K-(DABCYL)

Fig. 1. TNFα-based glycosylated bioprobe of ADAM17 exosites.

Targeting secondary substrate binding sites (exosites) can potentially work as an alternative strategy for drug discovery; however, there are only a few reports of potential exosites in ADAM protease structures. In this study we utilized a series of TNFα-based glycosylated, α-helical, and linear substrates to probe ADAM protease interactions to identify the structural features that determine substrate specificity. We found that non-catalytic domains of ADAM17 did not directly bind the substrates used in the study, but affected the binding nevertheless, most likely due to steric hindrance. Non-catalytic domains of ADAM17 affected the size/shape of the carbohydrate binding pocket contained within catalytic domain of ADAM17 suggesting that non-catalytic domains of ADAM17 play a role in substrate specificity and might help explain differences in substrate repertoires of ADAM17 and its closest homologue, ADAM10. We also addressed the question of which substrate features can affect ADAM protease specificity. We found that all ADAM proteases tested significantly decreased activity when the TNFα-derived sequence was induced into α-helical conformation, suggesting that conformation plays role in determining ADAM protease substrate specificity. These findings can help in the discovery of ADAM isoform- and substrate-specific inhibitors.

Results and Discussion

The cleavage of TNFα by ADAM17 on the cell surface occurs in its juxtamembrane region 20 amino acids away from transmembrane domain [1]. The juxtamembrane region of TNFα is disordered according to Uniprot database entry based on the primary structure composition [2], whereas the transmembrane region is predicted to be α-helical. We have previously shown that TNFα-based substrate can form an α-helix under conditions emulating the intramembrane environment [3].

Table 1. Kinetic parameters for hydrolysis of random coil and α-helical TNFα-based substrates and by ADAM10, 12, and 17. Results are reported as Ave ± SD (n=3).

Enzyme	k_{cat}, s^{-1}		K_M, μM	
	Random coil	*α-helical*	*Random coil*	*α-helical*
ADAM17	0.34±0.03	0.003±0.001	24±6.2	2.0±0.1
ADAM10	0.06±0.01	0.004±0.001	13±0.5	5.5±1.5
ADAM12	0.006±0.01	ND	18±5	ND

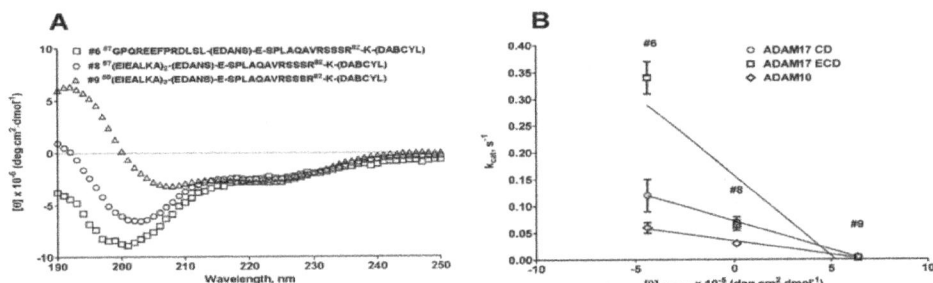

Fig. 2. Correlation of circular dichroism spectra of juxtamembrane (#6), substrate with 2 α-helical heptads (#8) and α-helical (#9) substrates.

We were interested to see whether activity of ADAM17 can be modulated by a substrate possessing a distinct secondary structure. The α-helical substrate exhibited significant improvement of affinity towards ADAM17 as compared to the random coil substrates. K_M values were 24 ± 6.2 µM and 2.0 ± 0.1 µM for random coil and α-helical substrates, respectively (Table 1). The α-helical substrate demonstrated greater than 100- and 20-fold decrease of k_{cat} as compared to random coil substrate. Such a dramatic effect of α-helical structure on enzyme activity is likely a consequence of tight binding exhibited by this substrate. It is, therefore, entirely possible that membrane-bound ADAM17's repertoire of substrates can be limited to sequences located within disordered regions of cell surface proteins. By the same token, it is conceivable that ADAM17 activity can be regulated by α-helical structure. Additionally, the linear regression analysis of a plot of molar absorptivity at λ = 192 nm versus k_{cat} exhibited excellent R^2 values for ADAM17 CD and ADAM10 ECD (Figure 2B, R^2 = 0.999 and 0.982, for ADAM17 CD and ADAM10 ECD, respectively) and a very good R^2 value of 0.815 for ADAM17 ECD. This suggests a strong correlation between the secondary structure of TNFα-based substrates and activity of ADAM10 and 17.

Acknowledgments

This work was supported by the James and Esther King Biomedical Research Program (2KN05 to DM), the National Institutes of Health (DA033985 to DM, CA098799 to GBF, DA031370 to RAH), the Multiple Sclerosis National Research Institute (to GBF), and the State of Florida, Executive Office of the Governor's Office of Tourism, Trade, and Economic Development.

References

1. Black, R.A., Rauch, C.T., Kozlosky, C.J., Peschon, J.J., Slack, J.L., Wolfson, M.F., Castner, B.J., Stocking, K.L., Reddy, P., Srinivasan, S., Nelson, N., Boiani, N., Schooley, K. A., Gerhart, M., Davis, R., Fitzner, J.N., Johnson, R.S., Paxton, R.J., March, C.J., Cerretti, D.P. *Nature* **385**, 729-733 (1997).
2. Uniprot. (2012) P01375 (TNFA_HUMAN).
3. Minond, D., Cudic, M., Bionda, N., Giulianotti, M., Maida, L., Houghten, R.A., Fields, G.B. *J. Biol. Chem.* **287**, 36473-36487 (2012).

Proceedings of the 23rd American Peptide Symposium
Michal Lebl (Editor)
American Peptide Society, 2013

Identification and Structure-Activity Studies of Novel Cryptides Hidden in Mitochondrial Proteins

Hidehito Mukai[1,2], Yoshinori Hokari[2,3], Tetsuo Seki[2,3], Akiyoshi Fukamizu[3], and Yoshiaki Kiso[1]

[1]*Laboratory of Peptide Science, Graduate School of Bio-Science, Nagahama Institute of Bio-Science and Technology, Nagahama, Shiga, 526-0829, Japan;* [2]*Institute of Applied Biochemistry, University of Tsukuba, Tsukuba, Ibaraki, 305-8572, Japan;* [3]*Graduate School of Life and Environmental Sciences, University of Tsukuba, Tsukuba, Ibaraki, 305-8577, Japan*

Introduction

Neutrophils are known to involve innate immunity, but endogenous factors which induce acute transmigration and activation of them have not been well elucidated so far. We therefore explored such unknown factors and recently identified two novel neutrophil-activating peptides, mitocryptide-1 (MCT-1) and -2 (MCT-2), which were derived from mitochondrial cytochrome c oxidase and cytochrome b, respectively [1-5]. We also found the presence of many neutrophil-activating peptides presumably derived from various mitochondrial proteins other than MCT-1 and MCT-2, indicating that neutrophils are regulated by many unidentified mitochondrial protein derived peptides [4,5]. We named such functional peptides hidden in protein structures "cryptides" [3-7]. After our discovery, the evidence has been accumulated that fragmented peptides derived from various functional proteins have variety of biological functions that are distinct from the roles of their mother proteins, and the presence of non-classical bioactive peptides derived from functional proteins, including cytosolic ones, is widely recognized [6-10]. In addition, specific receptor molecules and cellular signaling mechanisms for those cryptides are going to be characterized [6-9].

Here, we report our attempt to identify novel neutrophil-activating cryptides hidden in various mitochondrial proteins. We also communicate structure-activity relationships of these cryptides on the activation of neutrophilic differentiated HL-60 cells.

Results and Discussion

We extracted and purified neutrophil-activating peptides from porcine hearts using cation-exchange chromatography, gel filtration, preparative RP-HPLC, cation-exchange HPLC, and micro-analytical RP-HPLC based on the activity to induce β-hexosaminidase release from the neutrophilic differentiated HL-60 cells [4,5,11]. The primary structure of the substances in the purified fractions was analyzed by automated Edman degradation and MALDI-TOF-MS. The obtained peptide sequence of a substance was Leu-Glu-Asn-Pro-Lys-Lys-Tyr-Ile-Pro-Gly-Thr-X-Met-Ile-Phe-Ala-Gly-Ile-OH, where X is an unidentified amino acid. The amino acid sequence of this octadecapeptide was identical to that of porcine mitochondrial cytochrome c_{68-85}. In addition, the presence of derivative peptides of the octadecapeptide having different chain-length was indicated. We synthesized the octadecapeptide based on the sequence of human cytochrome c_{68-85} and confirmed that it induced β-hexosaminidase release from neutrophilic differentiated HL-60 cells, demonstrating that the peptide was a neutrophil-activating cryptide produced from cytochrome c. We named this neutrophil-activating cryptide as mitocryptide-CYC (MCT-CYC).

Since we found the presence of various neutrophil-activating peptides derived from mitochondrial cytochrome c, we investigated structure-activity relationships of human cytochrome c and its related derivativepeptides on β-hexosaminidase release from neutrophilic differentiated HL-60 cells. Although cytochrome c itself (cytochrome c_{1-104}) did not stimulate the differentiated HL-60 cells, cytochrome c_{68-104} with a MCT-CYC sequence at its N-terminus induced β-hexosaminidase release (EC$_{50}$: 7×10^{-6} M). MCT-CYC (cytochrome c_{68-85}) exhibited higher maximum response than cytochrome c_{68-104} on the stimulation of β-hexosaminidase release but required 30 times higher concentration for the induction. In case of cytochrome c_{70-85}, it induced β-hexosaminidase release at hundred times lower concentrations than MCT-CYC (cytochrome c_{68-85}) with the same maximum response

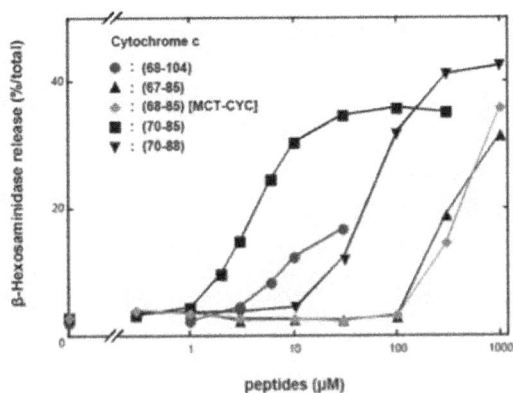

Fig. 1. Effects of mitocryptide-CYC and its derivatives on β-hexosaminidase release from HL-60 cells differentiated into neutrophilic cells.

(EC_{50}: 4×10^{-6} M), but cytochrome c_{70-88} required ten times higher concentrations than cytochrome c_{70-85}. These results indicate that peptides derived from the C-terminal portion of cytochrome c could activate neutrophilic cells. The present study suggest that cryptides produced from cytochrome c may play an important role in scavenging toxic debris from apoptotic cells by neutrophils, because cytochrome c is involved in the apoptotic process.

Recently, we are attempting to identify neutrophil-activating peptides systematically, i.e., we are predicting neutrophil-activating cryptides from mitochondrial proteins according to their common features of the distribution of positive charged residues and aromatic or aliphatic residues. In those studies, the C-terminal peptides of cytochrome c including cytochrome c_{70-85} have been predicted as neutrophil-activating ones. The actual existence of MCT-CYC and its related peptides in the tissues demonstrate that our strategy is effective for the systematic identification of neutrophil-activating cryptides derived from mitochondrial proteins.

Acknowledgments

The present study was supported by research grants from the Ministry of Education, Culture, Sports, Science and Technology, Japan (No. 21603014; 40089107) and Nagase Science and Technology Foundation. The authors thank the special technical assistance in the present study by Maruhachi Muramatsu Inc., Yaizu, Shizuoka, Japan and Boehringer Ingelheim Japan Inc., Tokyo, Japan.

References

1. Mukai, H., Hokari, Y., Seki, T., Nakano, H., Takao, T., Shimonishi, Y., Nishi, Y., Munekata, E., In Lebl, M. and Houghten, R.A. (Eds.) *Peptides; The Wave of the Future (Proceedings of the Second International and the Seventeenth American Peptide Symposium)*, American Peptide Society, San Diego, 2001, p. 1014-1015.
2. Mukai, H., Matsuo, Y., Kamijo, R., Wakamatsu, K., In Chorev, M. and Sawyer, T. K. (Eds.) *Peptide Revolution: Genomics, Proteomics & Therapeutics (Proceedings of the Eighteenth American Peptide Symposium)*, American Peptide Society, San Diego, 2003, p. 553-555.
3. Ueki, N., Someya, K., Matsuo, Y., Wakamatsu, K., Mukai, H., *Biopolymers (Pep. Sci.)* **88**, 190-198 (2007).
4. Mukai, H., Hokari, Y., Seki, T., Takao, T., Kubota, M., Matsuo, Y., Tsukagoshi, H., Kato, M., Kimura, H., Shimonishi, Y., Kiso, Y., Nishi, Y., Wakamatsu, K., Munekata, E. *J. Biol. Chem.* **283**, 30596-30605 (2008).
5. Mukai, H., Seki, T., Nakano, H., Hokari, Y., Takao, T., Kawanami, M., Tsukagoshi, H., Kimura, H., Kiso, Y., Shimonishi, Y., Nishi, Y., Munekata, E. *J. Immunol.* **182**, 5072-5080, (2009).
6. Mukai, H., Kiso, Y., Wakamatsu, K. *Seikagaku* **84**, 524-532 (2010).
7. Mukai, H., Seki, T., Hokari, Y., Fukamizu, A., Kiso, Y., In Fujii, N. and Kiso, Y. (Eds.) *Peptide Science 2010 (Proceedings of Fifth International Peptide Symposium)*, Japanese Peptide Society, Osaka, 2011, p. 29.
8. Seki, T., Fukamizu, A., Kiso, Y., Mukai, H. *Biochem. Biophys. Res. Commun.* **404**, 482-487 (2011).
9. Heimann, A.S., Gomes, I., Dale, C.S., Pagno, R.L., Gupta, A., de Souza, L.L., Luchessi, A.D., Castro, L.M., Giorgi, R., Rioli, V., Ferro, E.S., Devi, L.A. *Proc. Nat. Acad. Sci. USA* **104**, 20588-20593 (2007).
10. Samir, P., Link, J. *AAPS J.* **13**, 152-158 (2011).
11. Hokari, Y., Seki, T., Nakano, H., Matsuo, Y., Fukamizu, A., Munekata, E., Kiso, Y., Mukai, H., *Prot. Pept. Lett.* **19**, 680-687 (2012).

Proceedings of the 23rd American Peptide Symposium
Michal Lebl (Editor)
American Peptide Society, 2013

The Structure of the Central Side Chain is Crucial for Anoplin Hemolytic Activity

Jens K. Munk[1], Lea Thøgersen[2], Lars Erik Uggerhøj[3], Reinhard Wimmer[3], Niels Frimodt-Møller[4], Kresten Lindorff-Larsen[5], and Paul R. Hansen[1]

[1]Dept. of Drug Design and Pharmacology, Faculty of Health and Medical Sciences, Univ. of Copenhagen, Copenhagen Ø, 2100, Denmark; [2]Bioinformatics Research Center, Faculty of Science, Univ. of Aarhus, Aarhus C, 8000, Denmark; [3]Dept. of Biotechnology, Chemistry and Environmental Engineering, Univ. of Aalborg, Aalborg, 9000, Denmark; [4]Dept. of Clinical Microbiology, Hvidovre Hospital, Hvidovre, 2650, Denmark; [5]Dept. of Biology, Faculty of Science, Univ. of Copenhagen, Copenhagen N, 2200, Denmark

Introduction

Anoplin, H-GLLKRIKTLL-NH$_2$, is a decamer amide from the venom of the Japanese solitary wasp, *Anoplius samariensis* [1]. Anoplin has good antimicrobial properties, very low hemolytic activity [2,3] and forms an amphipathic α-helix in the presence of membranes and membrane mimetics. When α-helical, its apolar face is a central isoleucine (position 6) flanked by four leucines (positions 2, 3, 9 and 10). Central apolar residues strongly affect hemolytic activity of peptides [4]. Here we investigate position 6 of anoplin in that context. We have synthesized several anoplin analogs with modifications at position 6. Using molecular dynamics on anoplin and three of the analogs, we observe that reduction of position 6 side chain rotamer freedom increases hemolytic activity of the peptide.

Results and Discussion

The hemolytic activities of anoplin and analogs with different position 6 amino acid side chains (Figure 1) are shown in Table 1. Figure 2 shows the rotamer distributions and rotational freedoms of the side chains of Ile, Ail, Epa and Cha as observed by molecular dynamics. From 100 observations, only those with position 6 in a helical conformation were included. It can be seen that anoplin has

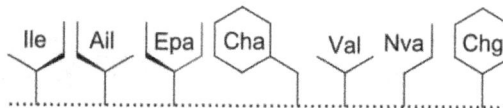

Fig. 1. Position 6 side chains used in this study. Isoleucine (Ile), allo-isoleucine (Ail), 2-amino-3-ethylpentanoic acid (Epa), cyclohexylalanine (Cha), valine (Val), norvaline (Nva) and cyclo-hexylglycine (Chg).

low hemolytic activity; its EC$_{10}$ (the concentration needed to lyse 10% of red blood cells suspended in PBS) is 1052 μM. When helical, the Ile side chain readily flips between rotamer states, making 53 changes. The ethyl branch of Ile mainly occupies states *m* (58% of the time) and *t* (33% of the time). The Ail side chain conveys 4.4 times more hemolytic activity than Ile. When helical, Ail rarely changes rotamer state, making only 12 flips, and its ethyl branch occupies the *m* state 91% of the time. Epa conveys even more hemolytic activity than Ail, making the peptide 50 times more hemolytic than anoplin. When helical, Epa rotation is also restricted more than Ail rotation is; we observe only 6 changes between rotamer states, and one of the ethyl branches in Epa (the one equivalent to the ethyl branch in Ail, and in black in Figure 1) occupies the *m* state 78% of the time. When helical, the Cha side chain rotates more freely than Ail and Epa, but less so than Ile, making 17 rotamer state changes. Its C$^\gamma$ occupies the *m* and *t* states 1:1, but it never occupies the *p* state. The rotational freedom therefore puts it in the same range as Ail, and the rotamer distribution is similar to Ile.

Table 1. Hemolytic activities (EC$_{10}$ values) for peptides with different amino acids in position 6.

Amino Acid	Ile	Ail	Epa	Cha	Val	Nva	Chg
EC$_{10}$, μM	1052	237	21	40	736	1659	175

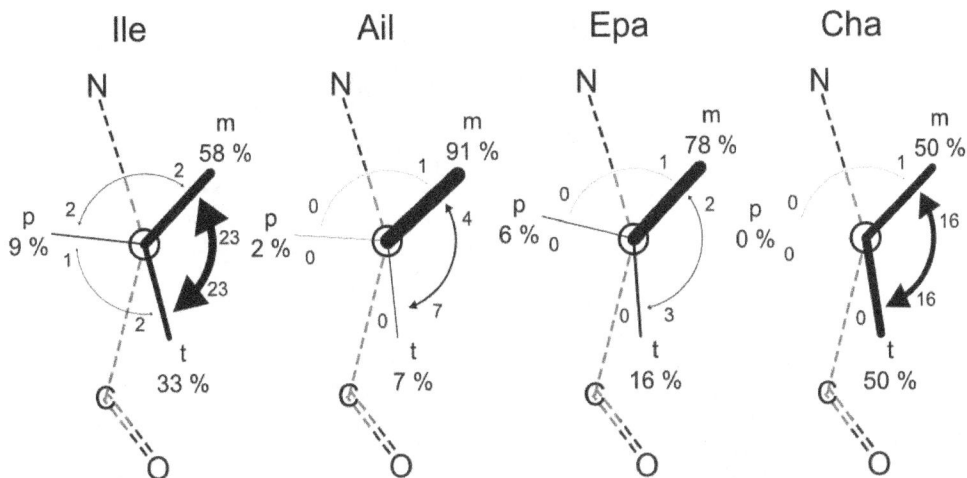

Fig. 2. Side chain rotation patterns of Ile, Ail, Epa and Cha positioned as amino acid #6 in anoplin and analogs. Curved arrows denote number of rotamer state changes observed during 100 observations; percentages are rotamer distributions in states p, m *and* t.

However, Cha conveys 26 times more hemolytic activity than Ile, more than can be explained by restricted rotation alone. We speculate this hemolytic activity is furthered by the size of the cyclohexyl group, since large hydrophobes are known to cause hemolysis in their own right [5]. This may also contribute to the hemolytic activity conveyed by Epa. Hence, for side chains such as Ile, Ail and Epa, which are similar in size and hydrophobicity, restriction of conformational freedom conveys hemolytic activity. We speculate that this relates to observations made for other peptides wherein increased conformational freedom, achieved by disruption of α-helical stability, has been shown to alleviate hemolysis [4].

We can expand this line of thinking to the three side chains not investigated using molecular dynamics. The mobility of the Val side chain is known to be similar to that of Ile [6], and Val can be considered as a midpoint between Ile and Ail. The hemolytic activity matches this view. Nva is the least restricted side chain since it is not β-branched, and smaller than Cha. This combination makes Nva the best choice in terms of hemolysis. Chg resembles Epa, but the conformational rigidity imposed by the ring seems to cause about $175/21 \approx 8$ fold less hemolytic activity than Epa, despite the slight increase in size and thus hydrophobicity. Compared to Val, Chg is about twice as large and thus causes more hemolysis. Chg is also larger than Ile and Ail, which is probably why it conveys more hemolytic activity.

In conclusion, we have shown that increased hemolytic activity could well be caused by decreased conformational freedom of the central apolar side chain.

Acknowledgments

We thank technician Sabaheta Babajic for peptide synthesis and purification, and Prof. Dr. med. Kaj Winther for providing blood. The present work is performed as a part of the Danish Center for Antibiotic Research and Development financed by The Danish Council for Strategic Research (#09-067075). KL-L is supported by a Hallas-Møller Stipend from the Novo Nordisk Foundation.

References

1. Konno, K., et al. *Biochim. Biophys. Acta* **1550**(1), 70-80 (2001).
2. Ifrah, D., et al. *J. Pept. Sci.* **11**(2), 113-121 (2005).
3. Munk, J.K., et al. Submitted.
4. Zhu, W.L., et al. *J. Biochem. Mol. Biol.* **40**, 1090-1094 (2007).
5. Chen, Y., et al. *J. Biol. Chem.* **280**, 12316-12329 (2005).
6. Lovell, S.C., et al. *Proteins* **40**, 389-408 (2000).

Proceedings of the 23rd American Peptide Symposium
Michal Lebl (Editor)
American Peptide Society, 2013

Development of Potent Dynorphin A Analogs as Kappa Opioid Receptor Antagonists

Cyf N. Ramos Colon[1,2], Yeon Sun Lee[2], Sara M. Hall[2], David R. Rankin[3], Josephine Lai[3], Frank Porreca[3], and Victor J. Hruby[2]

[1]College of Pharmacy, Departments of [2]Chemistry and Biochemistry, [3]Pharmacology, University of Arizona, Tucson, AZ, 85721, U.S.A.

Introduction

Chronic pain is the most ubiquitous disease with an incidence of 100 million people in the U.S. Opiate therapy is the mainly prescribed treatment for chronic neuropathic pain. However opioids do not address the mechanisms of neuropathic pain and thus have limited efficacy against this type of pain [1]. While opioids may reduce the pain states experienced by the patients, they have adverse effects such as tolerance, addiction, and medication overuse with long-term administration. In addition, other side effects can manifest, including constipation and severe toxicity [1,2]. It has been shown that these adverse effects are mediated through the κ opioid receptor (KOR), while nociception is mediated mainly through the μ opioid receptor (MOR). Based on this, there is a need to develop potent KOR antagonists for the treatment of chronic neuropathic pain without serious side effects.

Dynorphin A (Dyn A, Tyr-Gly-Gly-Phe-Leu-Arg-Arg-Ile-Arg-Pro-Lys-Leu-Lys-Trp-Asp-Asn-Gln) is one of three endogenous opioid peptides with high affinities for the μ, δ (DOR), and κ opioid receptors, with a preference for the KOR. Dyn A mediates an inhibitory effect through the opioid receptors resulting in nociception. In contrast, [des-Tyr[1]]-Dyn A has been shown to induce an excitatory effect instead through non-opioid mechanism on the bradykinin 2 receptor (B2R) [3].

For our development of KOR antagonists, we utilized our previous Dyn A structure-activity relationship (SAR) results for the B2R and performed further modifications on Dyn A to increase the biological activities and selectivity for the KOR. We hypothesize that through the modification of the Dyn A structure we will find analogs with high binding affinities and selectivity as antagonist for the KOR.

Results and Discussion

Dyn A analogs were synthesized by standard solid phase peptide synthesis using Fmoc-chemistry and cleaved by a 95% TFA cocktail solution containing 2.5% TIS, 2.5% water. Crude peptides were purified by RP-HPLC using 10-50% of acetonitrile in water containing 0.1% TFA. Radioligand competition binding assays were done for synthesized analogs using [^{3}H] DAMGO for MOR, [^{3}H] DPDPE for DOR, and [^{3}H] U69,593 for KOR using transfected cell membranes.

As shown in Table 1, we have identified that Arg at position 7 is not important nor necessary for binding at KOR as previously published [4]. We found that a modification of the *C*-terminal acid to amide in **CYF110, CYF111, and CYF112** increases binding affinity at all three opioid receptors. Substitutions of the *N*-terminal amine group with (2S)-2-methyl-3-(2,6-dimethyl-4-hydroxyphenyl)propanoic acid (Mdp) or Acetyl (Ac) groups increases selectivity for KOR over MOR and DOR (**CYF111, CYF112, CYF113**). It has been previously shown that substitution at the *N*-terminal amine of agonist opioid peptides can change their functions from agonists to antagonists at the receptors [6]. Therefore, we expect that ligands **CYF111, CYF112** and **CYF113** will have antagonist functions at the KOR. From this series of ligands we obtained the highest selectivity for KOR over the other opioid receptors from compound **CYF112**. As a lead compound, we will utilize **CYF112** to improve affinity, and selectivity and also obtain antagonist function at the KOR.

Table 1. Binding affinities of Dyn A analogs at KOR, MOR, and DOR in radioligand competition binding assays.

Dyn A	1	2	3	4	5	6	7	8	9	10	11	12	13	KOR K_i (nM)	MOR K_i (nM)	DOR K_i (nM)	κ/μ/δ
(1-13)[4]	Y	G	G	F	L	R	R	I	R	P	K	L	K	0.33			
(1-13)-H₂[5]													NH₂	0.015			
CYF104														0.3	4	4	1/13/13
CYF107														0.2	6	6	1/30/30
CYF110												NH₂		0.08	0.94	3	1/12/38
CYF111	Mdp											NH₂		62	1600	210	1/26/3
CYF112	Ac-											NH₂		74	820	2200	1/11/30
CYF113	Mdp													222	1800	580	1/8/3

Mdp- (2S)-2-methyl-3-(2,6-dimethyl-4-hydroxyphenyl)propanoic acid; Ac- acetyl group.

Future studies will focus on the modifications of the Dyn A structure such as cyclizations and non-natural amino acid substitution and functional assays will be followed to determine functional activities of the analogs.

Acknowledgments

Funding was provided by grants from the US Public Health Service, NIH, and NIDA (P01DA006248).

References

1. Hanlon, K.E., Herman, D.S., Agnes, R.S., Largent-Milnes, T.M., Kumarasinghe, I.R., Ma, S.W., Guo, W., Lee, Y.S., Ossipov, M.H., Hruby, V.J., Lai, J., Porreca, F., Vanderah, T.W. *Brain Res.* **1395**, 1-11 (2011).
2. Jensen, T.S., Gottrup, H., Sindrup, S.H., Bach, F.W. *Eur. J. Pharmacol.* **429**, 1-11 (2001).
3. Lai, J., Luo, M.-C., Chen, Q., Ma, S., Gardell, L.R., Ossipov, M.H., Porreca, F. *Nat. Neurosci.* **9**, 1534-1540 (2006).
4. Chavkin, A.G. *Proc. Nat. Acad. Sci. USA* **78**, 6543 (1981).
5. Lung, F.D.T., Chen, C.H., Liu, J.H. *J. Pept. Res.* **66**, 263-276 (2005).
6. Schiller, P.W., Weltrowska, G., Nguyen, T.M.-D., Lemieux, C., Chung, N.N., Lu, Y. *Life Sci.* **73**, 691-698 (2003).

Proceedings of the 23rd American Peptide Symposium
Michal Lebl (Editor)
American Peptide Society, 2013

Connecting Loop Effects on Beta-Strand Association Rates

Brandon L. Kier, Jordan M. Anderson, and Niels H. Andersen

Department of Chemistry, University of Washington, Seattle, WA, 98195, U.S.A.

Introduction

Features that form tertiary interactions favoring the folded states of proteins are frequently sequence remote. The assembly of β sheets by association of additional β strands to an initially-formed nucleating hairpin is an example. When there are long loops between the growing β sheet and the newly added strand, loop search could be a fundamental limiting factor in protein folding. In the more general case, loop search times may also dictate folding barrier heights, as all transition states must involve the formation of at least some long-range interactions.

Previous studies of loop contact times have involved fluorescent probe quenching or FRET pairs [1]. This technique has some issues; transient "contacts" at 5Å distances could effect quenching. We suspected that tertiary structure formation at the ends of loops might require closer contact and longer search times and that "loop contact times" might not be a suitable basis for modeling the protein folding speed limit. Herein we report a series of model peptides constructed for the determination of the loop search times required for long-range beta-strand association. Folding *via* a turn-zipper mechanism is unlikely past a 6 amino acid loop length, but competing hydrophobic collapse/assembly after a loop search still applies. Key features of our test constructs include the use of long, flexible loops (to preclude folding *via* a competing zippering from the turn locus mechanism), efficient β strands (rich in high beta-propensity β-branched amino acids) and a terminal capping motif which provides both exceptional stability [2] and useful probes for CD, NMR, and fluorescence studies of fold stability and folding dynamics.

Data presented in this account indicate that the loop search times required to create protein-like tertiary structure appear to be as much as 100 times greater than loop "contact" times associated with chromophore interactions at the ends of loops in the random-coil state.

Results and Discussion

The designed loop systems have the following sequences: RWITVTI-(GGGGKK)$_n$-IRVWE with n = 1, 2, 3, and 4. They correspond to RWITVTI-loop-KKIRVWE, with the highly flexible loop made up of glycines with some lysines added for solubility. The loop length, given by 6n − 2, varies from 4 to 22 residues. The general features of these constructs are illustrated in Figure 1.

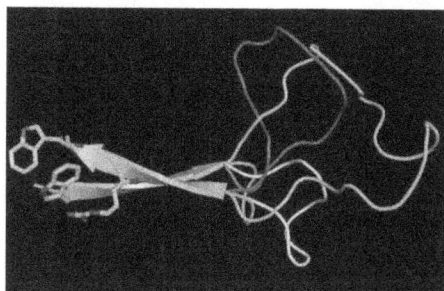

Fig. 1. The loop constructs, at loop lengths 4, 10, 16, and 22 are shown. For the shortest loop there is a single turn conformation; all of the longer loops are flexible, only one of many possible conformations is shown. The alignment of an Arg and two Trp sidechain within the β-cap is also shown.

Of course, most proteins aren't as simple as these minimal constructs; unlike our (GGGGKK)$_n$-containing sequences, most long loops in proteins are less flexible and have some innate propensity for structure. We chose to compare our long-loop "hairpins" to a circularly permuted WW domain of equivalent "loop" length, and incorporating an identical Trp/Trp β-cap motif: cp-WW = RWFYFNRITGKRQFERPKGLVKGWEKRWD. While the crucial end groups (the beta cap and two β strands) are the same, the intervening sequence is not; the unstructured poly-Gly is replaced here with a segment that includes the 3rd strand of the WW domain. NMR studies have revealed that it forms a fully-native fold. The stabilities, as melting temperatures, and folding dynamics are collected in Table 1.

Table 1. Stability and folding dynamics data.

Construct	Tm (°C)	Folding rates as 1/k_F (μsec)			
		290 K	300 K	310 K	320 K
Loop length (#residues) = 2	90	--		~2	
= 4	69	29	16		
= 10	48	83	70		
= 16	36	160	110	80	~45
= 22	10	430	~280		
cp-WW	**55**	--		19	37

All dynamics data were based on the exchange-broadening observed for proton NMR resonances for sites with large chemical shift differences in the folded versus unfolded state [3]. The edge indole in the edge-to-face Trp/Trp pair provided sites for rate comparisons. Some of Arrhenius plots appear in Figure 2. There is a clear dependence between folding rate and loop length for the otherwise-identically folded peptide loop models. Unsurprisingly, folding rates increase monotonically with *decreasing* loop length. Perhaps more surprising are the length-dependent changes in the unfolding rates, which also decrease with longer loops (though this effect is less pronounced). For the loop models, both the folding and unfolding rates increase as the temperature is increased.

The WW domain circular permutant exhibited similar unfolding rates to the loop model with a comparable loop length but faster folding rates. In addition, the temperature dependence for the folding rate was clearly inverted with faster folding at lower temperature as has been observed for non-permuted WW domains and many other proteins.

We have shown that hairpin folding can be decoupled from potential turn-nucleating pathways in order to directly probe strand-association times, as a function of contact order. In the β-capped loop models no basis for structuring is present in the long flexible loops. In the case of the circularly permuted WW domain, the sequence between the terminal β strands is folded in the native state and alternative folding mechanisms besides a simple loop search to find the β-cap are available and result in folding acceleration.

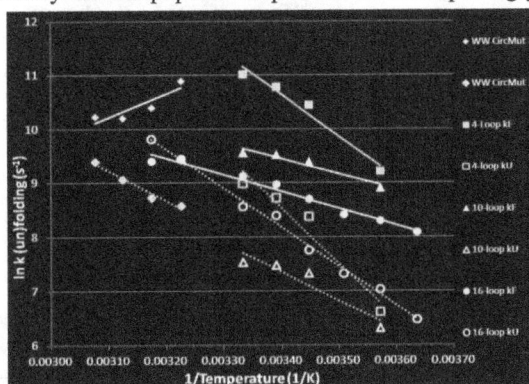

Fig. 2. Arrhenius plots of folding (solid lines) and unfolding rates (dotted lines, open symbols) for β-capped loops and cp-WW. The data for cp-WW could only be obtained at higher temperatures where there is a significant unfolded population.

Acknowledgments

Supported by grant CHE-1152218 (from the National Science Foundation) and NIH grant # GM-099899 (for studies of the WW domain circular permutant).

References

1. a) Fierz, B., Satzger, H., Root, C., Gilch, P., Zinth, W., Kiefhaber, T. *Proc. Nat. Acad. Sci. USA* **104**, 2163-2168 (2007); b) Fierz, B., Kiefhaber, T. *J. Am. Chem. Soc.* **129**, 672-679 (2007).
2. a) Kier, B.L., Andersen, N.H. *J. Am. Chem. Soc.* **130**, 14675-14683 (2008). b) Kier, B.L., Shu, I., Eidenschink, L.A., Andersen, N.H. *Proc. Nat. Acad. Sci. USA* **107**, 2657-2667 (2010).
3. Scian, M., Shu, I., Olsen, K.A., Hassam, K., Andersen, N.H. *Biochemistry* **52**, 2556-2564 (2013).

Proceedings of the 23rd American Peptide Symposium
Michal Lebl (Editor)
American Peptide Society, 2013

Mode of Action of the Sulfhydryl Group in Virolytic Peptide Triazole Thiol Inhibitors of HIV-1

Lauren D. Bailey, Caitlin Duffy, Huiyuan Li, Arangassery Rosemary Bastian, Ali Emileh, and Irwin Chaiken

Department of Biochemistry and Molecular Biology, Drexel University College of Medicine, Philadelphia, PA, 19102, U.S.A.

Introduction

HIV-1 entry is mediated by the interaction of the trimeric envelope glycoprotein (Env) on the virus membrane surface with host cell receptors. However, Env is the only virus-specific protein on the virion surface and is essential for cell receptor interactions and subsequent virus-cell fusion. Therefore, HIV-1 Env is an important target to directly inhibit and thus block the initial steps leading to host cell infection. Our lab has synthesized peptide triazoles, a class of novel entry inhibitors. These peptides contain a substituted triazole derivative formed from a synthetically introduced azido-proline amino acid and bind to gp120 with close to nanomolar affinity [1,2]. Site-directed mutagenesis and molecular dynamics simulation have shown that peptide triazole binding overlaps the CD4 binding pocket [3,4]. Peptide triazoles cause cell-independent gp120 shedding, and variants containing C-terminal cysteines cause cell-free virolysis as evidenced by internal p24 capsid release [5]. We are investigating the mode of action by which the sulfhydryl group causes irreversible inactivation. We hypothesize that the thiol interferes with conserved disulfides clustered proximal to the CD4 binding site in gp120 through "disulfide exchange", which could deform the Env protein spike, and subsequently the viral membrane, leading to p24 release. The process of disulfide exchange has been found to be necessary for HIV viral infection [6].

Results and Discussion

We previously identified KR13, a peptide triazole containing a free Cys sulfhydryl group at the C-terminal amide. As with other members of its peptide triazole family, KR13 inhibits both CD4 and co-receptor surrogate mAb 17b binding to gp120 with similar potencies [5], exhibits specific antiviral activity against HIV-1 BaL pseudotype and causes gp120 shedding [7]. Strikingly, KR13 and other peptide triazoles containing the C-terminal CysSH group have the capability to cause virolysis of HIV-1 as evidenced by release of p24 and formation of a shrunken residual virion particle [7]. The unique process of virolysis led us to investigate the role of the C-terminal cysteine sulfhydryl. We hypothesize that the free SH group of KR13 may function by triggering disulfide exchange between the thiol of the peptide triazole and one or several gp120 disulfides. These disulfides are found in a cluster proximal to the CD4 binding site of gp120 (Figure 1). The disulfide cluster in gp120 is composed of C296-C331 (V3 loop) disulfide, C385-C418 (C4) disulfide, and C378-C445 (C3) disulfide.

Based on a prior molecular dynamics simulated model of peptide triazole-gp120 binding [4] followed by an extrapolation of the likely trajectory of the C-terminus, the KR13 CysSH was predicted to be able to reach the disulfide cluster (Figure 1)

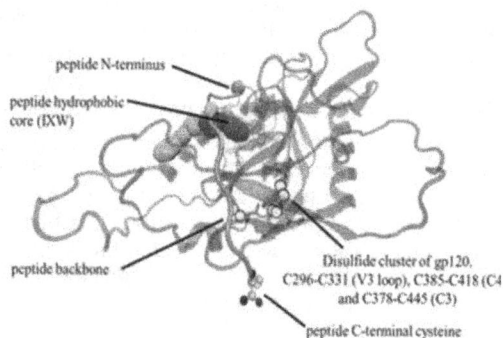

Fig. 1. Peptide triazole thiol bound to gp120. The binding model shows peptide triazole thiol KR13 bound in a site overlapping the CD4 pocket, with the C-terminal CysSH trajectory to (and possibly beyond) the conserved disulfide cluster, the latter of which encompasses possible sites of disulfide exchange. Trp residue of IXW is in red; and sulfur atoms of 3 clustered gp120 disulfides and peptide CysSH are represented by yellow balls.

Table 1. Peptide triazole thiol library characterization. Amino acid sequences of peptide triazole thiol truncates are listed, along with measured antiviral, virolytic, and relative virolytic (with respect to binding as observed by antiviral) activities.

Peptide Triazole Thiols	Sequence	Linker Atoms	Antiviral (μM)	p24 Release (μM)	p24 EC50/ Antiviral EC50
KR13	RINNIXWS-EA-M-M-βA-Q-βA-C	26	0.074±0.003	0.866±0.55	11.7
Peptide 1	RINNIXWS-EA-βA-Q-βA-βA-C	24	0.143±0.045	1.8±0.9	12.6
Peptide 2	RINNIXWS-EA-βA-Q-βA-C	20	0.114±0.040	1.32±0.26	11.6
Peptide 3	RINNIXWS-EA-βA-Q-C	16	0.197±0.038	2.2±0.6	11.2
Peptide 4	RINNIXWS-EA-βA-C	13	0.286±0.013	4.76±0.73	16.6
Peptide 5	RINNIXWS-EA-C	9	0.469±0.019	26.7±2.8	56.9
Peptide 6	RINNIXWS-E-C	6	0.554±0.007	33.2±0.9	59.9
Peptide 7	RINNIXWS-C	3	0.550±0.063	38.4±4.9	69.8
Peptide 8	RINNIXW-C	0	7.03±2.2	>>500	>>71.1
Peptide 9	NNIXW-C	0	26.3±0.78	>>500	>>19.0

making these disulfides possible sites of thiol-disulfide exchange. We hypothesize that there must be a minimal length of C-terminal linker needed for the CysSH to contact with the disulfides for efficient virolysis. Therefore, decreasing the length of the peptide triazole thiol linker should suppress lysis but proportionally retain binding affinity, antiviral potency and gp120 shedding activity, all of which are caused by active core binding but not the free SH. Therefore, we synthesized and characterized a set of C-terminal serially truncated peptide triazole thiols, based on the parent peptide KR13 binding core structure (Table 1). The serially truncated peptide triazole thiols displayed a strong dependence of lysis activity on length of the linker between the IXW pharmacophore and SH group (Table 1 and Figure 2), which argues that a minimal length is required for efficient virolysis. These observations support our hypothesis that the virolytic effect of peptide triazole thiols is promoted by a disulfide exchange between the peptide C-terminal thiol group and a specific disulfide cluster in gp120. Future studies in this project will include refining the peptide triazole pharmacophore, investigating the disulfide exchange process caused by the peptide triazole thiols, and identifying the specific Env protein disulfides involved.

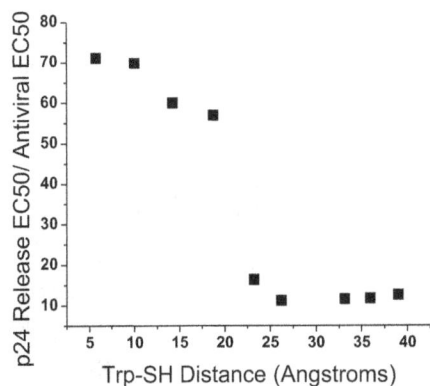

Fig. 2. Relative p24 lysis activity vs. pharmacophore-CysSH distance (based on cell infection). C-terminally truncated peptide triazole thiols lose virolytic potency as the distance between the thiol group and Trp residue of the IXW pharmacophore shortens.

Acknowledgments

We thank the National Institutes of Health for supporting this project with funds from grant 5P01GM056550 and LDB 1F31AI108485-01 from NIGMS.

References

1. Gopi, H., et al. *J. Mol. Recognition* **22**, 169-174 (2009).
2. Umashankara, M., et al. *ChemMedChem* **5**, 1871-1879 (2010).
3. Tuzer, F., et al. *Proteins* **81**, 271-290 (2013).
4. Emileh, A., et al. *Biochemistry* **52**, 2245-2261 (2013).
5. Bastian, A., et al. *ChemMedChem* **6**, 1335-1339 (2011).
6. Hugues, L., et al. *Proc. Nat. Acad. Sci. USA* **41**, 4559-4563 (1994).
7. Bastian, A., et al. *unpublished results* (2013).

Proceedings of the 23rd American Peptide Symposium
Michal Lebl (Editor)
American Peptide Society, 2013

Design, Synthesis and Structural Analysis of Cyclopeptides Against VEGF-Receptor

A. Caporale[1*], G. Focà[1,2*], N. Doti[1,2], A. Sandomenico[1,2], and M. Ruvo[1,2]

*[1]CIRPeB, Via Mezzocannone, 16, 80134 Napoli, Italy; [2]CNR-IBB, Via Mezzocannone, 16, 80134 Napoli, Italy; *These authors contributed equally to this work.*

Introduction

Angiogenesis is a physiological process mostly regulated by activation of VEGF receptors VEGF-R1 (Flt-1) and VEGF-R2 (KDR) by a set of structurally conserved ligands of the *Vascular Endothelial Growth Factor* (VEGF) family. Inappropriate receptor activation leads to pathological processes strictly correlated to tumor growth and their inhibition is a well-assessed strategy to stop or slow down cancer growth [1]. Both ligands and receptors have been targeted to block pathological angiogenesis, however a broad structural similarity amongst the different ligands and the two receptors makes difficult the design of new selective inhibitors. Several antiangiogenic compounds have been so far reported and developed at various stages [2]. A comparative analysis of known VEGFRs-binding peptides suggests that a minimum structural requirement for binding to VEGFR2 could be the motif XPR, where X is a hydrophobic residue crucial for receptor recognition. To identify new selective receptor inhibitors, we started from known VEGFR2-binding peptides [3-5] and designed and synthesized a small set of D-cyclopeptides containing this motif. A preliminary screening of their ability to bind to VEGFR1 and VEGFR2 by SPR was performed, showing that some of them bind both receptors with an affinity in the low micromolar range, whereas one is able to bind VEGFR1 in a selective manner.

Results and Discussion

The library components were designed using as template two previously reported VEGFR binding peptides: the linear peptide ATWLPPR [3], denoted as "CTRL", and the cysteine-bridged CPQPRPLC peptide [4,5]. Peptide C1 was the retro-inverso of the ATWLPPR lacking one Proline, to put the hydrophobic Leucine adjacent to the Arginine. Peptide C2 was the retro-inverso of ATWLPPR. The peptide C3 was designed to exchange the hydrophobic patch before the PPR motif with a charged residue. Peptides from C5 to C7 were designed as variants of peptide C3, by moving the "rp" motif along the sequence. The retro-inversion was introduced to increase proteolytic stability, maintaining the overall side chain topology, whereas cyclization was introduced to restrict the conformational freedom in solution.

The synthesis of the library was carried out on the solid phase, following the Sheppard's protocol using Fmoc protected Amino Acids and HATU as coupling reagent. The cyclization was carried out in slightly alkaline buffers (pH 8.0) on crude peptides. Purification was accomplished after reaction completion and acidification.

The interaction between VEGF mimetic cyclopeptides and VEGF-R1 and VEGF-R2 receptors was assessed by SPR assays using a BIACORE 3000 system. A CM5 sensor chip was covalently derivatized with recombinant KDR and Flt-1 proteins following an amino coupling strategy. The screening was performed simultaneously on both receptors, starting from a concentration of 20 nM until 20 μM.

NAME	M.W. Calc. (g/mol)	M.W. Found (g/mol)	K_D vs KDR receptor (M⁻¹)	K_D vs Flt-1 receptor (M⁻¹)
CTRL	838,5	839.4	//	9,92E-04
C1	1002,5	1003.2	9.36 E-06	3.57 E-06
C2	1099,5	1100.4	1.65 E-05	1.86 E-05
C3	857,4	858.3	//	1.78 E-05
C4	966,5	484.5 (2+)	//	//
C5	850,4	851.5	//	//
C6	850,4	851.4	//	//
C7	850,4	851.3	//	//

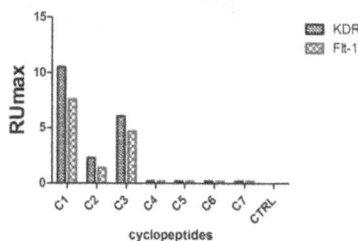

Fig. 1. a) Names, MW and K_Ds for the two VEGFRs determined by dose-response binding assays; b) SPR screening: plot of RUmax achieved at the concentration of 20 μM for the binding of the different cyclopeptides to the two receptors.

Data (Figure 1b) showed that C1 and C2 bound to both VEGF-R1 and VEGF-R2 receptors while the C3 cyclopeptide seemed specific for Flt-1. To further investigate the binding in terms of affinity and specificity on both receptors, we carried out a dose-dependent assay for three positive cyclopeptides C1, C2, and C3. By fitting data using the BIAevaluation analysis package (version 4.1, Pharmacia Biosensor) we extrapolated the K_D values (Figure 1a). The C2 peptide bound with a similar K_D to both receptors, while C1 recognized Flt-1 with a slightly increased affinity compared to KDR. In addition, we found a weak interaction of the linear and C3 peptides with Flt-1 receptor only.

In conclusion, we confirmed that the motif PR, preceded often by a hydrophobic residue, occurring in several VEGF receptor binding peptides, such as the sequence of PlGF-derived peptides [4-6], might be an important scaffold for binding to KDR and to Flt-1 [3-5]. We found that some of the library components bound both receptors (C1 and C2) with moderate affinity, whereas one of them, C3, selectively bound to Flt-1. Preliminary CD analyses (data not shown) suggest that the peptides adopt very different conformations in solutions and this, together with the different arrays of side chains can affect receptor recognition. NMR analyses are currently underway to investigate the contribution to binding deriving from side chains and/or backbone conformation.

Acknowledgments

We gratefully acknowledge the support from Regione Campania for PROGETTO "CAMPUS (Progetti di Ricerca industriale e sviluppo sperimentale") and MIUR for project FIRB-MERIT N RBNE08NKH7 to M. Ruvo. We also thank Consiglio Nazionale delle Ricerche for project CCNC.

References

1. Folkman, J. *N. Engl. J. Med.* **285**, 1182-1186 (1971).
2. National Cancer Institute, *http://www.cancer.gov/cancertopics/factsheet/Therapy/angiogenesis-inhibitors.*
3. Binétruy-Tounaire, R., et al. *EMBO J.* **19**, 1525-1533 (2000).
4. Giordano, R.J., et al. *Proc. Nat. Acad. Sci. USA* **107**, 5112-5117 (2010).
5. Pasqualini, et al. US2012/0028880 A1.
6. Diana, D., et al. *J. Biol. Chem.* **286**, 41680-41691 (2011).

Proceedings of the 23rd American Peptide Symposium
Michal Lebl (Editor)
American Peptide Society, 2013

Molecular Modeling Studies of Peptide Based Lectinomimics

Michael Cudic, Maria C. Rodriguez, Austin Yongye, Karina Martinez, and Predrag Cudic

Torrey Pines Institute for Molecular Studies, Port St. Lucie, FL, 34987, U.S.A.

Introduction

Carbohydrates mediate numerous biological processes involving cell-cell recognition. Concurrently, establishing connections between glycan structures and their functions has fuelled strong interest in mimicking the action of lectins, carbohydrate binding proteins, for potential bioanalytical and/or biomedical applications.

Cyclic peptides represent a particularly attractive approach for the design and preparation of artificial carbohydrate receptors [1]. Peptide cyclization lowers the conformational flexibility that might lead to the formation of less entropically unfavorable conformations in order to adopt the optimal complex geometry. It has been reported recently that small cyclic peptides possessing a *β*-turn/*β*-sheet conformation display lectin-like properties. The simplest of these is odorranalectin, a 17-mer cyclic peptide containing one disulfide bond isolated from skin secretions of the frog *Odorrana grahami* (Figure 1) [2].

Odorranalectin natural product was found to selectively bind to L-fucose with a binding affinity (K_d) of 55 μM [2]. Selectivity and affinity of peptide based lectinomimic can be tuned by the sequence modification. In particular, modification using a combinatorial

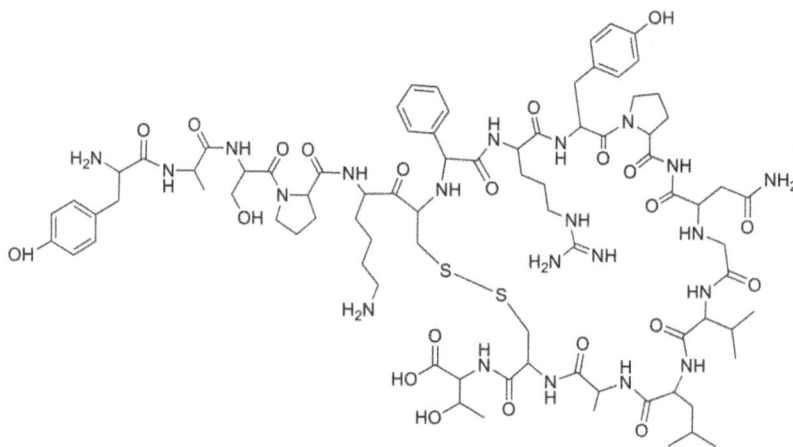

Fig. 1. Natural product odorranalectin.

chemistry approach offers enormous potential for synthesis of new and more selective peptide based lectinomimics. In the case of odorranalectin combinatorial library, Cys residues have to be omitted from the sequence in order to avoid potential problems with the thiol group oxidation during the library synthesis. However, substitution of disulphide bond in odorranalectin with other linkers may affect its conformation and thus its ability to form a hydrophobic pocket required for binding of carbohydrate ligands. To assess the effect of the disulphide bond substitution with different linkers on odorranalectin's conformation, we performed a conformational analysis of the natural product as well as a series of analogs with amide, diazine, diselenium, ethyne, ethane and thioether linkers (Figure 2a).

Results and Discussion

The Cartesian coordinates of odorranalectin were downloaded from the protein databank, pdbid: 2JQW. This file contained 20 NMR structures of odorranalectin, of which one model was selected as the starting structure. The geometry of each structure was optimized using Macromodel v9.9 [3], and conformational sampling was performed by employing the

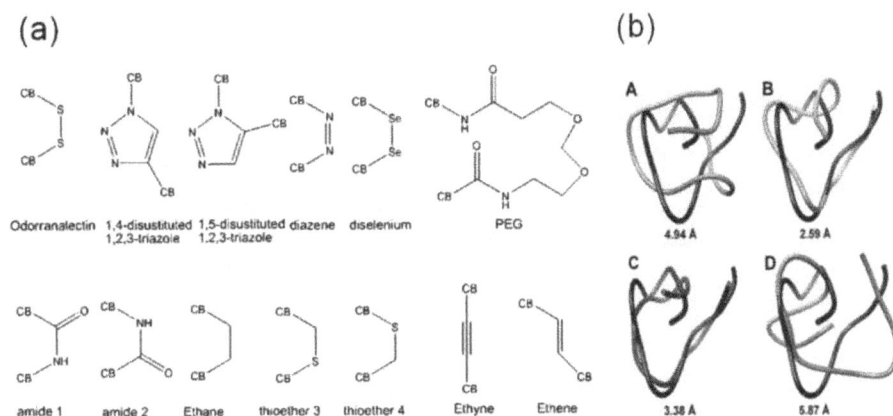

Fig. 2. Bridges employed computationally (a) and backbone overlays (b). (A) amide 1, (B) amide 2, (C) thioether 3 and (D) thioether 4. The weighted-average RMSD values are shown underneath each overlay. CB stands for C_β atom in the sidechain of Cys mimetics.

Macrocycle Conformational Sampling option of Macromodel v9.9 with the OPLS 2005 force field. A distance-dependent dielectric was used with the enhanced sampling mode. For each structure conformers above 5.0 kcal/mol of the identified global minimum were excluded, as were conformers that were less than 1.5 Å RMSD with a previously sampled conformer. The large scale low-mode method was applied to 10000 simulation cycles. To further reduce the number of conformers of each compound sampled by Macromodel, the backbone atoms of its conformers were superimposed and clustered using the trjconv and g_cluster modules of GROMACS [4] respectively. A pairwise RMSD distance matrix was computed between the centroids of all the compounds using a python script in Chimera v1.62. [5].

The best conformational overlay was 2.38 Å exhibited by amide analog **2** with a percent population of 80%, followed by thioether analog **3** with a value of 2.42 Å and percent population 42.3% (Figure 2b). In summary, we have shown that the disulfide bond in the odorranalectin natural product can be substituted with an amide bond without *significant conformational* change, thus making this analog ideally suited for modification using a combinatorial chemistry approach.

Acknowledgments

We would like to thank TPIMS summer internship program for support of M.C.

References

1. Rodriguez, M.C., Cudic, P. *Chimica Oggi/Chemistry Today* **29**, 36-42 (2011).
2. Li, J., et al. *PLosOne* **3**, e2381 (2008).
3. Macromodel 9.9, Schrödinger, LLC, New York, NY (2012).
4. Hess, B., Kutzner, C., van der Spoel, D., Lindahl, E.J. *Chem. Theory Comput.* **4**, 435-447 (2008).
5. Pettersen, E.F., Goddard, T.D., Huang, C.C., Couch, G.S., Greenblatt, D.M., Meng, E.C., Ferrin, T.E. *J. Comput. Chem.* **25**, 1605-1612 (2004).

Proceedings of the 23rd American Peptide Symposium
Michal Lebl (Editor)
American Peptide Society, 2013

Inhibition of Insulin Aggregation Through Supramolecular Host-Guest Interactions

Fiona C. Christie[1], Athene M. Donald[2], and Oren A. Scherman[1]

[1]Melville Laboratory for Polymer Synthesis, Department of Chemistry, Lensfield Road, Cambridge, UK, CB2 1EW; [2]Department of Physics, J.J. Thomson Avenue, Cambridge, UK, CB3 0HE

Introduction

The stability of protein-based pharmaceuticals, such as insulin, is important for their production, storage, delivery and administration. Insulin is a peptide hormone capable of forming amyloid-like structures and as such can be problematic during the manufacture and delivery processes. The formation of ordered, amyloid-like structures in proteins has been associated with several diseases such as Alzheimer's, Parkinson's and type II diabetes, and is now thought to be a generic feature of protein aggregation. As such, there is considerable interest in controlling their formation and gain better understanding of the aggregation process. Cucurbit[n]uril (CB[n]) homologues are a series of macrocyclic host molecules of which the larger CB[7] and CB[8] molecules are capable of binding guests containing an aromatic moiety, such as phenylalanine [1] (Figure 1). Herein, we report the effects of their complexation in aqueous media with human insulin and its aggregation-prone fragments. We show that the N-terminal phenylalanine residue of human insulin binds to CB[8] in a 2:1 molar ratio, resulting in an inhibition of the previously reported aggregation process [2]. Removal of the insulin from the CB host results in aggregation behaviour not observed in the complexed form, suggesting CB as a potential inhibitor of this type of phenomena.

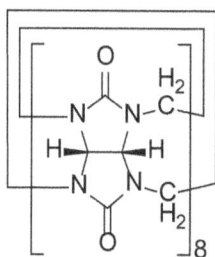

Fig. 1. The CB[8] macrocycle is capable of accepting two Phe residues simultaneously.

Results and Discussion

The effect of complexation between human insulin and CB[8] was monitored by dynamic light scattering (DLS) over multiple temperature cycles. It was found that, under the protein denaturing conditions used (20% acetic acid, 0.2M NaCl) at 25°C, addition of the CB[8] to the protein resulted in the formation of an insoluble aggregate. Upon heating above 37°C, the complex resolubilised and the light scattering intensity was greatly reduced. As the solution cooled back to room temperature, the aggregate precipitated out once again – a fully reversible process. Studies by Chinai, et al. [1] showed that the smaller homologue CB[7] binds solely to the N-terminal Phe residue on the B-chain of human insulin with an association constant of 1.5×10^6 M^{-1}. This work revealed that the larger CB[8] cavity was capable of binding the N-terminal Phe of two insulin molecules simultaneously (Figure 2) with a K_a of 1×10^6 M^{-2}. Competitive replacement experiments were performed to determine whether the binding of the insulin to the CB[8] was inducing the aggregation.

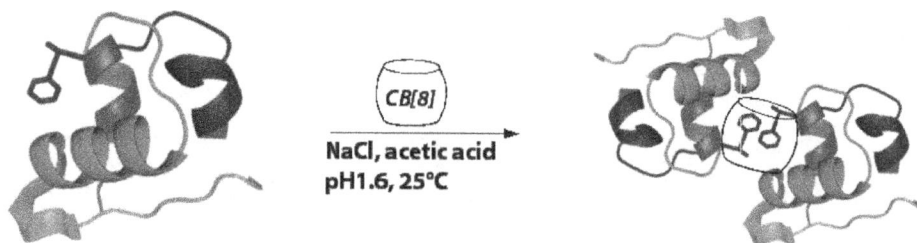

Fig. 2. Addition of insulin to CB[8] under protein denaturing conditions resulted in a 2:1 complex which precipitated out of solution.

Adamantylamine is a suitable guest molecule known to have a strong association constant of 8.2×10^8 M^{-1} for binding with CB[8] [3]. The addition of 1 equivalent of adamantylamine to a solution of 2:1 insulin-CB[8] did not displace the insulin immediately; however after one heating and cooling cycle, the components of the solution were fully soluble, indicating displacement of the insulin from the CB[8] host.

Centrifugation of the aggregate removed any non-complexed insulin and CB[8] from the precipitate. Resuspension of the resultant solid in a number of solvents (water; acetic acid; 1% insulin/acetic acid; 100µM CB[8]; 100µM adamantylamine) showed that both the insulin and the CB[8] were present in the precipitate. Resuspension in adamantylamine resulted in a colorless solution, while addition of the solid to any of the other solvents resulted in a precipitated aggregate.

Insulin is known to form spherulite structures [4] during the aggregation process and the mechanism of growth has been widely characterized. The presence of spherulites is confirmed by microscopy; a sample containing spherulite structures is typified by a "Maltese cross" pattern when viewed through crossed polarizers. Incubation of a 1% insulin sample both at room temperature and at 70°C in 20% acetic acid resulted in spherulite formation over 72 hours. In the presence of CB[8], however, only the sample heated at 70°C displayed spherulite formation. The sample incubated at room temperature did not show any evidence of spherulite behavior.

Complexation of insulin with the macrocyclic host molecule CB[8] is achieved in a 2:1 molar ratio in 20% acetic acid in the presence of 0.2M NaCl. Addition of CB[8] to an insulin solution results in the formation of an insoluble complex. Inhibition of the aggregation of human insulin at room temperature has been achieved by complexation of the protein with the macrocyclic host molecule CB[8]. The inhibition of protein amyloid formation is desirable for the design of protein based therapeutic agents and could have potential application in the treatment of other amyloid-based diseases.

Acknowledgments

FCC gratefully acknowledges the EPSRC for funding through the Nanoscience and Technology Doctoral Training Centre at the University of Cambridge.

References

1. Chinai, J.M., Taylor, A.B., Ryno, L.M., Hargreaves, N.D., Morris, C.A., Hart, P.J., Urbach, A.R. *J. Am. Chem. Soc.* **133**, 8810-8813 (2011).
2. Krebs, M.R.H., Bromley, E.H.C., Rogers, S.S., Donald, A.M. *Biophys. J.* **88**, 2013-2021 (2005).
3. Day, A., Arnold, A.P., Blanch, R.J., Snushall, B. *J. Org. Chem.* **66**, 8094-8100 (2001).
4. Krebs, M.R.H., MacPhee, C.E., Miller, A.F., Dunlop, I.E., Dobson, C.M., Donald, A.M. *Proc. Nat. Acad. Sci. USA* **101**, 14420-14424 (2004).

Proceedings of the 23rd American Peptide Symposium
Michal Lebl (Editor)
American Peptide Society, 2013

Self-Assembled Peptide Materials for Prevention of HIV-1 Transmission

John T. M. DiMaio[1], David Easterhoff[2], Annah M. Moore[1], Stephen Dewhurst[2], and Bradley L. Nilsson[1]

[1]Department of Chemistry, University of Rochester, Rochester, NY, 14627, U.S.A.; [2]Department of Microbiology and Immunology, University of Rochester, Rochester, NY, 14627, U.S.A.

Introduction

Development of microbicides for prevention of sexual transmission of HIV-1 is of critical importance. Molecules that perturb entry of the virus into host cells are promising components of microbicide strategies since these so-called "entry inhibitors" are less prone to resistance than other classes of HIV therapeutics [1]. We hypothesized that attachment of an HIV entry inhibitor to a self-assembling peptide hydrogelator would facilitate a multivalent display of the inhibitor at the surface of the resulting fibril assembly. These hydrogel materials have potential to be applied as intravaginal/intrarectal anti-HIV microbicides. Herein we report efforts to affect multivalent display of aplaviroc (APV) [1], an entry inhibitor that targets CCR5, on peptide nanofibrils derived from the amphipathic $(FKFE)_2$ peptide [2-4] (Figure 1) in order to create materials that can prevent interaction of HIV with target cells, thus abrogating HIV transmission.

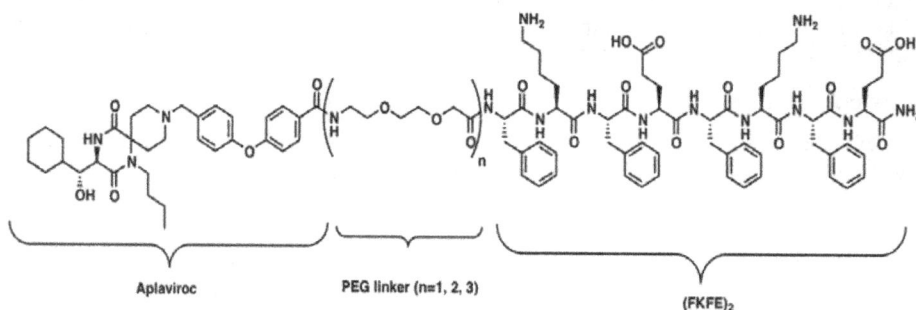

Fig. 1. APV attached to $(FKFE)_2$, an amphipathic self-assembling peptide.

Results and Discussion

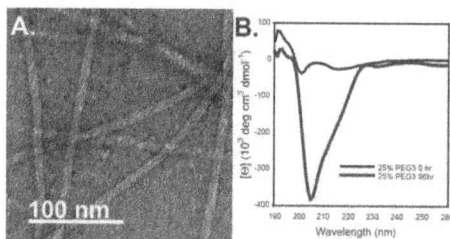

Fig. 2. A. TEM of PEG3 nanoribbons; B. CD spectrum of PEG3 nanoribbons.

Synthetic APV was attached to $(PEG)_n$-$(FKFE)_2$-NH_2 peptides (n = 1, 2, or 3). Representative data for APV-$(PEG)_n(FKFE)_2$-NH_2 with 3 PEG units (PEG3) is shown. PEG3 successfully assembles into nanoribbon structures 9.86 ± 0.49 nm in diameter (Figure 2A) that display CD signatures similar Ac-$(FKFE)_2$-NH_2 fibrils (Figure 2B) [2]. Co-assembly of APV-bearing peptides at varying ratios with Ac-$(FKFE)_2$-NH_2 was also assessed. It was found that PEG3 co-assembled at varying ratios with Ac-$(FKFE)_2$-NH_2 resulted in fibrils with identical morphological and spectroscopic properties as PEG3 assemblies. Varying the ratio of PEG3 in these assemblies is an effective strategy to tune the density of APV displayed on these materials.

We next conducted experiments to characterize whether multivalent display on amyloid-inspired materials imparted an advantage to binding to CCR5-bearing BC7 cells relative to monomeric APV. Two antibodies were used to determine the efficiency of binding: (1) 2D7, which binds to CCR5 noncompetitively with APV; and (2) 45531, which binds to the same region of CCR5 as APV. While PEG3/Ac-(FKFE)$_2$-NH$_2$ assemblies with 25% PEG3 bind to CCR5-bearing BC7 cells, they did so less effectively than monomeric APV (Figure 3A). This result was unexpected in light of literature precedent in which advantages for multivalent display of APV were observed [1]. We hypothesize that this decrease in binding efficiency of the APV-fibril display occurs because APV is, to some extent, sequestered to the hydrophobic bilayer core of the (FKFE)$_2$-derived materials [2-4], reducing availability for binding to CCR5. This burying is because of the high hydrophobicity of APV, and fibril formation of (FKFE)$_2$ is dictated by desolvation of the hydrophobic face of the peptide. 25% PEG3 remains bound to CCR5 in higher concentration than APV; however, not significantly (Figure 3B).

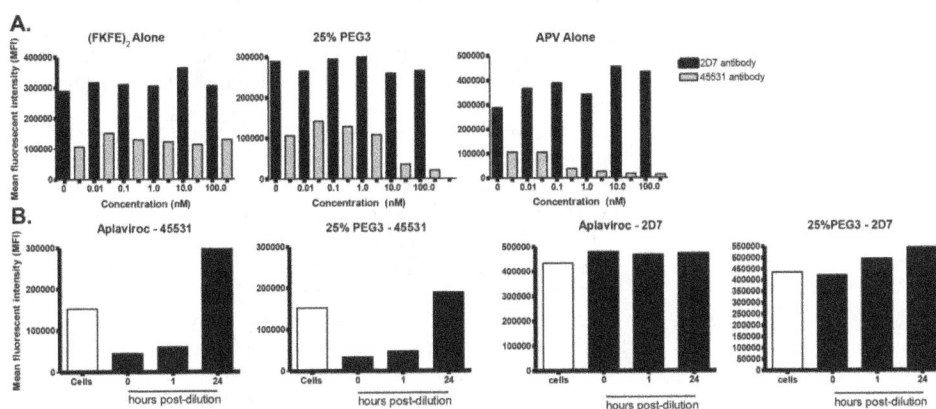

Fig. 3. *Binding of CCR5 and competitive binding of CCR5 with APV or PEG3 nanoribbons; A. BC7 cells were treated with 2D7 to assess overall CCR5 expression (high MFI indicates high CCR5 expression), and then compared with 45531 to determine the overall amount of APV or PEG3 binding to CCR5 (low MFI indicates binding of APV or PEG3); B. BC7 cells were treated with 45531 and fluorescence was measured over time to determine if PEG3 would remain bound to CCR5 longer than APV.*

These studies establish proof of principle for the use of amyloid-inspired peptide materials as components of anti-HIV microbicides. The display of entry inhibitors on peptide nanofibrils is a promising strategy for the creation of functional anti-HIV materials, although challenges remain in the design and implementation of these materials. Specifically, strategies to ensure that anti-HIV functionality is available at the surface of fibril assemblies, and not sequestered in the interior of the fibril where binding to target is inhibited. These challenges may be overcome by modifying the peptide display and by appropriate selection of HIV targeting ligands.

Acknowledgments

This work was supported by a Creative and Novel Ideas in HIV Research (CNIHR) grant and a Provost's Multidisciplinary grant (University of Rochester).

References

1. Gavrilyuk, J. et al. *ChemBioChem* **11**, 2113-2118 (2010).
2. Bowerman, C.J., et al. *Mol. BioSyst.* **5**, 1058-1069 (2009).
3. Easterhoff, D. et al. *Biophys. J.* **100**, 1325-1334 (2011).
4. Bowerman, C.J., et al. *Biopolymers* **98**, 169-184 (2012).

Proceedings of the 23rd American Peptide Symposium
Michal Lebl (Editor)
American Peptide Society, 2013

Analysis of Supramolecular Assemblies via Electrostatic Force Microscopy

Sha Li[1], Anton Sidorov[2], Anil K. Mehta[1], Dibyendu Das[1], Zhigang Jiang[3], Thomas M. Orlando[2], and David G. Lynn[1]

[1]*Emory University, Departments of Chemistry and Biology, Atlanta, GA, 30322, U.S.A.;*
Georgia Institute of Technology; [2]*School of Chemistry and Biochemistry, and*
[3]*School of Physics, Atlanta, GA, 30332, U.S.A.*

Introduction

The nucleating core of the Amyloid β peptide implicated in Alzheimer's disease, when formulated as Ac-KLVFFAE-NH$_2$ (Aβ(16-22)), self-assembles in aqueous environments as peptide bilayers maintaining a thickness similar to biological phospholipid membrane [2]. The hydrogen-bonded peptides are arranged in antiparallel out-of-register β-sheets and stacks of these sheets make up one of the leaflets of the bilayer membrane[2]. The *N*-terminal lysine residues are exposed on the surface of each leaflet with its backbone amide not incorporated into the hydrogen-bonding network [3].

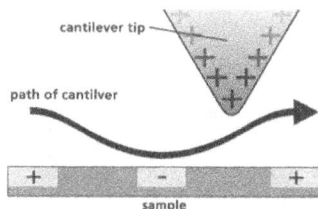

Fig. 1. Anticipated displacement during electrostatic force microscopy [1].

These structural models [3] for Aβ(16-22) and its congener E22L present ordered crystalline structures with a high density of alkyl ammonium ions located at precise positions across a nanoscale grid on a hollow nanotube surface. Methods for mapping these charges in aqueous solutions to define positions of macromolecular adsorbents, surface imperfections, domain size, and even surface dynamics are limited [4-6]. However, extensions of Atomic Force Microscopy (AFM), by applying a bias on the probe tip, have been developed to measure the electric field gradient distribution above dry surfaces, including mineral crystal faces, graphene layers and other solid materials [7]. Here we demonstrate electrostatic force microscopy (EFM) analyses (Figure 1) that map the charge distribution on these self-assembled peptide membranes in a partially dried state. Our results are consistent with homogeneously charged peptide membrane surfaces that achieve microscale order and open new opportunities for characterizing more complex and dynamic self-assembled engineered materials.

Results and Discussion

Peptides were synthesized using standard FMOC chemistry on a Liberty CEM Microwave Automated Peptide Synthesizer, purified (> 99%) on a C18-reverse phase column with an acetonitrile-water gradient containing 0.1% trifluoroacetic acid, and their molecular weights confirmed by MALDI-TOF using a 2,5-dihydroxybenzoic acid matrix [2,3]. The purified peptides were allowed to assemble under acidic or neutral conditions in 40% acetonitrile/water. Maturation of peptide assembly was determined by monitoring circular dichroism elipticity at 217 nm.

Electrostatic Force Microscopy was performed on a Park System XE-100 AFM. Pt-Ir coated electrically conductive cantilevers were used to map electrical properties on the samples. A series of measurements with a positive tip bias were taken on a standard sample with P, N and neutral coatings (Figure 2). Topography images (Figure 2, right) showed repeated steps with two different heights. However, EFM amplitude images (Figure 2, left) indicated three different coating areas with positive, neutral and negative charges.

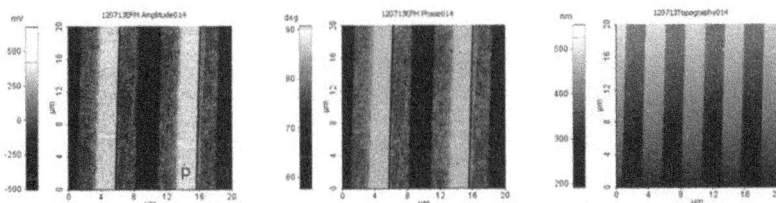

Fig. 2. Standard sample with P, N and neutral coatings, from left to right: EFM amplitude, EFM phase, topography.

Aqueous samples were deposited on Si/SiO_2 substrates and air-dried over 12 hours. Using these same EFM methods, the NH_4Cl and Na_2HPO_4 samples showed the existence of crystalline or aggregates forms of the salts without any specific patterns (Figure 3a,b). EFM measurements on self-assembled Ac-KLVFFAL-NH_2 nanotubes showed homogeneous

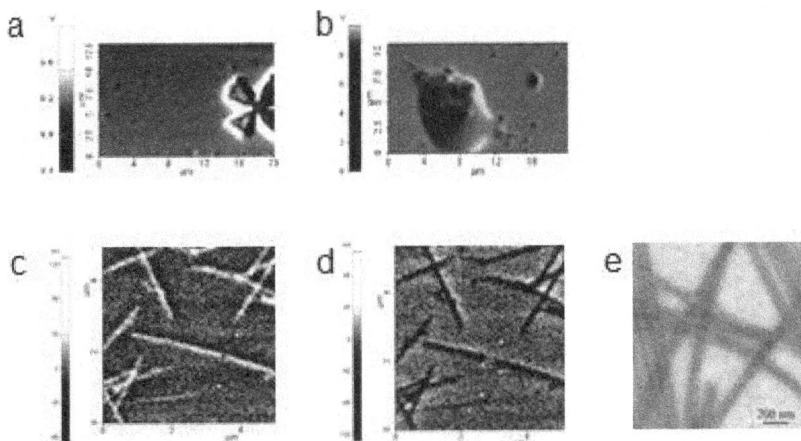

Fig. 3. EFM amplitude images of dried salt solutions a) NH_4Cl and b) Na_2HPO_4; Ac-KLVFFAL-NH_2 assemblies c) positive bias and d) negative bias EFM and e) TEM under acidic conditions. Red line indicated the site for electrochemical potential measurements.

positively-charged micrometer-long nanotubes. With a positive tip bias (Figure 3c), the positively-charged surfaces are shown in bright colored lines and the electrochemical potential was ~55 millivolts. With a negative tip bias (Figure 3d), the dark lines gave a potential of ~40 millivolts. While the absolute potential voltage changes depend on salt and pH, the images clearly reflect the fixed charge on the assembly surface and confirm both the highly ordered surfaces and the use of EFM to map these dynamic supramolecular surfaces.

Acknowledgments

This work was jointly supported by NSF and the NASA Astrobiology Program, under the NSF Center for Chemical Evolution, CHE-1004570; The Division of Chemical Sciences, Geosciences, and Biosciences, Office of Basic Energy Sciences of the U.S. Department of Energy through Grant DE-ER15377. We thank Robert P. Apkarian Microscopy Core, Emory University, for TEM.

References

1. http://www.parkAFM.com.
2. Childers, W.S., et al. *Angewandte Chem. Int. Ed.* **49,** 4104-4107 (2010).
3. Mehta, A.K., et al. *J. Amer. Chem. Soc.* **130,** 9829-9835 (2008).
4. Clausen, C.H., et al. *Scanning* **33,** 201-207 (2011).
5. Lee, G., et al. *Applied Phys. Lett.* **101,** 043703-043704 (2012).
6. Gramse, G., et al. *Applied Phys. Lett.* **101,** 213108-21312 (2012).
7. Burnett, T., Yakimova, R., Kazakova, O. *Nano Letters* **11,** 2324-2328 (2011).

Proceedings of the 23rd American Peptide Symposium
Michal Lebl (Editor)
American Peptide Society, 2013

Amino Acid and Peptide-Derived Co-Assembled Hydrogels for Cell Culture Applications

Wathsala Liyanage and Bradley L. Nilsson

Department of Chemistry, University of Rochester, Rochester, NY, 14627, U.S.A.

Introduction

The extracellular matrix (ECM), a complex biologically active environment, provides mechanical and biochemical cues to cells [1]. Design of biomaterials that can mimic the dynamic native ECM by supporting cell-material interaction, cell viability, and proliferation is a rapidly expanding field of study. Hydrogels derived from self-assembled peptides have proven to be versatile materials that closely mimic the 3D extracellular environment [2]. Fmoc-Phe-derived self-assembled hydrogel materials [3,4] (Figure 1) have potential as economical alternative biomimetic materials for cell culture and tissue engineering applications [5]. The utility of Fmoc-Phe-based gels as ideal hydrogels for cell culture applications is yet limited due to the need for organic co-solvents and less efficient hydrogelation of complex cell culture media mixtures [6]. Herein we report initial efforts to overcome these limitations and to establish the biochemical suitability of Fmoc-Phe-derived hydrogels as 3D cell culture matrices.

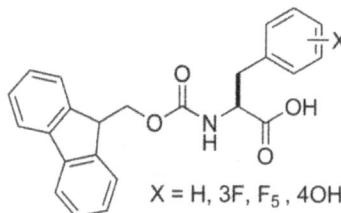

Fig. 1. Chemical structure of Fmoc-X-Phe derivatives.

Results and Discussion

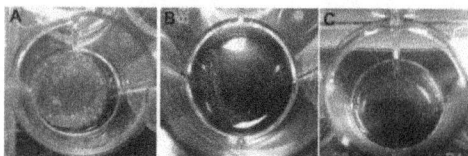

Fig. 2. Digital images of Fmoc-X-Phe hydrogels after perfusion with DMEM for 72 h. A) Fmoc-F$_5$-Phe; B) Fmoc-3F-Phe; C) Fmoc-Tyr.

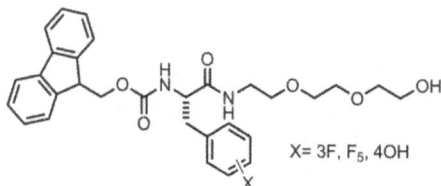

Fig. 3. Chemical structure of Fmoc-X-Phe-PEG derivatives.

The use of Fmoc-Phe-based hydrogel materials for tissue culture applications has been limited by several factors. First, the need to induce gelation by dilution of Fmoc-Phe derivatives into water from an organic co-solvent (most frequently DMSO) limits compatibility with living cells. Second, the hydrogelation of complex cell culture media has been challenging relative to gelation of simple aqueous solutions. We attempted to overcome these limitations by forming hydrogels in water by dilution of Fmoc-Phe derivatives (Figure 1) from DMSO into water followed by perfusion of the resulting hydrogels with cell culture media (DMEM) in order to remove the organic co-solvent and enrich the gel with the materials necessary for cell growth. Fmoc-X-Phe derivatives were sterilized by UV exposure and hydrogelation was initiated by dilution of sterile filtered Fmoc-X-Phe DMSO stock solutions into water (4.9 mM Fmoc-Phe derivative, 2% DMSO/water, *v/v*). The resulting hydro-gels were perfused with cell culture media (DMEM) at 37°C. After 72 h of perfusion, it was found that these hydrogels were of limited stability; each hydrogel exhibited precipitation of the gel fibril network and loss of mechanical integrity (Figure 2).

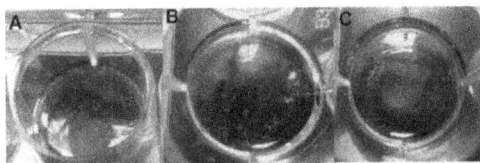

Fig. 4. Digital images of Fmoc-X-Phe/Fmoc-X-Phe-PEG hydrogels after DMEM perfusion. A) Fmoc-3F-Phe:Fmoc-3F-Phe-PEG (8:2) B) Fmoc-Tyr:Fmoc-Tyr-PEG (8:2) C) Fmoc-F5-Phe:Fmoc-F5-Phe-PEG (8:2).

Coassembly of Fmoc-Phe derivatives with Fmoc-Phe-PEG derivatives (Figure 3) was explored as a strategy to stabilize the hydrogel network to precipitation. This strategy has been used to stabilize Fmoc-Phe-derived gels in simple aqueous solutions [3]. Co-assembled Fmoc-X-Phe/Fmoc-X-Phe-PEG hydrogels were formed in water by dilution from DMSO as described above and perfused with DMEM for 72 h (Figure 4). The Fmoc-F5-Phe/Fmoc-F5-Phe-PEG (8:2) hydrogels showed excellent solvolytic stability and shear responsive behavior even after 4 days (Figure 4). A rheological oscillatory frequency sweep was conducted before perfusion and after several exchanges of cell culture media to test the rigidity of hydrogel (Figure 5A); the resulting gel exhibited stable rheological properties. Transmission electron microscopy (TEM) imaging of the hydrogel fibril before (Figure 5B) and after (Figure 5C) DMEM perfusion confirmed that the fibril network remained intact and unaltered in morphology.

These studies indicate that co-assembly of Fmoc-Phe and Fmoc-Phe-PEG derivatives provide hydrogels that are stable under conditions necessary for *ex vivo* tissue culture. While these hydrogels are formed by dilution from DMSO solutions into water, perfusion with DMEM cell culture media ultimately provides hydrogel materials that stable and that have appropriate rheological properties to support cell growth. Initial cell culture experiments provide validation that cells are viable and can proliferate in these gels. Thus, Fmoc-Phe-derived hydrogels have great potential as cost-effective ECM-mimetics for cell culture applications. Further cell culture assessments will be conducted to test the biocompatibility of the gel materials in terms of cell

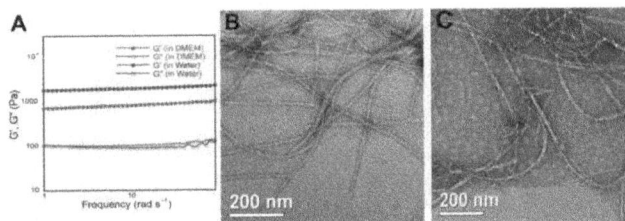

Fig. 5. A) Oscillatory frequency sweep of Fmoc-F5-Phe-OH: Fmoc-F5-Phe-PEG B-C) TEM images of Fmoc-F5-Phe-OH: Fmoc-F5-Phe-PEG before and after perfusion with culture media.

viability, adhesion, proliferation, migration, and differentiation. Future efforts will focus on engineering appropriate biochemical characteristics of these materials as a function of cell type.

Acknowledgments

We gratefully acknowledge Karen Bentley (University of Rochester Electron Microscopy Core) for her assistance with TEM imaging, and Dr. Kanika Vats (Department of Biomedical Engineering) for assistance with cell culture experiments. We also acknowledge the NSF (DMR-1148836) for support of this work.

References

1. Geckil, H., Xu, F., Zhang, X., Moon, S., Demirci, U. *Nanomedicine* **5**, 469-484 (2010).
2. Kretsinger, J.K., Haines, L.A., Ozbas, B., Pochan, D.J., Schneider, J.P. *Biomaterials* **26**, 5177- 5186 (2005).
3. Ryan, D.M., Doran, T.M., Nilsson, B.L. *Chem. Commun.* **47**, 475-477 (2011).
4. Ryan, D.M., Doran, T.M., Anderson, S.B., Nilsson, B.L. *Langmuir* **27**, 4029-4039 (2011).
5. Jayawarna, V., Richardson, S.M., Hirst, A.R., Hodson, N.W., Saiani, A., Gough, J.E., Ulijn, R.V. *Acta Biomaterialia* **5**, 934-943 (2009).
6. Ryan, D.M., Nilsson, B.L. *Polym. Chem.* **3**, 18-33 (2012).

Proceedings of the 23rd American Peptide Symposium
Michal Lebl (Editor)
American Peptide Society, 2013

Nanoparticles for Peptide Synthesis and Biological Applications

Victoria Machtey[1], R. Khandadash[1], Yuval Ebenstein[3], Aryeh Weiss[2], and Gerardo Byk[1]

[1]Dept. of Chemistry, Laboratory of Nano-Biotechnology; [2]School of Engineering, Bar Ilan University, 52900-Ramat-Gan; [3]School of Chemistry, NanoBioPhotonics, Tel Aviv University, Tel Aviv, Israel

Introduction

Developing new methods for *in vivo* targeting and tracking of small molecules in live cells is crucial for extending the scope of a variety of biological applications such as drug discovery, drug targeting, medical diagnostics and tissue engineering.

To develop advanced systems for live cell and *in vivo* screening based on the use of small molecules anchored to inert vehicle particles, we have synthesized a new type of bio-compatible, non-toxic nanoparticles (NPs) suited both for synthesis of peptides on their surface and for live cells and *in vivo* applications. These new NPs can be used for novel generations of screening systems where a small molecule is synthesized on or conjugated to bio-compatible and properly labeled NPs so that it can be further screened/tracked in live-cell assays or even *in vivo* without toxicity risks originated from the vehicle particle.

Results and Discussion

Self-assemblies were formed for three different concentrations of block copolymer mixtures (exp. 1-3 Table 1). The highest concentration gave a self-assembly of about 200 nm with no significant effect of temperature on size. On the other hand, decreasing the concentration of the block copolymers mixture resulted in self-assemblies of about 120 nm, which swelled upon heating. NIPAAM, the monomer that generates the thermo-sensitive polymer PNIPAAM, forms hydrogen bonds with water at room temperature. However, upon heating, NIPAAM forms NPs of about 0.6 nm (exp. 4 in Table 1), corresponding to a micelle state of the molecules. We envisaged the use of the NIPAAM monomer in combination with PPO-PEO block copolymers for tuning the size of the self-assembled NPs upon heating.

Table 1. Self-assembly studies of nanoparticles formation under different conditions.

Exp	[a]Block (mg)	NIPAAM (mg)	H_2O (ml)	PVP (mg)	BIS (mg)	KPS (mg)	Size at 25°C (nm)	Size at 73°C (nm)
1	256	-	4	2	2	-	201.3±3.13	209.7±11.12
2	120	-	4	2	2	-	127.9±7.28	168.69±5.91
3	80	-	4	2	2	-	120.3±4.24	153.1±1.21
4	-	75	4	2	2	-	-	0.643±0.026
5	256	75	4	2	2	-	283.5±6.24	30.34±1.88
6	256	75	4	2	2	2	**17.92±2.31**	**18.09±3.14**
7	120	75	4	2	2	-	124±11.64	84.86±9.488
8	120	75	4	2	2	2	**183.6±6.84**	**76.52±9.488**
9	80	75	4	2	2	-	121.1±8.77	159.2±5.241
10	80	75	4	2	2	2	**300.5±5.09**	**160.4±4.2**

Interestingly, at room temperature, the mixtures of both monomers at different ratios in the presence of PVP generated self-assemblies similar to those obtained with the block copolymers alone. For higher concentration of block copolymer we observed an increase in size from 201 to 283 nm (exp. 1 and 5 Table 1), while for lower concentrations of copolymer there was no significant change in size in the presence of NIPAAM (exp. 7 and 9 as compared to exp. 2 and 3 Table 1). On the other hand, at 73°C there were substantial changes: the highest concentration of copolymer with a given quantity of NIPAAM results in a self-assembly of about 30 nm, which represents a substantial change from room temperature to 73°C (from 283 to 30 nm, exp. 5 Table 1). In addition, formulations with lower concentrations of co-polymers also underwent significant size changes at 73°C, from 124 to 85 nm and from 121 to 159 nm (exp. 7, 9 Table 1).

The addition of initiator and heating at 73°C transformed the self-assemblies into cross-linked NPs with similar sizes to those obtained at the same temperature with no polymerization (exp. 6, 8 and 10 Table 1 with particle sizes of 18, 76 and 160 nm respectively at 73°C) in all formulations. The highest ratio of block-copolymer (exp. 6, Table 1) produced particles with no significant sensitivity to temperature. On the other hand, at lower concentrations of block-copolymer (exp. 8 and 10, Table 1), the highest thermo-sensitivity effect was observed (exp. 10). Particles of 300 nm at 25°C shrink to 160 nm at 73°C. This effect was caused by the different content of PNIPAAM within the polymeric network.

Results suggest that self-assemblies are formed at room temperature by the block-copolymer monomers. At room temperature, monomeric NIPAAM interacts with the hydrophilic shell of these assemblies. Upon heating, the NIPAAM, together with the relatively hydrophobic PPO segments of the block-copolymer collapse to generate a core shell of self-assembly with a hydrophobic core formed by PPO and NIPAAM, and a corona formed by PEO chains. The concentration of NIPAAM in the initial formulation will affect the volume of the hydrophobic core so that they will be filled with NIPAAM. Since part of the block-copolymer is a di-block copolymer composed of acrylated-PPO-PEO-PPO-NH$_3^+$, the ammonium groups are exposed at the outer layer of the shell with a concomitant stabilization of the assemblies by electrostatic repulsion between particles. The addition of initiator and cross-linker brings about the generation of stable cross-linked NPs (see Figure 1 for the proposed mechanism).

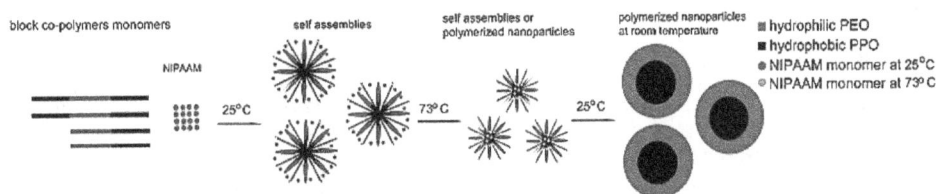

Fig. 1. Proposed mechanism for generation of nanoparticles.

NPs were characterized by TEM, DLS and NMR. NPs were non-toxic at up to 5mg/ml in PC3 and NIH3T3 cells and *in vivo* in the zebra fish model. Finally, NPs were tracked in live cells using confocal microscopy and there intracellular fate was quantified.

Overall, we developed a new type of NP with improved *in vitro* and *in vivo* toxicity profiles. The mechanism of NP formation seems to be self-assembly assisted NP formation prior to polymerization. The size of the preformed self-assemblies can be tuned by mixing different ratios of hydrophilic and hydrophobic precursors prior to polymerization. The NPs were nontoxic, as demonstrated *in vitro* and *in vivo* studies. The NPs can be modified through amino groups present at their surface. These promising results warrant the development of novel bio-medical applications [1,2].

Acknowledgments

This work was financed by the ISF grant number 830/11 and "The Marcus Centre for Medicinal Chemistry of Bar Ilan University". RK and VM are indebted to the BIU President Scholarships and RK is also indebted to Israel Council for Higher Education for the Converging Technologies Fellowship.

References

1. Byk, G., Partouche, S., Weiss, A., et al. *J. Comb. Chem.* **12**, 332-345 (2010).
2. Khandadash, R., Partouche, S., Weiss, A., et al. *Open. Opt. J.* **5**, 17-27 (2011).

Proceedings of the 23rd American Peptide Symposium
Michal Lebl (Editor)
American Peptide Society, 2013

Control of Site-Specific Silica Precipitation Using PNA Peptides and DNAs

Kenji Usui[1], Hiroto Nishiyama[1], Kazuma Nagai[1], Takaaki Tsuruoka[1], Satoshi Fujii[1], and Kin-ya Tomizaki[2]

[1]Faculty of Frontiers of Innovative Research in Science and Technology (FIRST), Konan University, Kobe, 650-0047, Japan; [2]Innovative Materials and Processing Research Center and Department of Materials Chemistry, Ryukoku University, Otsu, 520-2194, Japan

Introduction

Biomineralization, precipitation of inorganic compounds, can be modulated by natural proteins. These proteins can control both the shape and the size of the precipitate with high reproducibility and accuracy during the biomineralization processes. Short peptides derived from the sequences of the natural mineralization proteins have been found for precipitation of inorganic compounds [1,2]. Because these peptides are easy to design and synthesize compared to proteins, they are attractive molecules for the construction of organic-inorganic nanostructures. However, there have been few studies to date using these peptides for well-controlled site-specific precipitation yielding homogeneously-shaped inorganic deposits. In this study, we focused on silica precipitation and demonstrated a site-specific precipitation using DNA and designed peptides with peptide nucleic acids (PNA) [3]. Site-specific precipitation would provide the controlled distribution of organic and inorganic molecules, contributing to bottom-up nanotechnology.

Results and Discussion

We first designed PNA (peptide nucleic acid) -conjugated peptide (Pspp) for this purpose. This peptide consists of two parts. The first part was composed of a silica-precipitating sequence (spp) described previously [4]. The other part was a PNA sequence as a binding module for a complementary sequence of termini of a DNA prepared for the site-specific precipitation. Consequently, Pspp could interact with the DNA site-specifically and then, the silica precipitation could occur only on the peptide at both termini of the DNA, resulting in silica-addressable precipitation.

After the design and synthesis of Pspp [5], we checked its ability to precipitate silica by itself using atomic force microscopy (AFM). Pspp alone (Figure 1c) provided better control of the shape and size of the silica particles compared to a peptide containing only a silica-precipitating sequence (spp).

The addressable precipitation of silica using Pspp and long-chain DNA with the complementary DNA sequences to the PNA sequence at both termini was then demonstrated. AFM images showed DNA and silica precipitates as long chains and nanometer-sized spheres (Figure 1a), indicating the successful site-specific precipitation of silica on the DNA. Additionally, without Pspp, DNA alone could not precipitate silica (Figure 1b). These results suggested that Pspp can bind to DNA specifically and precipitate silica site-specifically at both termini of DNA.

DLS (dynamic light scattering) was also conducted to check the organic-inorganic structures. As a result, these sizes corresponded to the AFM results.

Currently we are optimizing the site-specific precipitation of silica using Pspp and long-chain or short-chain DNA and analyzing the precipitates by DLS, AFM, SEM (scanning electron microscopy), TEM (transmission electron microscopy) and EDX (energy dispersive X-ray spectroscopy).

Our results provide good examples of the controlled localization of inorganic compounds and organic compounds. Such systems would be a promising powerful tool for the nanobiotechnology and material fields.

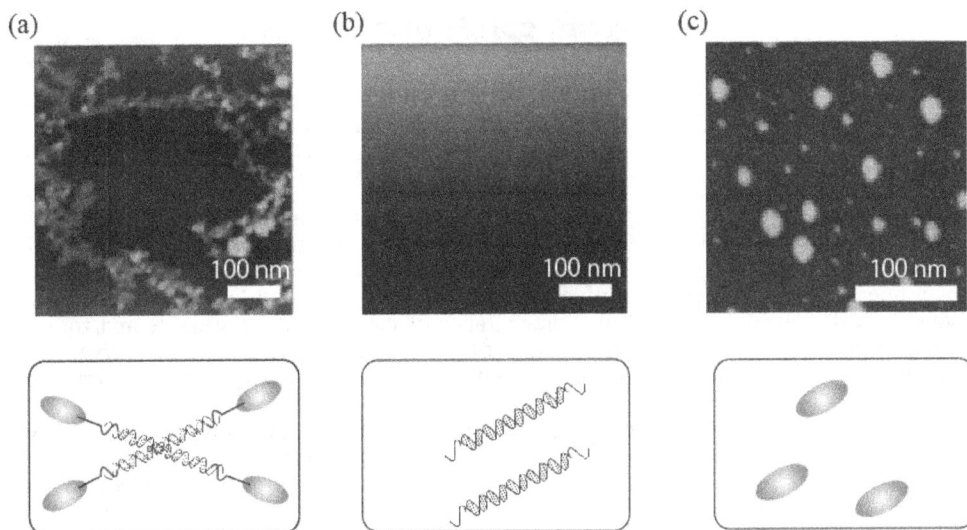

Fig. 1. Atomic force microscopy (AFM) images and illustrations of silica precipitation using (a) Pspp peptide and long-chain DNA, (b) long-chain DNA alone (without Pspp), (c) Pspp alone (without long-chain DNA).

Acknowledgments

This study was supported in part by grants from The Murata Science Foundation, and The CASIO Science Promotion Foundation.

References

1. Acar, H., Garifullin, R., Guler, M.O. *Langmuir* **27**, 1079-1084 (2011).
2. Banerjee, I.A., Yu, L., Matsui, H. *Proc. Nat. Acad. Sci. USA* **100**, 14678-14682 (2003).
3. Nielsen, P.E., Egholm, M., Berg, R.H., Buchardt, O. *Science* **254**, 1497-1500 (1991).
4. Knecht, M.R., Wright, D.W. *Chem. Commun.* 3038-3039 (2003).
5. Sano, S., Tomizaki, K.-Y., Usui, K., Mihara, H. *Bioorg. Med. Chem. Lett.* **16**, 503-506 (2006).

Proceedings of the 23rd American Peptide Symposium
Michal Lebl (Editor)
American Peptide Society, 2013

Metal-Ligand Interactions to Promote the Assembly of a Collagen Peptide into Fibrils

Jeremy Gleaton and Jean Chmielewski

Department of Chemistry, Purdue University, West Lafayette, IN, 47906, U.S.A.

Introduction

Collagen is one of the most abundant proteins that accounts for approximately one third of the total protein content within humans [1]. The primary structure of collagen is composed of a trimeric repeating unit, most commonly Xaa-Yaa-Gly, where Xaa and Yaa are most often L-proline and 4(R)-hydroxy-L-proline [2]. This sequence forms a left-handed polyproline type II (PPII) helical conformation. Three strands of the PPII helix assemble and form the overall right-handed triple helical structure of collagen. This rigid structure serves many functions throughout the body, such as in tendons, ligaments, bones, and the extracellular matrix.

Because of the versatility of collagen, many researchers have sought ways to mimic natural collagen. Such mimics could play a crucial role in regenerative medicine in addition to offering novel biomaterials. To date, a number of research groups have investigated new collagen materials using short collagen mimetic peptides (CMPs) [3]. One such method for triggering CMP assembly is to incorporate ligands for metal ions within the sequence [4]. Metal-ligand interactions between triple helices promote the assembly of larger collagen based structures.

Our designed system (Figure 1) was modeled after the metal-ligand interactions found within a variety of bacterial siderophores that contain a catechol moiety. Siderophores have an exquisite ability to chelate iron (III) [5]. With this knowledge, we aimed to promote the radial assembly of the CMP into a collagen mimetic. We designed a CMP with repeating units of Pro-Hyp-Gly, with the catechols attached *via* lysine side chains to have the attachment of the catechol moieties positioned along the backbone of the peptide thus allowing the ligands to extend from the collagen triple helix. We hypothesized that addition of Fe^{3+} ions would elicit a radial assembly through the interaction of the catechol moieties between neighboring collagen triple helices.

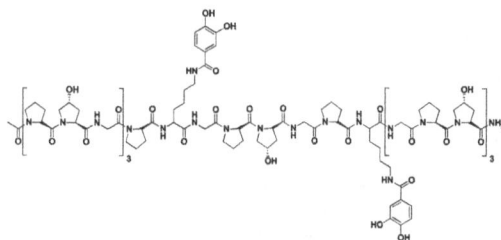

Fig. 1. Molecular structure of designed CMP **H(3,4-DHBA)2**.

Results and Discussion

The designed peptide, **H(3,4-DHBA)2** was synthesized on ChemMatrix Rink Amide resin using an Fmoc-based strategy with HBTU as a coupling agent. Lysine residues were incorporated with the acid labile 4-methyltrityl (Mtt) protecting group at the ε-amino group. The Mtt groups were removed on resin with 1.8% TFA, and the desired carboxylic acid, 3,4-dihydroxy benzoic acid was coupled. After synthesis of the peptide, it was cleaved from resin and purified to homogeneity using reverse phase-HPLC. The structure of **H(3,4-DHBA)2** was confirmed using MALDI mass spectrometry.

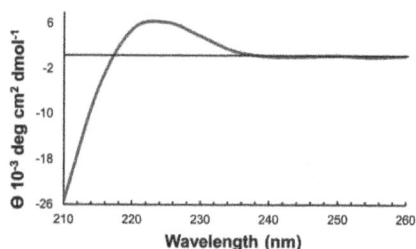

Fig. 2. CD Spectrum of **H(3,4-DHBA)2** *(180 μM) in 50 mM CAPS, pH 10.*

We first evaluated **H(3,4-DHBA)2** by circular dichroism (CD) to determine if the peptide adopted a collagen triple helical structure. The data showed a distinctive maximum at 225 nm that is characteristic of the PPII helical structure that is a signature of collagen

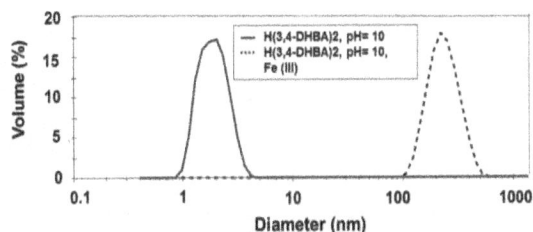

Fig. 3. DLS data of **H(3,4-DHBA)2** (1 mM) (solid green, 50 mM CAPS pH 10) and with Fe(ClO₄)₃ (0.33 mM) (dashed red. 50 mM CAPS).

peptides (Figure 2). We next determined the melting temperature (T_m) of the peptide. Monitoring at 225 nm over a temperature range of 4 to 90°C revealed that the peptide had a T_m of 47°C, as compared to the T_m for POG₉ of 67°C [6].

Prior to assembly, all samples were thermally annealed to facilitate the formation of uniform triple helices. Formation of fibrils were exhibited over a variety of peptide to metal ratios, here we present data for a ratio of 2:1. The assembly of **H(3,4-DHBA)2** was studied using dynamic light scattering (DLS) (Figure 3). In the absence of metal, a signal at ~3 nm was observed, a value previously reported for individual triple helices of collagen peptides [4]. In the presence of the metal ion, Fe(III) (2:1 ratio of peptide to metal ion), DLS suggested that larger aggregates had formed with an average size of ~285 nm. Upon the addition of Fe^{3+} to the peptide, the solution turned to a deep green-blue color. This color change is indication of metal-ligand charge transfer interaction that occurs for 3,4-dihydroxybenzoic acid and Fe (III) [7].

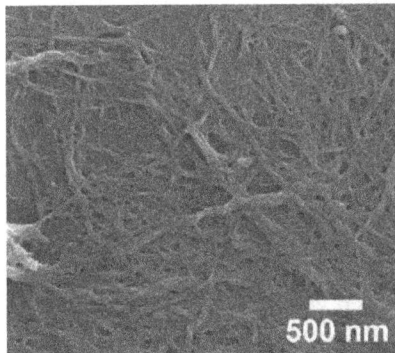

Fig. 4. SEM micrograph of **H(3,4-DHBA)2** (1.00 mM) incubated with Fe(ClO₄)₃ (0.33 mM).

We next visualized the morphology of the structures using scanning electron microscopy (SEM). An intricate network of fibrils with substantial branching, having a variety of lengths and widths was observed (Figure 4). It may be possible for fibrils to form due to flexibility between the metal ligand linkages; analogous to a system previously reported [1]. Here the possible formation of sticky ends was proposed to explain fibril formation via linear growth.

In summary we have demonstrated the use of a catecholate moiety within a collagen-based peptide to promote the assembly of a CMP. Specifically using **H(3,4-DHBA)2**, we demonstrated the formation of fibrils upon the addition of Fe^{3+} at a metal to peptide ratio of 2:1. Further research will investigate variants of this system by making use of different ligands, in an effort to tune the system for a variety of drug delivery and biomaterial applications.

Acknowledgements

We are grateful to NSF (1213948-CHE) for their support of this research and Charles Rubert Pérez for the assistance with SEM imaging.

References

1. Brinckmann, J. *Topics Curr. Chem.* **247**, 1-6 (2005).
2. Ramshaw, J.A., Shah, N.K., Brodsky, B. *J. Struct. Biol.* **122**, 86-91 (1998).
3. Shoulders, M.D., Raines, R.T. *Annu. Rev. Biochem.* **78**, 929-958 (2009).
4. a) Przbyla, D.E., Chmielewski, J. *Biochemistry* **49**, 4411-4419 (2010); b) Pires, M.M., Przbyla, D.E., Rubert-Perez, C.M., Chmielewski, J. *J. Am. Chem. Soc.* **133**, 14469-14471 (2011); c) Przybyla, D.E., Rubert-Perez, C.M., Gleaton, J., Nandwana, V.,Chmielewski, J. *J. Am. Chem. Soc.* **135**, 3418-3422 (2013).
5. Raymond, K.M., Dertz, E.A., Kim, S.S. *Proc. Nat. Acad. Sci. USA* **100**, 3584-3588 (2003).
6. a) Holmgren, S.K., Bretscher, L.E., Taylor, K.M., Raines, R.T. *Nature* **392**, 666-667 (1998); b) Holmgren, S.K., Bretscher, L.E., Raines, R.T. *Chem. Biol.* **6**, 63-70 (1998).
7. Hider, R.C., Mohd-Nor, A.R., Silver, J. *J. Chem. Soc. Dalton* 609-622 (1981).

Proceedings of the 23rd American Peptide Symposium
Michal Lebl (Editor)
American Peptide Society, 2013

Functional Illumination in Living Cells

Sara Ahadi, Gaurav Bhardwaj, Tuong Pham, Eduardo Sanchez, David Olivos, Mary Saunders, Lin Tian, and Kit S. Lam

Department of Biochemistry and Molecular Medicine, University of California, Davis, Sacramento, CA, 95817, U.S.A.

Introduction

In the last decade, application of genetically encoded fluorescent probes and sensors in molecular imaging has greatly improved our understanding about how specific molecules orchestrate cellular functions and how errant cells cause diseases. Green fluorescent protein and its relatives of color palette have been successfully employed in a broad range of biological disciplines, reporting the distribution, abundance, dynamics, interaction and conformational changes of essential signaling molecules in time and space by engineering fluorescent protein (FP) chimeras [1]. However, engineering FP chimeras has long been limited by the large size (27KDa) and biophysical properties of FPs. Here, we use highly efficient one-bead-one-compound (OBOC) combinatorial library technique to rapidly discover eukaryotic amino acid containing peptides (~1500 daltons) that can bind to and activate the fluorescence of malachite green (MG) with high affinity, specificity and superior signal-to-noise ratio. We believe some of these peptides can be developed into highly sensitive and specific genetic encoded small illuminants (GESIs) for probing a wide range of biochemical and cellular functions, microscopically, in living cells.

Results and Discussion

We designed a 12-mer linear OBOC peptide library [2] based on the calcium binding EF hand structure [3], in which the first amino acid is Asp, the sixth amino acid is Gly, and the 12th amino acid is Glu. The rest of the positions contain all 19 natural amino acids except Cys to avoid peptide cyclization. The chemical structure of such library is as follows: $DX_dX_dX_dX_dGXXXXXE$—Bead. In the X position, the resin beads were distributed evenly for all 19 amino acids, whereas in the X_d position, 50% of the beads were reacted with Asp and the remaining 50% of the beads were distributed evenly for all the remaining 18 amino

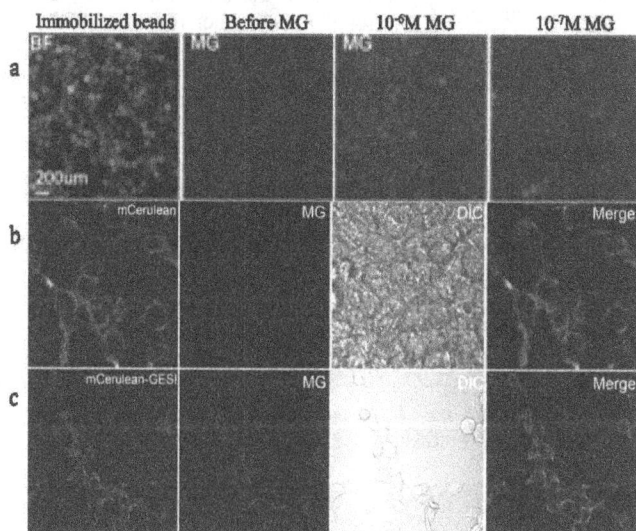

Fig. 1. (a) Automated high-throughput screening system. Beads were immobilized on 100mm Petri dish, followed by incubation with MG; (b and c) Characterization of genetically encoded MG-activating peptides in 293T cells.

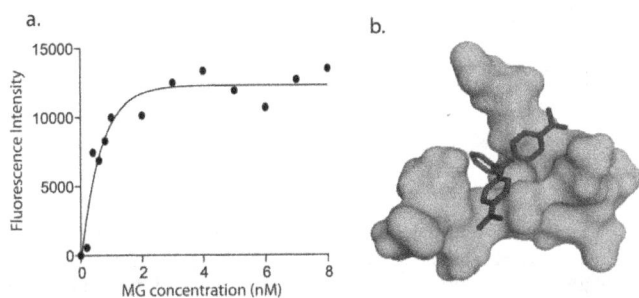

Fig. 2. (a) Fluorescence titration curve for binding of MG to Peptide 12 in solution, (b) Simulated structure of peptide-MG complex.

acids. In this way, Asp is preferred in the first 5 residues of every peptide. This library design assures high probability of Ca^{2+} binding to provide binding pocket for MG. Such flexibility of library design is not possible with phage-display or yeast-display peptide libraries, which are limited by genetic code and expression preferences of different proteins. For library screening, OBOC library beads were immo-bilized in Petri dish for real-time tracking and infrared fluorescence of the beads was inspected with a wide-field fluorescent microscope in the presence and absence of calcium 5 mins after MG addition. The latter was achieved with addition of EDTA. Those beads with strongest fluorescence at lowest concentration of MG were selected for sequencing and further characterization in living cells (Figure 1a).

To evaluate the utility of lead peptides as GESI in mammalian cells, we fused individual peptide tags to the C-terminus of a cyan fluorescence protein (Cerulean 3) anchored on the plasma membrane. We used a -GGSGSGGS- linker between C-terminus of Cerulean and the peptide to increase the solubility and stability of the recombinant fusion proteins. The fusion protein allowed us to see the labeling ability of the peptides by co-localization of both cyan and infrared fluorescence. Four out of 11 lead peptides fused proteins showed clear co-localization of infrared and cyan fluorescence upon MG addition. The fluorescence intensity reached plateau within seconds and remains stable during the 10 minutes period of imaging window. In contrast, no apparent infrared fluorescence was observed when membrane-anchored Cerulean control (without grafted peptide) was expressed upon MG addition (Figure 1b and c).

For detail characterization, we selected one of the peptides that can activate MG with calcium. Solution form of the positive peptide was synthesized and titrated with MG in the presence of 2mM Ca^{2+}. Fluorescent signal at 650nm was measured. The binding curve for peptide 12 (DWWDWGNHGYTE) was found to follow one-site specific binding, with high affinity (Kd, ~410 pM) (Figure 2a). Computational modeling predicted stably folded structure of peptide 12 upon MG binding, constraining intramolecular motions of MG (Figure 2b).

In summary, we successfully used OBOC combinatorial technology to discover novel small peptide illuminants that can be genetically encoded in living cells. Using our new high throughput method, we could screen thousands of beads in less than 10 min and readily identify those beads that bind to MG only in presence of calcium.

Our results show that GESI can be used as a powerful platform for sensor design and engineering in live cell imaging. One unique feature of GESI is that the fluorescent signal of the exogenously added organic dye is activated only upon binding to the peptide, thus, making the background staining very low. Beside MG, our peptide library can be also screened against other fluorophores such as 4-hydroxybenzlidene imidazolinone (HBI) and its derivatives [4] to discover GESIs that bind to a broad color spectrum of organic dyes and expand the spectrum to infrared, making multiplex and whole animal imaging possible.

References

1. Van Roessel, P., et al. *Nature Cell Biol.* **4,** 15-20 (2002).
2. Lam, S.K., et al. *Nature* **354**, 82-84 (1991).
3. Zhou, Y., et al. *Cell Calcium* **46.1**, 1-17 (2009).
4. Paige, J.S., et al. *Science* **333**, 642-646 (2011).

Proceedings of the 23rd American Peptide Symposium
Michal Lebl (Editor)
American Peptide Society, 2013

Quantitation of MT1-MMP Activity at the Cell Surface

S. Pahwa and G.B. Fields

Torrey Pines Institute for Molecular Studies, Port St. Lucie, FL, U.S.A.
E-mail: gfields@tpims.org

Introduction

MT1-MMP plays critical roles during tumor malignancy and is one of the best-validated proteolytic enzyme targets on cancer cells. It is up-regulated in several tumor types, including breast, cervical, and ovarian cancer and is significantly associated with adverse outcome [1]. In spite of the large proteolytic repertoire and of the robust proteolytic activity of engineered soluble forms of the enzyme, several studies indicate that soluble MT1-MMP mediated proteolysis is not sufficient per se for efficient penetration of ECM barriers. Thus, localization of the enzyme at the cell surface is essential to translate its proteolytic activity into a modification of cell function. Overall, the transmembrane nature of MT1-MMP allows it to influence extracellular remodeling of the matrix surrounding tumor cells as well as intracellular signaling events involved in cell invasion. Like virtually all cell surface proteases, quantitative assessment of MT1-MMP in its native environment has not been achieved. A mechanistic examination of MT1-MMP at the cell surface would unravel the influences of binding partners on activities, and set the stage for the development of unique, non-active site inhibitors. Such inhibitors may act only on selective functions of MT1-MMP, reducing potential toxicities. MMP activity assays are typically based on fluorescence resonance energy transfer utilizing synthetic peptides [2]. Fluorescence can be detected with outstanding sensitivity, and these continuous assays allow for rapid, convenient kinetic evaluation of proteases and thus greatly aid mechanistic studies of these enzymes. In this study, MT1-MMP and several MT1-MMP mutants were stably transfected and a cell-based FRET assay utilized to quantify protease activity. Activity comparisons were made for soluble MT1-MMP and surface-bound MT1-MMPs to evaluate the importance of the cell surface and specific MT1-MMP domains on activity.

Results and Discussion

ΔTM-MT1-MMP (a deletion mutant lacking the transmembrane domain) was transfected transiently in COS-1 cells to generate soluble MT1-MMP. WT-MT1-MMP (wild type) and different constructs were stably transfected in MCF-7 cells. MCF-7 cells are normally deficient in MT1-MMP [3]. The constructs were WT-MT1-MMP, MT1-MMP(ΔCT) (enzyme with the cytoplasmic tail deleted), MT1-MMP(ΔHPX) (enzyme with the hemopexin-like domain deleted), and MT1-MMP(MMP-1 CAT) (enzyme with the catalytic domain of MMP-1) (Figure 1). Estimation of the amount of active enzyme was obtained by TIMP-2 titration. Kinetic parameters (K_M, k_{cat}, k_{cat}/K_M) were determined by calculation of rates of hydrolysis of the triple-helical substrate fTHP-9 [(Gly-Pro-Hyp)$_5$-Gly-Pro-Lys(Mca)-Gly-Pro-Gln-Gly~Cys(Mob)-Arg-Gly-Gln-Lys(Dnp)-Gly-Val-Arg-(Gly-Pro-Hyp)$_5$-NH$_2$], which is selective for MT1-MMP compared with MMP-1 [2].

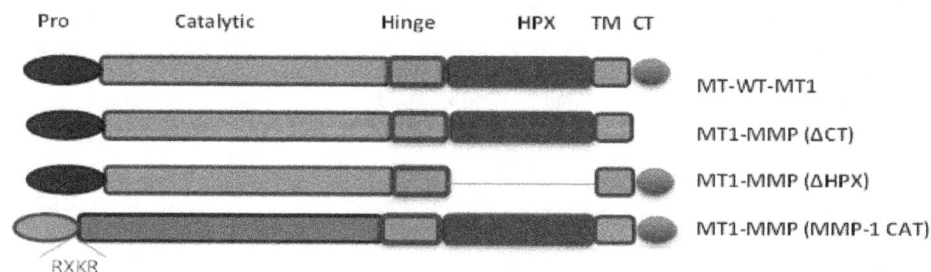

Fig. 1. WT-MT1-MMP and mutants.

High levels of MT1-MMP expression in transfected cells relative to mock transfected cells were verified by western blotting. The *in situ* cell based assays were run in a 384-well format on a tissue-culture treated opaque microplate. The cells were seeded in OptiMEM Medium at 6,000 cells/well density and then incubated overnight at 37°C. The assays were carried out in serum-free OptiMEM with fTHP-9 dissolved in the same media. Proteolytic activity was determined by calculating the percentage increase in fluorescence compared to the background signal provided by the corresponding dilution of the substrates with no cells using a multiwell plate fluorimeter at λ_{ex} = 324 nm and λ_{em} = 405 nm [2]. K_M for soluble MT1-MMP was 18.6±1.4 μM and 15.1±2.2 μM for WT-MT1-MMP, whereas k_{cat}/K_M for the soluble enzyme is 3- to 4-fold higher as compared to the membrane-tethered enzyme (Figure 2). Thus, the catalytic efficiency of the enzyme for fTHP-9 is less in the cell surface environment. Catalytic efficiency and TIMP-2 titration data suggested that MT1-MMP (ΔCT) transfected cells had enhanced (2-fold) MT1-MMP expression as compared to wild type probably due to inhibition of internalization at the cell surface.

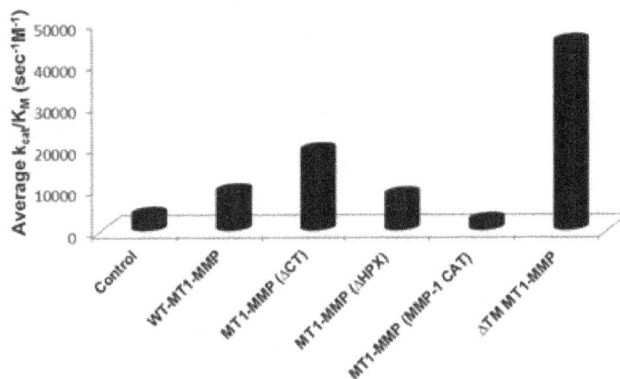

Fig. 2. Comparison of catalytic efficiency of different membrane anchored MT1-MMP constructs with soluble MT1-MMP.

Overall, we have developed a cell-based assay useful for the mechanistic and quantitative evaluation of MT1-MMP in its native environment.

References

1. Itoh, Y., Seiki, M. *J. Cell. Physiol.* **206**, 1-8 (2006).
2. Minond, D., Lauer-Fields, J., Cudic, M. Overall, C.M., Pei, D., Fields, G.B. *J. Biol. Chem.* **281**, 38302- 38313 (2006).
3. Rozanov, D.V., Deryugina, E.I, Ratnikov, B.I., Quigley, J.P., Strongin, A.Y. *J. Biol. Chem.* **276**, 25705-25714 (2001).

Proceedings of the 23rd American Peptide Symposium
Michal Lebl (Editor)
American Peptide Society, 2013

Synthesis and Characterization of ^{99m}Tc-Labeled Peptides as a Potential Amyloid Plaques Imaging Agent

Danielle V. Sobral[1], Ana C.C. Miranda[2], Marycel F. Barboza[2], Fábio L.N. Marques[3], Clóvis R. Nakaie[4], and Luciana Malavolta[1]

[1]Faculdade de Ciências Médicas da Santa Casa de Sao Paulo, Sao Paulo, 01221-020, Brazil;
[2]Instituto Israelita de Pesquisa Albert Einstein, Sao Paulo, Brazil; [3]Faculdade de Medicina,
Universidade de Sao Paulo, Sao Paulo, Brazil; [4]Escola Paulista de Medicina,
Universidade Federal de Sao Paulo, Sao Paulo, Brazil

Introduction

One of the major pathological landmarks of Alzheimer's disease (AD) is the presence of amyloid deposits in the brain. In this regard, recent findings demonstrate that the development of molecular imaging biomarkers with binding affinity to amyloid in the brain is therefore in the forefront of imaging biomarker and radiochemistry research [1,2]. Following with previous report, in vivo amyloid plaques imaging with radiolabeled peptides are under investigation and have emerged as an important diagnostic tool for the AD treatment [3,4]. In this work, we present a new approach to establish the labeling and the quality control procedures of the amyloid peptide fragments as potential biomarkers by using [99m]Tc(I)–carbonyl-peptide compounds.

Results and Discussion

Based upon the (1-42) Aβ-amyloid peptide present in Alzheimer's disease, the VHHQKLVFFAED (12-24) fragment and also the (36-42) and (16-20) but attaching a His residue at their C-terminal positions (VGGVVIAH and KLVFFH, respectively), were synthesized, purified and tested as potential biomarkers.

By using a very convenient method [5] of the organometallic aqua-ion $[^{99m}Tc(H_2O)_3(CO)_3]^+$, abbreviated TcCO, directly from $[^{99m}TcO_4]^-$ under 1 atm of CO we obtained high specific activities and very stable complexes (Figure 1A). No intermediates could be detected, and moreover the Tc (I) center, once has been formed upon reduction, was trapped efficiently by the three CO molecules. Even at high temperature and extended reaction time, TcCO is stable in high pH.

Employing the [99m]Tc(I)–carbonyl compound recently prepared, we obtained highly specific activities and extremely stable complexes to the His-(1-42) β-amyloid peptide fragments, which may be one of the most convenient attachment sites because Tc (I) is stabilized as a tricarbonyl complex in aqueous solution, but readily exchanges its water ligand for the nitrogen of imidazoles. Labeling studies of KLVFFH (16-20), VGGVVIAH (36-42) and VHHQKLVFFAEDV (12-24) amyloid fragments were investigated between pH 5-8, time reaction of 30, 60 and 90 minutes and peptides concentration of 10^{-4}M.

The results showed that stable metal-peptide complexes were obtained in the range of 60 to 80% according to the peptide under study. The purity and radiolabelling efficiencies were checked by Whatman 3MM chromatography paper, TLC-SG (Al), solid phase extraction (Sep-Pak C18 filter) in different mobile phases. The best labeling for all peptides were observed at pH 7.0 and 90 min using acetone as solvent (Figure 1B and Table 1).

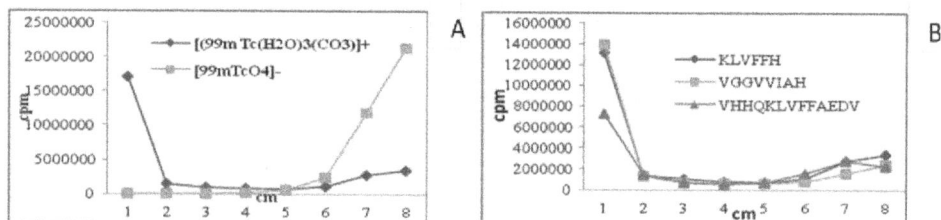

Fig. 1. Chromatographic profile and purity of (A) $[^{99m}Tc(H_2O)_3(CO_3)]^+$ from $[^{99m}TcO_4]^-$ and (B) $^{99m}Tc(CO_3)$-peptides in Whatmann 3MM (acetone).

Table 1. Radiolabelling efficiencies and purity of $^{99m}Tc(CO_3)$-Peptides.

Peptides	pH	$^{99m}TcCO_3$ (mCi)	Radiolabeling Efficiencies (%)			
			Whatman 3MM Acetone		Sep-Pak C-18	
			60min	90min	60 min	90min
	5.0	2.8	42	48	52	55
VGGVVIAH	7.0	3.0	75	80	78	81
	8.0	3.4	68	71	73	76
KLVFFH	7.0	3.1	61	70	63	72
VHHQKLVFFAEDV	7.0	3.5	57	62	59	62

We also tested the influence of the position of histidine on its complexation efficiency and the potential importance of the *C*-terminal carboxyl group. After 90 min at pH 7.0 about 80% of the TcCO was complexed by *C*-terminal His-peptide and the most inefficiently labeled was the endo-peptide which reached 62% (Table 1). Noteworthy, higher metal labeling yield was achieved with peptide segments bearing His residues at peptide *C*-terminal position, thus pointing to a position-dependent effect for the ^{99m}Tc coupling reaction.

In conclusion, the VGGVVIAH, KLVFFH and VHHQKLVFFAED peptides were efficiently synthesized and showed potentiality for the diagnostic assays of the Alzheimer's disease. The tested radiolabeling strategies showed successful results. TcCO radioisotope is easily prepared from a Tc generator and CO gas at atmospheric pressure. In regard to the radiolabeling approach, the $[Tc(H_2O)_3(CO)_3]^+$ complex is an extremely convenient reagent for labeling His residue in a peptide structure. We believe that this technology might be useful for a broad range of *in vitro* and *in vivo* diagnostics using this type of peptides.

Acknowledgments

Grants from the Brazilian FAPESP (Proc. 2010/20197-5) and CNPq agencies are gratefully acknowledged.

References

1. Brockschnieder, D., et al. *J. Nucl. Med.* **53**, 1794-1801 (2012).
2. Morais, G.R., Paulo, A., Santos, I. *Eur. J. Org. Chem.* **7**, 1279-1293 (2012).
3. Cabral, F.R., et al. In Kokotos, G., Constantinou-Kokotou, V. and Matsoukas, J. (Eds.) *Peptides 2012: (Proceedings of the 32th European Peptide Symposium)*, Athens, Greece, 2012, p. 222.
4. Guerrero, S., et al. *Bioconj. Chem.* **23**, 399-408 (2012).
5. Waibel, R., et al. *Nat. Biotechnol.* **17**, 897-901 (1999).

Proceedings of the 23rd American Peptide Symposium
Michal Lebl (Editor)
American Peptide Society, 2013

Targeting GLP-1R for Non-Invasive Assessment of Pancreatic β-Cell Mass

Bikash Manandhar[1], Su-Tang Lo[2], Eunice Murage[1], Mai Lin[2], Xiankai Sun[2], and Jung-Mo Ahn[1*]

[1]Department of Chemistry, University of Texas at Dallas, Richardson, TX, 75080, U.S.A.; [2]Department of Radiology, University of Texas Southwestern Medical Center, Dallas, TX, 75390, U.S.A.

Introduction

Pancreatic β-cell mass (BCM) is significantly reduced in patients with both type 1 and type 2 diabetes. Although early diagnosis is highly beneficial, the clinical diagnostic tools currently available reveal the disease status only after over the half of BCM is lost. Thus, reliable and non-invasive methods to determine BCM are of great value in early detection of the disease and evaluation of therapeutic interventions. As a β-cell biomarker, glucagon-like peptide-1 receptor (GLP-1R) is found to be specifically localized in the pancreatic β-cells and can serve as a suitable target for the development of molecular imaging probes [1]. Its endogenous ligand glucagon-like peptide-1 (GLP-1) is an incretin and interacts with G-protein coupled GLP-1R with nanomolar affinity. However, it has not been successfully used to determine BCM due to its rapid degradation *in vivo* by proteases like dipeptidyl peptidase-IV (DPP-IV) and neutral endopeptidase 24.11 (NEP 24.11), significantly lowering plasma half-life ($t_{1/2}$ ~2 min) [2]. In our prior studies, we have demonstrated that strategically positioned lactam bridges at the *N*- and *C*-terminal regions of GLP-1 can enhance receptor interaction by stabilizing two helical structures in the peptide and provide proteolytic stability by shielding the cleavage sites from the enzymes [3,4].

Results and Discussion

Based on the findings, we have designed a bicyclic GLP-1 analog EM2198 that has two lactam bridges between residues 18-22 and 30-34. It has D-Ala at position 8 that completely

Fig. 1. Primary structure of EM2198 and cleavage sites of DPP-IV and NEP 24.11.

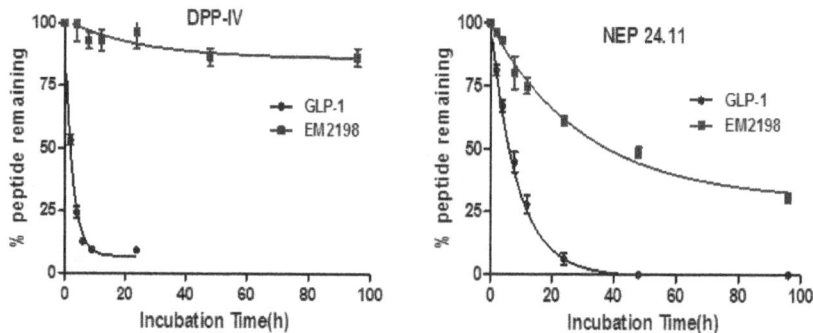

Fig. 2. In vitro stability of GLP-1 and EM2198 to DPP-IV and NEP-24.11.

protected the peptide from DPP-IV degradation (Figure 2). The introduced lactam bridges were effective to deter NEP 24.11 degradation, leaving more than 60% intact after 24h incubation. To conjugate a metal chelator DOTA by using maleimide-thiol coupling chemistry, a cysteine was introduced at the *C*-terminus of the peptide with a 6-aminohexanoic acid (Ahx) as a spacer. The *C*-terminal modification was found not to compromise its high affinity to the receptor (EC_{50} = 1.2 nM, EC_{50} of GLP-1 = 4.7 nM).

After labeling the peptide with a positron-emitting radiotracer ^{64}Cu, it was evaluated as a PET imaging agent (Figure 3). It is exciting that the PET imaging agent was preferentially taken up by the pancreas in healthy mice. To confirm the PET signal originated from the β-cells, we have examined it (1) on STZ-induced diabetic mice and (2) on healthy mice with a blocking dose of cold exendin-4 that would saturate the receptor. *In vivo* PET images were analyzed by normalizing the pancreas uptake to the liver uptake (Pancreas-to-Liver: P/L). The P/L ratio was 1.87 ± 0.28 in healthy mice and significantly reduced to 1.20 ± 0.14 and 1.11 ± 0.09 in the STZ-treated mice and the competitive blocking model, respectively. The *in vivo* PET imaging findings were also validated by *ex vivo* scans and by histological staining (data not shown). In summary, our study clearly demonstrates that BCM can be accurately and reliably measured by non-invasive PET imaging with a novel GLP-1-based peptide EM2198.

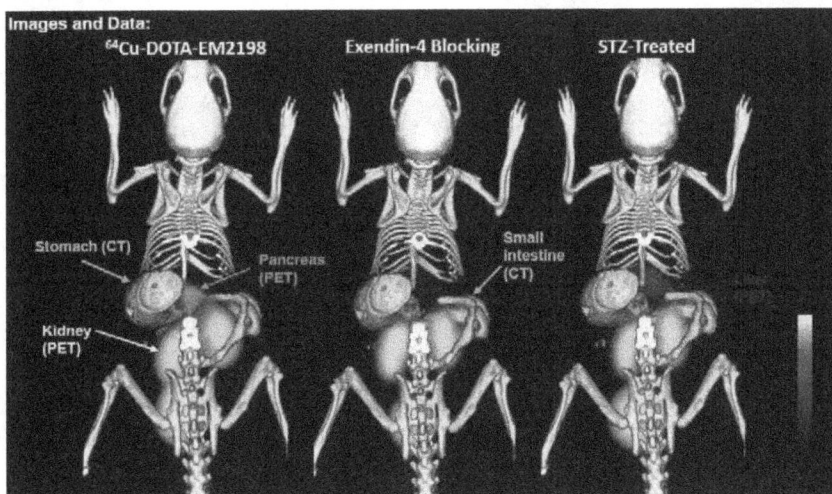

Fig. 3. In vivo PET images of pancreas on healthy mice, healthy mice co-injected with a blocking dose of cold exendin-4, and STZ-induced diabetic mice.

Acknowledgments

This work was supported by the Juvenile Diabetes Research Foundation (37-2009-103, 37-2011-20) and National Institute of Health (5R01DK092163-02).

References

1. Mukai, E., et al. *Biochem. Biophys. Res. Commun.* **389**, 523-526 (2009).
2. Kreymann, B., et al. *Lancet* **2**, 1300-1304 (1987).
3. Murage, E.N., et al. *Bioorg. Med. Chem.* **16**, 10106-10112 (2008).
4. Murage, E.N., et al. *J. Med. Chem.* **33**, 6412-6420 (2010).

Proceedings of the 23rd American Peptide Symposium
Michal Lebl (Editor)
American Peptide Society, 2013

Tag-Probe System for Imaging of Intracellular Proteins

Wataru Nomura, Nami Ohashi, Tetsuo Narumi, and Hirokazu Tamamura

*Institute of Biomaterials and Bioengineering, Tokyo Medical and Dental University,
Chiyoda-ku, Tokyo, 101-0062, Japan*

Introduction

Fluorescent imaging of proteins in living cells is a useful technique in chemical biology. A tag-probe system can label target proteins by fluorescent probes that bind specifically to tag peptides fused to the proteins. ZIP tag-probe pairs based on three α-helical antiparallel leucine zipper peptides have been developed. [1-4] A probe peptide is designed as an α-helical peptide, which possesses an environment sensitive fluorescent dye such as 4-nitrobenzo-2-oxa-1,3-diazole (NBD). Tag peptides are designed as two α-helical antiparallel peptides. The binding of the probe to its complement tag causes a remarkable change of fluorescent properties, blue shift of fluorescent spectra and an increase in fluorescent intensity, because the environment surrounding the fluorescent dye in the probe is changed to more hydrophobic. Since the labeled protein with strong fluorescence can be distinguished from the free probe with weak fluorescence, washing steps for excessed probes is not required for detection of target proteins. As the first application, the fluorescent imaging of a cell surface protein, chemokine receptor CXCR4, which was transiently expressed as tag-fusion on CHO-K1 cells was performed. In this study, as an example of intracellular proteins, protein kinase C (PKC) was utilized for detection of translocation by probe peptides having octaarginine sequence. Taken together, the ZIP tag-probe system involving three α-helical antiparallel leucine zipper peptides is a useful tool for fluorescent imaging of proteins.

Results and Discussion

Polyarginine peptides were utilized to add cell membrane permeability to the probe peptide. Futaki, et al. have developed the use of polyarginine peptides for transportation of macrobiomolecules to cytoplasm. An octa-arginine sequence (R8) was reported as a cell-penetrating peptide (CPP). Efficient delivery to the cytoplasm was expected to follow attachment of this CPP to the probe peptides, and in a first approach, three probe peptides with polyarginine sequences were synthesized (Figure 1). A probe α-helical peptide has an NBD moiety attached to the side chain of L-α,3-diaminopropionic acid and the resulting Dap(NBD) residue is situated at the X position in the probe peptide to locate the NBD dye in the hydrophobic region of the 3α-helical leucine zipper structure. In a fluorescent titration with the tag peptide, fluorescence spectra of the NBD-probe peptides with polyarginines showed a 13-17 fold increase of intensity as the concentration of the tag peptide is increased (Table 1). The emission maximum of the NBD dye shifted from 535 nm to 505 nm as the emission intensity increased. The three probes showed a reasonable increase of fluorescent intensity, indicating the formation of 3α helical bundles with the tag peptide, as the original polyarginine-free probe. The dissociation constants of the probes were markedly different. The apparent dissociation constants (K_d) of the probe peptides 1-3 with hepta- or octa-arginine peptides were determined as 181 nM, 894 nM, and 27 nM, respectively, by a nonlinear least-squares curve fitting method based on a 1:1 stoichiometry model (Table 1). Probe **3** with hepta-arginine (R7) showed strong affinity for the tag peptide. The value was slightly decreased compared to that of the parental probe polyarginine-free peptide. Probe **1** with R8 showed 5-fold decrease in binding affinity compared to that of the probe **3**. The decrease could be rationalized as a consequence of the cationic charges of polyarginine sequence.

Probe **1**: Ac-ALKKKLEALKKKXEALKKKLA-GS-RRRRRRRR-NH$_2$
Probe **2**: Ac-RRRRRRRR-GS-ALKKKLEALKKKXEALKKKLA-NH$_2$
Probe **3**: Ac-ALKKKLEALKKKXEALKKKLA-GS-RRRRRRR-NH$_2$

Fig. 1. Sequences of probe peptides with octa-arginine.

Table 1. The emission maxima, $\Delta I_{max}/I_0$ values (in parentheses) of the probe peptides and tag-probe complexes, and the dissociation constants (K_d) between the tag and the probe peptides.

	Probe 1	Probe 2	Probe 3
λ_{max} ($\Delta I_{max}/I_0$)	505 nm (17)	505 nm (13)	506 nm (15)
K_d[a]	181 nM[b]	894 nM[b]	27 nM[c]

[a] *Measurement conditions: 50 mM HEPES buffer solution (pH 7.2, 100 mM NaCl), at 25°C, [probe] = 0.5 μM.* [b],[c] *Determined by the fluorescent intensity change at 505 nm and 506 nm, respectively.*

The purpose of the development of tag-probe pairing system for protein imaging is to enable the visualization of protein dynamics by non-covalently attached probes. Ligand binding to the C1b domain in PKC leads to its membrane translocation. This translocation of PKC is of central importance for its function because the localization of PKC determines to which substrates it has access. As a ligand inducing translocation of PKC, phorbol 12-myristate-13-acetate (PMA) was utilized. A gene of PKCδ fused with tag and mKO was introduced to the mammalian expression vector and the plasmid DNA coding the fusion protein was transiently transfected to HeLa cells by electroporation. After 48-72 h of transfection, experiments for cell penetration and pairing formation with the tag peptide region were performed. As the first challenge, probe 1 at 1 μM was added to the cells and incubated for 30 min. Without washing, the dishes were mounted on a confocal microscope. Even in the presence of the protein expressed in cytosol, the distribution of probe 1 in cytosol was not changed. Instead of distribution in the cytosol, the fluorescence from the probe again showed formation of vesicles. To overcome this problem, two other experiments were performed in the cells expressing the PKC-mKO fusion protein with the tag peptide: (1) further reduction of the probe concentration and (2) treatment with 1-pyrenebutyrate before the addition of probes, the method to deliver R8 peptide efficiently to the cytosol was adopted by utilizing 1-pyrenebutyrate as its counteranion. Although the detailed function and mechanisms of this method have been studied in model systems based on liposomes, the mechanism of action in cell-penetration remains unclear. However, it has been reported that the effect of this compound brings successful distribution of R8 peptides to the cytosol. In the absence of PMA, PKCδ was distributed in the cytosol and the fluorescence from NBD indicated co-localization of PKCδ and the probe peptide by formation of the tag-probe complex. The concentration of probe 1 was 0.1 μM. Thirty minutes after the addition of 10 μM of PMA, the fluorescence from mKO showed translocation of PKCδ. It is known that the addition of PMA causes translocation of PKCδ to membrane compartments in cells such as nuclei and Golgi apparatus. Most of the fluorescence from NBD showed co-translocation as with PKCδ. These results suggest that the tag-probe complex is very stable and not disrupted by a dynamic change of localization of target proteins.

Our strategy could enhance the options of molecular imaging inside the cells because the turn-on fluorescent system can lower the signal/noise ratio and cell-permeability is controlled by probe peptides.

Acknowledgments

We thank to Professor Naoki Yamamoto, Dr. Kenji Ohba (Yong Loo Lin School of Medicine, National University of Singapore), and Professor Kazunari Akiyoshi (Graduate School of Engineering, Kyoto University).

References

1. Tsutsumi, H., Nomura, W., Abe, S., Mino, T., Masuda, A., Ohashi, N., Tanaka, T., Ohba, K., Yamamoto, N., Akiyoshi, K., Tamamura, H. *Angew. Chem., Int. Ed.* **48**, 9164-9166 (2009).
2. Tanaka, T., Nomura, W., Narumi, T., Tamamura, H. *J. Am. Chem. Soc.* **132**, 15899-15901 (2010).
3. Nomura, W., Mino, T., Narumi, T., Ohashi, N., Masuda, A., Hashimoto, C., Tsutsumi, H., Tamamura, H. *Biopolymers: Pep. Sci.* **94**, 843-852 (2010).
4. Tsutsumi, H., Abe, S., Mino, T., Nomura, W., Tamamura, H. *ChemBioChem* **12**, 691-694 (2011).

Proceedings of the 23rd American Peptide Symposium
Michal Lebl (Editor)
American Peptide Society, 2013

Biological Effects of Bivalent-Type CXCR4 Ligands with Rigid Linkers

Wataru Nomura, Tomohiro Tanaka, Toru Aoki, Tetsuo Narumi, and Hirokazu Tamamura

Institute of Biomaterials and Bioengineering, Tokyo Medical and Dental University,
Chiyoda-ku, Tokyo, 101-0062, Japan

Introduction

A chemokine receptor CXCR4 belongs to the G-protein coupled receptor (GPCR) family. Interaction with its endogenous ligand, stromal-cell derived factor-1α (SDF-1)/CXCL12, induces various physiological functions in an embryonic stage. Recent studies have indicated a pivotal role of homo- and hetero-oligomerization of CXCR4 in cancer metastasis. In the previous study, we have designed and synthesized novel CXCR4 bivalent ligands utilizing two FC131 analogues [*cyclo*(-D-Tyr-Arg-Arg-Nal-D-Cys-)](Nal = L-3-(2-naphthyl)alanine) as ligand units [1]. The units are connected by a polyproline or a PEGylated polyproline linker. A ligand with an optimum linker-length showed the strongest binding affinity. Thus the dimer-state of CXCR4 on the cell surface was estimated by the linker length. FACS analyses showed that the bivalent ligand can distinguish the amount of CXCR4 expression. As the next challenge, migration of cells was targeted by the bivalent ligands because it was unclear whether the binding of CXCR4 bivalent ligands can promote or suppress the migration induced by the chemotaxis depending on the concentration of SDF-1α.

Results and Discussion

The bivalent ligands with a T140 analog (Figure 1) were newly synthesized by standard Fmoc solid phase peptide synthesis. The binding affinity of the bivalent ligands were was evaluated in a competitive binding assay against [^{125}I]SDF-1α as reported previously [3]. In the case of bivalent ligands with cFC131, a *cyclo*-pentapeptide, the one with 20 proline linker showed highest binding affinity. For bivalent ligand with T140, the increase of binding affinity compared to monomer ligand (4F-benzoyl-TN14003) was not observed. However, in the case of cFC131 ligands, bivalent form with insufficient length of linkers showed great decrease in binding affinity (Table 1). For T140 ligands, the decrease in binding affinity was not observed. All the ligand showed less than 30 (nM) in IC$_{50}$. The ligand with 18 prolines showed the highest binding affinity among the linker variants. In this study, the longer linker than 18 prolines was not investigated. It is possible that the linker length with 18 prolines could be appropriate for binding of T140 ligands as shown in cFC131 ligands. Utilizing the

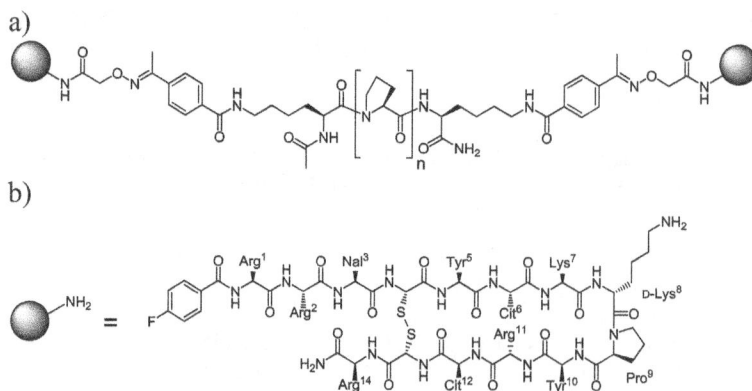

a)

b)

Fig. 1. a) Structure of polyproline linker for T140 dimers; b) The T140 unit (TZ14011) utilized in this study.

Table 1. Binding affinity of T140 bivalent ligands.

Compounds	IC50 (nM)
4F-benzoyl-TN14003	6.2
TZ14011/ 6 prolines	26.1
TZ14011/ 9 prolines	30.2
TZ14011/ 12 prolines	13.6
TZ14011/ 15 prolines	13.3
TZ14011/ 18 prolines	8.6

T140 ligands, the anti-chemotaxis activity in a chamber assay was conducted (Figure 2). The assay utilizing the series of the bivalent ligands with cFC131 did not show any anti-chemotaxis activity. The inhibition rates of the bivalent ligands were similar to the monomer ligands (T140 analog or FC131). For the dimers of TZ14011 [2], the increase of inhibition activity was observed as the linker length is getting longer. The numbers of prolines were 6, 9, 12, and 15. However, the ligand with 18 proline linker showed decrease in inhibition activity. As mentioned above, the binding affinity of the ligand with 18 proline linker was the highest. It is of interest that the anti-chemotaxis activity is not proportional to the length or binding affinity of the ligands. In addition, the reason for the bivalent ligands with cFC131 did not show anti-chemotaxis activity should be solved. The anti-chemotaxis activity is highly related with the agonism and antagonism of the ligands. In the previous study, the monomer unit, 4F-benzoyl-TN14003, was shown to inhibit the cell migration [3]. The T140 bivalent ligand showed stronger inhibition than the monomer units. Thus, the bivalent ligand could be a potent reagent against migration and invasion of cancer cells. The analysis of signaling pathways such as calcium mobilization or phosphorylation of the proteins in downstream processes will be conducted in the future study.

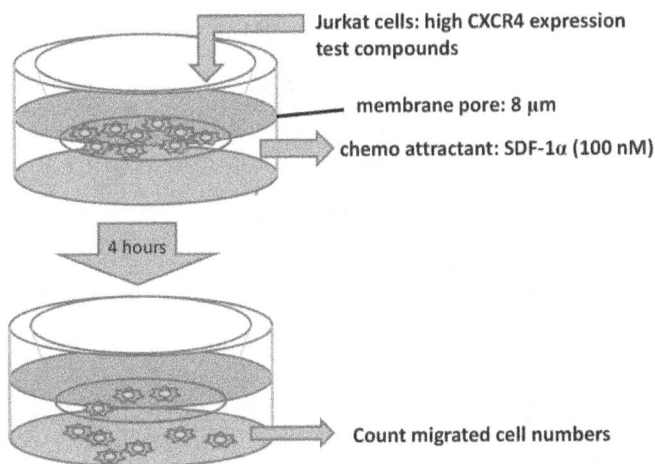

Jurkat cells: high CXCR4 expression
test compounds

membrane pore: 8 μm

chemo attractant: SDF-1α (100 nM)

4 hours

Count migrated cell numbers

Fig. 2. Experimental scheme for the migration assay by chemotaxis for SDF-1α.

The results indicate that our ligand design approach utilizing rigid polyproline linker would be useful for development of more effective anti-CXCR4 ligands.

Acknowledgments

We thank to Professor Naoki Yamamoto and Dr. Kenji Ohba (Yong Loo Lin School of Medicine, National University of Singapore), and Professor Kazunari Akiyoshi (Graduate School of Engineering, Kyoto University).

References

1. Tanaka, T., Nomura, W., Narumi, T., Tamamura, H. *J. Am. Chem. Soc.* **132**, 15899-15901 (2010).
2. Hanaoka, H., Mukai, T., Tamamura, H., Mori, T., Ishino, S., Ogawa, K., Iida, Y., Doi, R., Fujii, N., Saji, H. *Nucl. Med. Biol.* **33**, 489-494 (2006).
3. Mori, T., Doi, R., Koizumi, M., Toyoda, E., Ito, D., Kami, K., Masui, T., Fujimoto, K., Tamamura, H., Hiramatsu, K., Fujii, N., Imamura, N. *Mol. Cancer Ther.* **3**, 29-37 (2004).

Proceedings of the 23rd American Peptide Symposium
Michal Lebl (Editor)
American Peptide Society, 2013

Myristoylated Protein Kinase C Epsilon Peptide Inhibitor Exerts Cardioprotective Effects in Rat and Porcine Myocardial Ischemia/Reperfusion: A Translational Research Study

Matthew Montgomery, Jovan Adams, Jane Teng, Biruk Tekelehaymanot, Regina Ondrasik, Issachar Devine, Kerry-Anne Perkins, Qian Chen, Robert Barsotti, and Lindon H. Young

Department of Bio-Medical Sciences, Philadelphia College of Osteopathic Medicine (PCOM), Philadelphia, PA, 19131, U.S.A.

Introduction

Following an acute myocardial infarction, the rapid restoration of blood flow to the ischemic myocardium is the most effective method of limiting infarct size and improving patient outcomes. However, reperfusion leads to the loss of cardiomyocytes that were viable during the ischemic episode and is referred to as ischemia/reperfusion (I/R) injury [1]. This restoration of blood flow is known to result in the overproduction of reactive oxygen species (ROS), as incoming oxygen reacts with the damaged mitochondrial respiratory chain to produce superoxide that in turn leads to the enzymatic uncoupling of endothelial nitric oxide synthase (eNOS) and the production of more ROS and reduced nitric oxide (NO) bioavailability [2]. The reduction in NO has been shown to reduce the coupling efficiency of O_2 consumption and ATP synthesis in cardiac mitochondria, presumably leading to further ROS production during in I/R [3]. It is well established that protein kinase C epsilon (PKCε) activity can stimulate eNOS activity. During I/R, PKCε activity is activated via stimulation of cytokine receptors and enhances uncoupled eNOS activity [4]. We hypothesized that limiting PKCε activity during I/R using a cell permeable myristoylated PKCε peptide inhibitor (PKCε-) should limit ROS production in part by attenuating uncoupled eNOS activity and thereby improve heart function and reduce infarct size.

Results and Discussion

Untreated isolated rat (male Sprague-Dawley, 275-325g) hearts subjected to global I(30min)/R(45min) exhibited compromised cardiac function, while I/R hearts treated with 10μM PKCε- (N-myr-EAVSLKPT, MW=1054g/mol, Genemed Synthesis) given at the beginning of reperfusion had significant restoration for all cardiac function indices and coronary flow (Figure 1). Infarct size was significantly ($p < 0.01$) reduced from 48±2 (n=10) to 25±2% (n=11) in untreated compared to treated hearts (Figure 3). We also studied regional myocardial I(1hr)/R(3hr) injury in anesthetized pigs (castrated male Yorkshire Cross, 27-36kg) in which the left anterior descending coronary artery was occluded at the level of the second diagonal branch for 1hr using a fluoroscopically guided balloon catheter. Echocardiography was used to monitor ejection fraction during baseline, ischemia and post-reperfusion (1-3hr). PKCε- (0.8mg/kg, n=3) was given at reperfusion. At the end of the reperfusion period, cardiac ejection fraction in hearts treated with PKCε- recovered to 91±6% (n=3) of its baseline value (Figure 2), and infarct size was 13±0.3% (n=3) of the total area at risk (Figure 3). These parameters were significantly improved compared to saline treated myocardial I/R pigs that recovered to a final cardiac ejection fraction of only 70±3% (n =4, p<0.01) of baseline values and an infarct size of 34±4% (n= 3, p<0.01; Figures 2 and 3).

These results suggest that myristoylated PKCε- has profound cardioprotective effects that transcend species in both global (rat) and regional (porcine) I/R, which may be mediated by the inhibition of ROS production from uncoupled eNOS activity. Collectively these data suggest that PKCε- should be an effective therapeutic tool to ameliorate cardiac contractile dysfunction and tissue damage following acute myocardial infarction and subsequent primary coronary intervention treatment in humans.

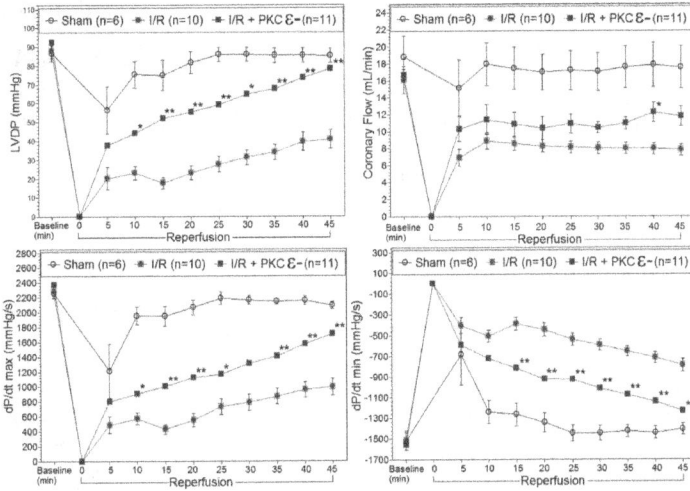

Fig. 1. Time course of left ventricular developed pressure (LVDP) (top left), coronary flow (top right), dP/dtmax (bottom left), and dP/dtmin (bottom right) in isolated perfused rat hearts subjected to 30 min ischemia prior to reperfusion. (*p<0.05,**p<0.01 vs. I/R control). Data were analyzed using ANOVA with the Student-Newman-Keuls test.

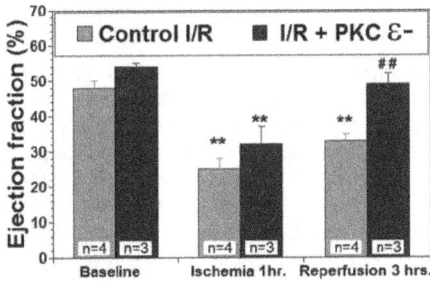

Fig 2. Ejection fraction of saline control and PKCε-treated I/R hearts during baseline, ischemia and 3hr reperfusion. Both groups exhibited a significant decrease in ejection fraction during ischemia compared to their respective baselines (**p<0.01 vs. hearts at baseline). Ejection fraction for PKCε-treated hearts was significantly increased compared to control hearts at 3hr reperfusion ($^{##}$p<0.01).

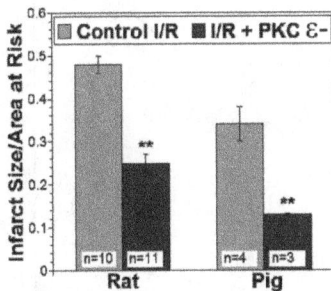

Fig 3. Ratio of infarcted tissue weight to that of area at risk. PKCε- treatment significantly decreased infarct size compared to untreated control I/R rat hearts (** p<0.01). PKCε-treated pig hearts had an infarct size of 13±0.3% , which was significantly reduced compared to saline control pigs with an infarct size of 34±4% (**p<0.01).

Acknowledgments

This study was supported by the Center for Chronic Disorders of Aging and the Department of Bio-Medical Sciences at PCOM.

References

1. Yellon, D.M., Hausenloy, D.J. *N. Engl. J. Med.* **357**, 1121-1135 (2007).
2. Perkins, K.A., et al. *Naunyn Schmiedebergs Arch Pharmacol.* **385**, 27-38 (2012).
3. Shen, W., et al. *Am. J Physiol. Heart Circ. Physiol.* **281**, H838-H846 (2001).
4. Teng, J.C., et al. *Naunyn Schmiedebergs Arch Pharmacol.* **378**, 1-15 (2008).

Proceedings of the 23rd American Peptide Symposium
Michal Lebl (Editor)
American Peptide Society, 2013

Peptide-DNA Hybrid G-Quadruplex Structures for Regulation of Protein Expression Depending on Protease Activity

Arisa Okada[1], Mai Taniguchi[1], Kenji Usui[1,2], and Naoki Sugimoto[1,2]

[1]Faculty of Frontiers of Innovative Research in Science and Technology (FIRST);
[2]Frontier Institute for Biomolecular Engineering Research (FIBER), Konan University,
7-1-20 Minatojima-Minamimachi, Chuo-ku, Kobe 650-0047, Japan

Introduction

Guanine (G) -rich sequences are found in the promoter or untranslated region of genes with high probability and these sequences in the genome may help regulate protein expression by forming a quadruplex structure (G-quadruplex) [1]. Consequently, a system capable of inducing G-rich sequences to form G-quadruplex structures would be useful to modulate protein expression. From this point of view, several studies for developing ligands such as RNAs [2], porphyrin derivatives [3] and others [4] to induce G-rich sequences to form a quadruplex structure toward regulating transcription have been recently reported. However, the promising ligands in the next generation would need ability of altering (switching) the structural thermodynamic properties, more G-rich sequence specificity and a greater degree of functionality such as delivering ability to cellular organelles. In this study, we attempted to construct a modulation system for protein expression using a designed peptide as the functional ligand with the on-to-off switching module for G-quadruplex forming depending on protease (calpain I) activity.

Results and Discussion

First of all, we designed and synthesized a peptide (GCalp) with G-rich peptide nucleic acids (PNAs) [5] for forming PNA-DNA hybrid G-quadruplex structure with a G-rich DNA. LLVY, a calpain I substrate sequence, was introduced to the center region of GCalp as the on-to-off switching module. Therefore, once GCalp would be digested by expressed calpain I, GCalp would lose the binding ability to the DNA and simultaneously a PNA-DNA quadruplex structure would collapse. Consequently, by using GCalp and a plasmid with a GCalp binding sequence at upstream of a target protein coding region, well-controlled system for protein expression could be achieved (Figure 1).

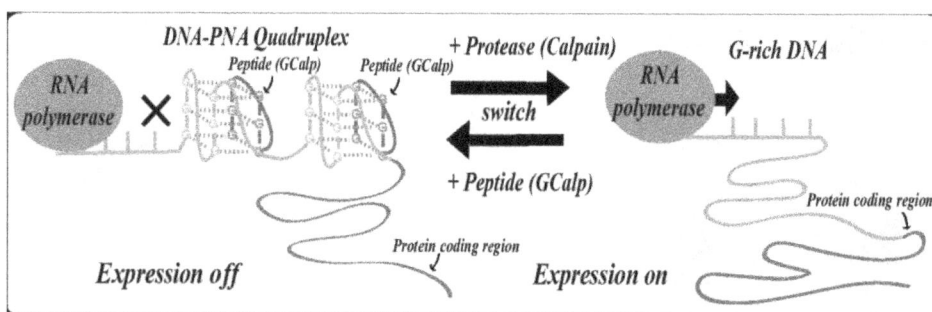

Fig. 1. Outline of this study.

After synthesis of GCalp by Fmoc chemistry [6], we checked GCalp binding property to a selected model G-rich DNA by CD spectroscopy and electrophoresis. Binding stoichiometry analysis showed 2:1 formation between GCalp and the model DNA using a fluorescent model DNA. Additionally, T_m values at 295 nm by UV spectroscopy were dependent on KCl concentration. Furthermore, we determined thermodynamic parameters for formation of a DNA G-quadruplex as well as a PNA-DNA G-quadruplex. The free energy of formation of the PNA-DNA G-quadruplex is greater than that of the DNA G-quadruplex. This emphasizes the fact that a very high affinity complex is formed between the model DNA and GCalp.

These results indicated that GCalp could form PNA-DNA quadruplex structures with the DNA.

Next, we demonstrated the on-to-off switching depending on calpain I activity using a fluorescent model DNA and GCalp. The fluorescent intensity of the PNA-DNA G-quadruplex was higher than that of the DNA G-quadruplex. And the intensity of digested GCalps by calpain I was almost same as that of the DNA G-quadruplex. Additionally, we checked by HPLC and MS whether calpain I could digest GCalp smoothly. These results indicated that GCalp functioned effectively as an on-to-off structural switching module as expected.

Finally, we demonstrated regulation of protein expression with GCalp using the cell-free protein synthesis system. A plasmid including GCalp binding sequences showed that protein expression efficiency with GCalp was ca. 20% of that without GCalp, whereas, the protein expression efficiency of a plasmid without GCalp binding sequences could not be influenced by the addition of GCalp.

Throughout this study, we constructed a control system for protein expression by regulation of forming peptide-DNA hybrid G-quadruplex structures depending on protease activity. With more improvement of binding-specificity, this system would be a promising tool for the well-controlled protein expression and regulation of important cellular events toward cell engineering and tissue engineering.

Acknowledgments

We thank Dr. Tamaki Endo (Frontier Institute for Biomolecular Engineering Research (FIBER), Konan University, Japan) for valuable discussions and generous support. K.U. and N.S. are also grateful for Grants-in-Aid for Scientific Research from MEXT for the Core Research project (2009-2014).

References

1. Kobayashi, K., Matsui, N., Usui, K. *J. Nucleic Acids Article* ID 572873 (2011).
2. Ito, K., Go, S., Komiyama, M., Xu, Y. *J. Am. Chem. Soc.* **47**, 19153-19159 (2011).
3. Jain, A., Grand, C., Bearss, D., Hurley, L. *Proc. Nat. Acad. Sci. USA* **99**, 11593-11598 (2002).
4. Yaku, H., Murashima, T., Miyoshi, D., Sugimoto, N. *Chem. Comm.* **46**, 5740-5742 (2010).
5. Neilsen, P., Egholm, M., Berg, R., Buchardt, O. *Science* **254**, 1497-1500 (1991).
6. Sano, S., Tomizaki, K., Usui, K., Mihara, H. *Bioorg. Med. Chem. Lett.* **16**, 503-506 (2006).

Proceedings of the 23rd American Peptide Symposium
Michal Lebl (Editor)
American Peptide Society, 2013

A Selective AIF/CypA Inhibitory Peptide Provides Protection in Neuronal Cells

N. Doti[1,2,3], P.L. Scognamiglio[1], A. Caporale[1], C. Reuther[2], A.M. Dolga[2], N. Plesnila[3,4], C. Culmsee[2], and M. Ruvo[1]

[1]*IBB-CNR, CIRPEB Via Mezzocannone, 16, 80134, Napoli, Italy;* [2]*Institute of Pharmacology and Clinical Pharmacy, Philipps University of Marburg, 35032, Germany;* [3]*RCSI, Dublin 2, Ireland;* [4]*ISD Großhadern Max-Lebsche Platz 30, D-81377, Munich, Germany*

Introduction

The *Apoptosis Inducing Factor* (AIF) is a phylogenetically ancient flavoprotein that contains a mitochondrial localization signal (MLS) on the *N*-terminus. Upon import into the mitochondria, the MLS is removed by a mitochondrial peptidase generating the mature form of the protein (62 kDa) [1], which is anchored to the Inner Mitochondrial Membrane (IMM), where it exerts NADH oxidase activity [1]. Based on findings with different knockout/knockdown models it was proposed that AIF might play a role in oxidative phosphorylation, mainly by modulating the structure and function of complex I of the respiratory chain [1,2].

In acute brain injury, AIF acquires a pro-death role upon translocation from the mitochondria to the nucleus, where it initiates chromatin condensation and large-scale DNA fragmentation. Inhibition of this translocation or reduction of AIF expression, such as in Hq mice, is neuroprotective [3].

Since nuclear translocation of AIF was not observed in cyclophilin A-depleted (CypA)[-/-] neurons exposed to hypoxia–ischemia, it was suggested that the translocation and DNA binding activities of AIF is mediated by the interaction with CypA [3]. Also, inhibiting AIF nuclear translocation prevents cell death, further suggesting that inhibiting AIF/CypA complex formation can be an effective strategy to block the lethal action of AIF.

Here we have investigated whether pharmacological inhibition of the AIF/CypA complex enables protection of neurons against NMDA receptor-independent and caspase-independent oxidative stress-induced apoptosis. The study has been performed with an AIF peptide targeting the AIF-binding site on CypA, using as model the hippocampal HT-22 neuronal cell line.

Results and Discussion

Starting from the reported molecular model of the AIF/CypA complex [4], we designed and prepared by chemical synthesis a set of AIF peptides, corresponding to the predicted interface between the two proteins. Peptides were tested by SPR competition binding assays for their ability to prevent AIF/CypA complex formation. Notably, one of the peptide tested, named Pospep, inhibited complex formation in a dose-dependent manner with an IC_{50} of about 3.0×10^{-6} M, while others tested in the same concentration range were essentially ineffective. Furthermore, SPR binding experiments with the selected peptide to the immobilized proteins indicated that it selectively bound CypA with a K_D of 1.15×10^{-5} M.

Then, we investigated whether Pospep could antagonize the lethal effect of the AIF/CypA complex in a hippocampal neuronal cell line (HT-22). In this model, treatment with high concentrations of glutamate (2÷5 mM) induces a caspase-indipendent mode of cell death mediated by AIF [5]. Noteworthy, the administration of the inhibitory peptide at concentrations ranging between 10 and 50 µM, as well as the silencing of CypA, significantly inhibited the neuronal loss by about 80% at the highest concentration tested by blocking the nuclear translocation of AIF (Figure 1).

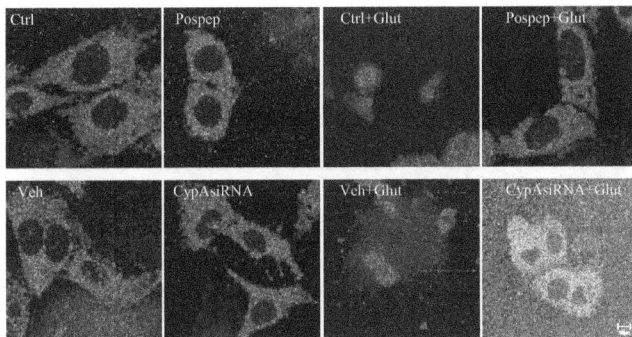

Fig. 1. Pospep and CypAsiRNA block glutamate-induced AIF nuclear translocation. Confocal images showed that the Pospep at 50 μM significantly blocked glutamate-induced nuclear translocation of AIF to an extent comparable to that observed after siRNA-mediated CypA-silencing in the HT-22 cells.

Remarkably, administration of the inhibitory AIF peptide to HT-22 cells also significantly eliminated most hallmarks of glutamate-mediated neuronal cell death, including AIF nuclear translocation, DNA condensation, mitochondria depolarization/fragmentation, and AIF perinuclear condensation, thus demonstrating that the selective targeting of the AIF/CypA complex affords a robust neuroprotection.

In summary, we investigated the possibility to block the lethal action of AIF by generating pharmacological inhibitors of the CypA/AIF interaction. Using an AIF-derived CypA-binding peptide, we show for the first time that this is a new therapeutic strategy to induce neuroprotection. In perspective, this is of particular relevance, because new inhibitors fulfilling the same task as the Pospep, potentially have no influence on the physiological functions of both CypA and AIF, and could therefore display advantageous pharmacological profiles.

References

1. Sevrioukova, I.F. *Antioxid. Redox Signal.* **14**, 2545-2579 (2011).
2. Polster, B.M. *Neurochem. Int.* **62**, 695-702 (2013).
3. Zhu, C., et al. *J. Exp. Med.* **204**, 1741-1748 (2007).
4. Candé, C., et al. *Oncogene* **23**, 1514-1521 (2004).
5. Tobaben, S., et al. *Cell Death Differ.* **18**, 282-292 (2011).

Proceedings of the 23rd American Peptide Symposium
Michal Lebl (Editor)
American Peptide Society, 2013

Design, Synthesis, Biophysical and Structure-Activity Properties of a Novel Dual MDM2 and MDMX Targeting Stapled α-Helical Peptide: ATSP-7041 Exhibits Potent *In Vitro* and *In Vivo* Efficacy in Xenograft Models of Human Cancer

V. Guerlavais[1], K. Darlak[1], B. Graves[2], C. Tovar[2], K. Packman[2], K. Olson[1], K. Kesavan[1], P. Gangurde[1], J. Horstick[1], A. Mukherjee[1], T. Baker[1], X.E. Shi[1], S. Lentini[1], K. Sun[1], S. Irwin[1], E. Feyfant[1], T. To[2], Z. Filipovic[2], C. Elkin[1], J. Pero[1], S. Santiago[1], T. Bruton[1], T. Sawyer[1], A. Annis[1], N. Fotouhi[2], T. Manning[1], H. Nash[1], L.T. Vassilev[2], Y.S. Chang[1], and T.K. Sawyer[1]

[1]Aileron Therapeutics Inc., Cambridge, MA, 02139, U.S.A.; [2]Roche Research Center, Hoffmann-La Roche, Inc., Nutley, NJ, 07110, U.S.A.

Introduction

Activation of the p53 tumor suppressor protein by disrupting its interactions with both MDM2 and MDMX provides an opportunity for the development of a novel cancer therapeutic to treat p53 wild-type tumors. Stapled Peptides are a breakthrough approach to create a new class of drugs that target intracellular protein-protein interactions [1,2]. Here, we describe the stapled α-helical peptide ATSP-7041, a novel dual inhibitor of MDM2 and MDMX that reactivates p53-dependent pathway in multiple cancer cell lines *in vitro* and *in vivo*.

Significant Structure-Activity Relationship Analysis Derived from Prototype Stapled Peptide Lead (ATSP-3900) and Ala-Scanning

Optimization of p53 Stapled Peptides was undertaken (Table 1) to improve both their biological and biophysical properties and optimization leveraged the sequence enhancements of a recently described phage display peptide (pDI), Ac-Leu[17]-Thr-Phe-Glu-His-Tyr-Trp-Ala-Gln-Leu-Thr-Ser[28]-NH2 [3]. The prototype molecule, Ac- Leu[17]-Thr-Phe-*cyclo*(R8-His-Tyr-Trp-Ala-Gln-Leu-S5)-Ser[28]-NH2 (ATSP-3900) achieved striking improvements in binding potency to MDM2 (Ki = 8nM) and MDMX (Ki = 12.7nM) compared to SAH-p53-8 [1,2].

As shown in Figure 1, the key amino acids (F[19], Y[22], W[23], L[26]) for the activity were confirmed relative to Ala-scanning and comparative analysis of both binding to MDM2 and MDMX as well as cellular viability in SJSA-1 cells (10% FBS).

Table 1. Ala-scanning on Stapled Peptide lead ATSP-3900.

Peptides		14	15	16	17	18	19	20	21	22	23	24	25	26	27	28	29		MDM2 binding Ki (nM)	MDMX binding Ki (nM)	SJSA-1,10% FBS Cellular Viability IC50 (uM)
SAH-p53-8	Ac-	Q	S	Q	Q	T	F	R8	N	L	W	R	L	L	S5	Q	N	-NH2	26	106	>30
ATSP-3900	Ac-				L	T	F	R8	H	Y	W	A	Q	L	S5	S		-NH2	8.0	12.7	15
Analog A[17]	Ac-				A	T	F	R8	H	Y	W	A	Q	L	S5	S		-NH2	3.0	7.3	12
Analog A[18]	Ac-				L	A	F	R8	H	Y	W	A	Q	L	S5	S		-NH2	24.1	105.2	23
Analog A[19]	Ac-				L	T	A	R8	H	Y	W	A	Q	L	S5	S		-NH2	228	13224	>30
ATSP-4641	Ac-				L	T	F	R8	A	Y	W	A	Q	L	S5	S		-NH2	5	34	5.9
Analog A[22]	Ac-				L	T	F	R8	H	A	W	A	Q	L	S5	S		-NH2	20.8	76.9	>30
Analog A[23]	Ac-				L	T	F	R8	H	Y	A	A	Q	L	S5	S		-NH2	196	8600	>30
Analog A[25]	Ac-				L	T	F	R8	H	Y	W	A	A	L	S5	S		-NH2	17.5	12.5	17.2
Analog A[26]	Ac-				L	T	F	R8	H	Y	W	A	Q	A	S5	S		-NH2	550	862	>30
Analog A[28]	Ac-				L	T	F	R8	H	Y	W	A	Q	L	S5	A		-NH2	5	11.2	8

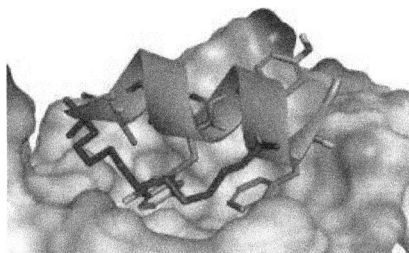

Fig. 1. X-Ray structure of complex ATSP-3900:MDMX.

The biophysical properties of SAH-p53-8, ATSP-3900 and ATSP-4641 are very different. In contrast to SAH-p53-8, ATSP-3900 and ATSP-4641 form idealized amphipathic helices with the hydrophobic amino acids F19, Y22, W23 and L26 forming a hydrophobic patch devoid of undesirable polar amino acids. Through iterative lead optimization, ALRN-3900 was further modified to enhance target binding, biophysical properties (e.g., solubility) and cell potency. This effort resulted in the identification of ATSP-7041 [4].

Solid-phase Synthesis and Purification of ATSP-7041

Synthesis as performed on-resin produced crude ATSP-7041 (E-isomer) with good yield (52%) and purity (64%). The "solid-phase ring-closing metathesis" reaction generated mostly E-isomer. Reversed-phase HPLC purification gave purified ATSP-7041 with a 10-15% yield.

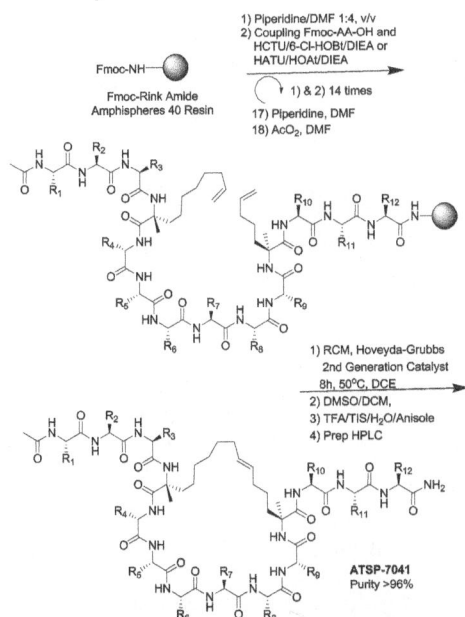

Fig. 2. Solid phase synthesis of ATSP-7041.

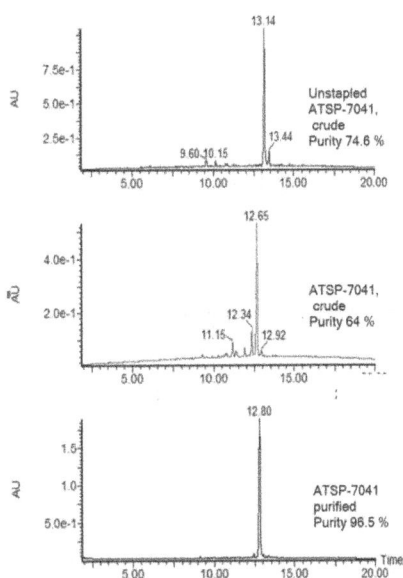

Fig. 3. HPLC traces of crude unstapled, stapled and purified ASTP-7041.

ATSP-7041 represents a potent and specific dual Stapled Peptide inhibitor of MDM2 (Kd= 0.91 nM) and MDMX (Kd= 2.31 nM) that induces p53-dependent apoptosis and inhibits cell proliferation in MDM2 and MDMX overexpressing tumor *in vitro* and *in vivo*.

References

1. Bernal, F., Tyler, A.F., Korsmeyer, S.J., Walensky, L.D., Verdine, G.L. *J. Am. Chem. Soc.* **129**, 2456-2457 (2007).
2. Bernal, F., Wade, M., Godes, M., Davis, T.N., Whitehead, D.G., Kung, A.L., Wahl, G.M., Walensky, L.D. *Cancer Cell* **18**, 411-422 (2010).
3. Hu, B., Gilkes, D.M., Chen, J. *Cancer Res.* **67**, 8810-8817 (2007).
4. Chang, Y.S., Graves, B., Guerlavais, V., Tovar, C., Packman, K., To, K.-H., Olson, K.A., Kesavan, K., Gangurde, P., Mukherjee, A., Baker, T., Darlak, K., Elkin, C., Filipovic, Z., Cai, H., Berry, P., Feyfant, E., Shi, E., Horstick, J., Annis, A., Manning, T., Fotouhi, N., Nash, H., Vassilev, L.T., Sawyer, T.K. *Proc. Nat. Acad. Sci. USA* Early Edition (Aug 14 2013), 1-10, 10 pp (2013).

Proceedings of the 23rd American Peptide Symposium
Michal Lebl (Editor)
American Peptide Society, 2013

Rational Design of Helix-Mimicking Small Molecules for Inhibiting BCL-2 Proteins in Prostate Cancer

Jung-Mo Ahn[1]*, Joongsoo Kim[1], Rakesh Kumar[1], Kajal Bhimani[1], Jaspal Singh[1], Pradeep Patil[1], Yuefeng Du[2], Timothy Dobin[2], Jonathan Nguyen[2], and Jer-Tsong Hsieh[2]

[1]Department of Chemistry, The University of Texas at Dallas, Richardson, TX, 75080, U.S.A.;
[2]Department of Urology, The University of Texas Southwestern Medical Center,
Dallas, TX, 75390, U.S.A.

Introduction

Cell death occurs for many reasons including physiological and pathological factors and plays a crucial role in the process of development, aging and diseases. The fate of a cell depends on an appropriate response to various environmental or intracellular stress stimuli and the mechanism of cell death can be broadly classified into two types, apoptotic and non-apoptotic. Apoptosis is largely regulated by protein-protein interactions between Bcl-2 family proteins that comprise of three groups. Anti-apoptotic proteins (e.g., Bcl-2, Bcl-xL, Bcl-w, Mcl-1, A1) are responsible for cell survival and have four conserved domains (BH1-BH4). Pro-apoptotic members induce cell death and are further classified to multi-domain proteins bearing BH1-BH3 (e.g., Bak, Bax) and BH3-only proteins (e.g., Bim, Bik, Bid, Puma, Noxa). The process of apoptosis is initiated by BH3-only proteins either activating multi-domain pro-apoptotic proteins or inhibiting anti-apoptotic members (direct or indirect activation model, respectively), resulting in release of cytochrome c through mitochondrial outer membrane permeability [1]. This event in turn begins caspase cascade that ultimately leads to cell death. Thus, small molecules that inhibit anti-apoptotic Bcl-2 proteins or activate pro-apoptotic ones would be of high interest in treating cancers.

Results and Discussion

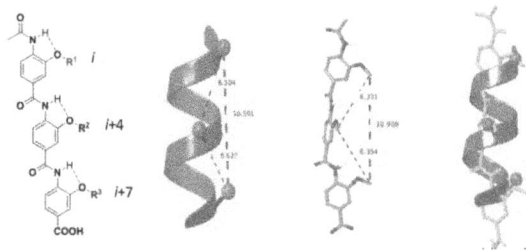

Fig. 1. Tris-benzamide scaffold and its confor-mation overlaid on a helix.

To the end, we have designed a series of small molecules based on a rigid tris-benzamide scaffold in an attempt to mimic BH3 domains of pro-apoptotic Bcl-2 proteins since the helical BH3 domains of pro-apoptotic proteins are found to strongly contribute to the interaction with anti-apoptotic proteins [2-4]. We have developed the tris-benzamide scaffold to reproduce a helical surface organized by three residues at the *i*, *i+3/i+4*, and *i+7* positions of an α-helix (Figure 1) [5].

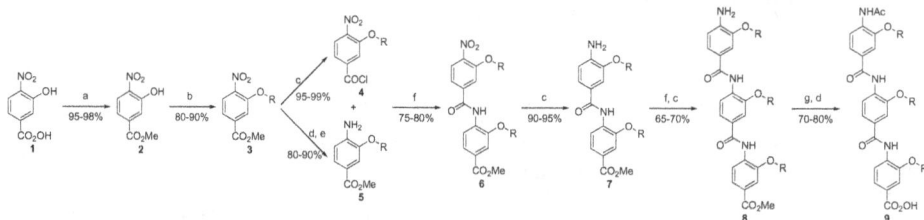

Scheme 1. Synthesis of tris-benzamides as peptidomimetics of BH3 domains.
Reagent and conditions: (a)H_2SO_4, MeOH, reflux, 8h (b)RX, K_2CO_3, DMF, 60°C, 10h (c)$SnCl_2$, THF/HCl/AcOH(4:1:1), rt, 15h (d)LiOH, THF/MeOH(4:1), rt, 15h (e)$(COCl)_2$, DMF(cat.), DCM, reflux, 1h (f)TEA, DCM, rt, 5h (g)Ac_2O, DCM, rt, 15h.

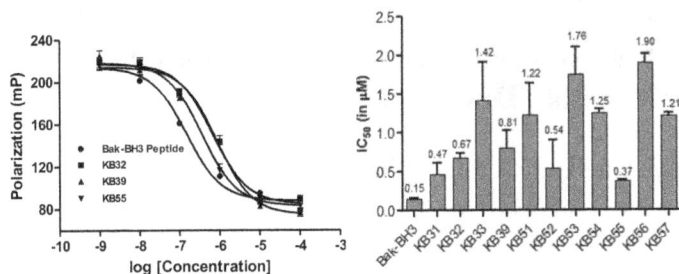

Fig. 2. Competitive binding assay of tris-benzamides.

Synthesis of tris-benz-amides [5-8] began with *O*-alkylation of methyl 3-hydroxy-4-nitrobenzo-ate **2** with various alkyl halides in presence of K_2CO_3 in DMF (Scheme 1). After hydrolysis of the methyl ester, acid chloride **4** was prepared with oxalyl chloride. It was then reacted with aniline **5** that was made by reducing the nitro group of **3** with tin (II) chloride, yielding bis-benzamide **6**. These steps were repeated to produce tris-benzamide **8** in high yields.

To examine whether the designed molecules can mimic the BH3 domains of pro-apoptotic Bcl-2 proteins, their binding affinities were determined by fluorescence polarization assays using Bcl-xL and a BH3 peptide derived from Bak (Figure 2). Several tris-benzamides were identified to have strong binding to the protein showing sub-micromolar IC_{50} values, indicating suitable mimicry of the BH3 domains of pro-apoptotic proteins.

Then, we have investigated the effects of the BH3 mimetics in human prostate cancer cell line (DU-145) that has a high expression level of Bcl-xL. MTT assays were carried out to determine cell proliferation in the presence of the tris-benzamides (Figure 3). Several compounds exhibited significant cytotoxicity towards the prostate cancer cell line. To determine the mechanism of cell death, induction of apoptosis was examined by fluorescence-activated cell sorting after DU-145 cells were treated with a range of concentrations of KB39 for 72h (data not shown). The results indicated that KB39 induced significant level of apoptosis in dose-dependent manner, suggesting the potential of rationally designed pepti-domimetics as therapeutic candidates for treating prostate cancer.

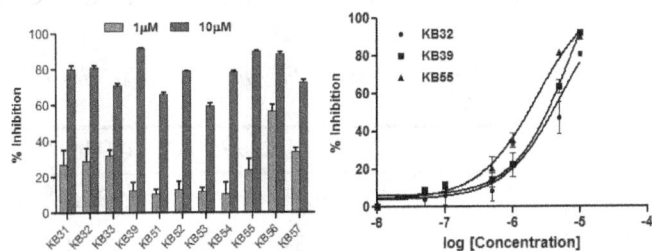

Fig. 3. Cytotoxicity assay of tris-benzamides on DU-145 prostate cancer cell line.

Acknowledgments

This work was supported in part by the Cancer Prevention and Research Institute of Texas (RP100718) and the Welch Foundation (AT-1595).

References

1. Kuwana, T., et al. *Mol. Cell* **17**, 525-535 (2005).
2. Kelekar, A., et al. *Mol. Cell Biol.* **17**, 7040-7046 (1997).
3. Lessene, G., Czabotar, P.E., Colman. P.M. *Nat. Rev. Drug Discov.* **7**, 989-1000 (2008).
4. Sattler, M., et al. *Science* **275**, 983-986 (1997).
5. Ahn, J.-M., Han, S.-Y. *Tetrahedron Lett.* **48**, 3543-3547 (2007).
6. Marimganti, S., et al. *Org. Lett.* **11**, 4418-4421 (2009).
7. Lee, T.-K., Ahn, J.-M. *ACS Comb. Sci.* **13**, 107-111 (2011).
8. Ravindranathan, P., et al. *Nat. Commun.* **4**, 1923, doi: 10.1038/ncomms2912 (2013).

Proceedings of the 23rd American Peptide Symposium
Michal Lebl (Editor)
American Peptide Society, 2013

Synthesis of Azapeptide Ligands of the CD36 Receptor Towards Potential Treatments of Age-Related Macular Degeneration

Yésica García-Ramos and William D. Lubell

*Département de Chimie Université de Montréal, C.P. 6128, Succursale Centre-Ville,
Montréal, Québec, H3C 3J7, Canada*

Introduction

The leading cause of blindness in North American adults, age-related macular degeneration (AMD) involves accumulation of sub-retinal deposits, degeneration of photoreceptors and choroidal neovascularization [1]. Targeting the CD36 receptor, because of its roles in oxidized lipid uptake and neovascularization [2], we have synthesized azapeptide analogues of growth hormone releasing peptide-6 (GHRP-6, His-D-Trp-Ala-Trp-D-Phe-Lys-NH$_2$), such as [azaY^4]- and [A^1, azaF^4]-GHRP-6 (**1** and **2**), which bind to this scavenger receptor without significant affinity for the growth hormone secretagogue receptor 1a (GHS-R1a, Table 1) [3,4]. Azapeptides **1** and **2**, both reduced inflammation; however, they exhibited opposite effects on neovascularization in a microvascular sprouting assay using mouse choroidal explants: the former suppressed angiogenesis and the latter exhibited angiogenic activity [3]. Considering subtle differences between the structures of **1** and **2** caused significant effects on neovascularization, we have now prepared a targeted library of azapeptide analogs to examine the influence of the 4-position aromatic residue. Employing the sub-monomer approach for solid-phase azapeptide synthesis [5], six analogs **7** were synthesized in which the phenol hydroxyl group of [azaY^4]-GHRP-6 was replaced by hydrogen and different electron withdrawing (F, CF$_3$, NO$_2$, I) and donating (OCH$_3$) substituent groups.

*Table 1. Affinities of GHRP-6 and aza-analogues **1** and **2** for the GHS-R1a and CD36 receptors.*

Compound	GHS-R1a IC$_{50}$	CD36 IC$_{50}$
GHRP-6	6.08x10^{-9} M	2.03x10^{-6} M
[azaTyr^4]-GHRP-6 (**1**)	1.57x10^{-5} M	2.80x10^{-5} M
[Ala1,azaPhe4]-GHRP-6 (**2**)	»10^{-5} M	7.58x10^{-6} M

Results and Discussion

On Rink amide resin, an Fmoc strategy was used to prepare dipeptide **3** employing HBTU and DIEA for amino acid couplings. The aza-glycinyl residue was added using *N,N'*-disuccinimidyl carbonate (DSC) to activate benzylidene hydrazone. Semicarbazone **4** was alkylated selectively with the respective benzyl bromide using Et$_4$NOH as base to provide substituted aza-phenylalanines **5** [6]. Hydrazone removal using hydroxylamine hydrochloride in pyridine and semicarbazide acylation with the symmetric anhydride of Fmoc-Ala provided azapeptides **6**, which were elongated and cleaved from the resin using standard Fmoc-based protocols to provide azapeptides **7**. Substituted [azaF^4]-GHRP-6 analogues **7** were thus effectively synthesized in 3-6% overall yields. Their CD36 receptor affinity and activity on neovascularization are under investigation and will be reported in due time.

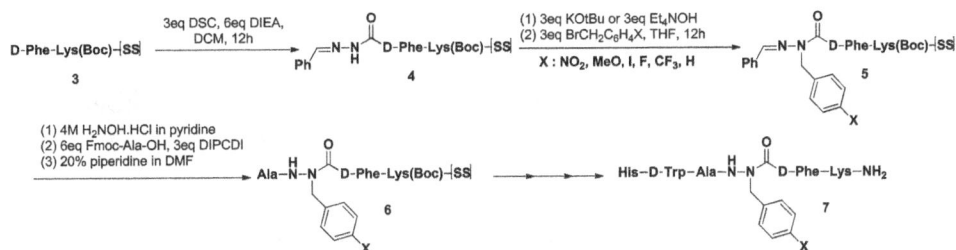

Scheme 1. Synthesis of [azaF⁴]-GHRP-6 analogues.

Table 1. Substituted [azaF⁴]-GHRP-6 analogues yields and characterization.

X	Yield(%)	HRMS [M+1]	
		m/z (calcd)	m/z (obsd)
H	3%	835.4362	835.4374
F	5%	853.4268	853.4269
CF$_3$	6%	903.4236	903.4238
OMe	6%	865.4468	865.4467
NO$_2$	5%	880.4213	880.4196
I	5%	481.1701	481.1706

Acknowledgments

We thank the Natural Sciences and Engineering Research Council of Canada, the Canadian Institutes of Health Research (grant #TGC-114046), the Ministère du développement économique de l'innovation et de l'exportation du Québec (#878-2012, Traitement de la dégénerescence maculaire), Amorchem and Boehringer Ingelheim for financial support.

References

1. Ho, Y-S., Poon, D.C-H., Chan, T-F., Chang, R. C-C. *Curr. Pharm. Design* **18**, 15-26 (2012).
2. Marleau, S., Harb, D., Bujold, K., Avallone, R., Iken, K., Wang, Y., Demers, A., Sirois. M.G., Febbraio, M., Silverstein, R.L., Tremblay, A., Ong, H. *Fed. Am. Soc. Exp. Biol. J.* **19**, 1869-1871 (2005).
3. Proulx, C., Picard, E., Boeglin, D., Pohankova, P., Chemtob, S., Ong, H., Lubell, W.D. *J. Med. Chem.* **55**, 6502-6511 (2012).
4. Sabatino, D., Proulx, C., Pohankova, P., Ong, H., Lubell, W.D. *J. Am. Chem. Soc.* **133**, 12493-12506 (2011).
5. Sabatino, D., Proulx, C., Klocek, S., Bourguet, C.B., Boeglin, D., Ong, H., Lubell, W.D. *Org. Lett.* **11**, 3650-3653 (2009).
6. Garcia-Ramos, Y., Proulx, C., Lubell, W.D. *Can. J. Chem.* **90**, 985-993 (2012).

Proceedings of the 23rd American Peptide Symposium
Michal Lebl (Editor)
American Peptide Society, 2013

Analysis of *N*-Amino-Imidazolinone Turn Mimic Geometry

Yésica García-Ramos, Caroline Proulx, and William D. Lubell

Département de Chimie, Université de Montréal, C.P. 6128, Succursale Centre-Ville,
Montréal, Québec, H3C 3J7, Canada

Introduction

Peptide turn conformations are essential elements for receptor recognition. In a program targeted on the CD36 receptor to develop treatments for age related macular degeneration, the leading cause of blindness in North American adults, we have been exploring *N*-amino imidazolin-2-one (Nai) peptide mimics. This novel heterocycle moiety combines elements of aza-amino acids and α-amino-γ-lactams inducing a rigid backbone geometry similar to ideal turn secondary structures [1,2]. Moreover, substituent groups may be added to the 4-position of the imidazolinone to mimic side chain functions. In pursuit of understanding of the influence of the 4-position substituent on the turn conformation, we have synthesized, crystallized and analyzed by X-ray crystallography a set of dipeptide models possessing 4-methyl and 4-benzyl groups [3,4].

Results and Discussion

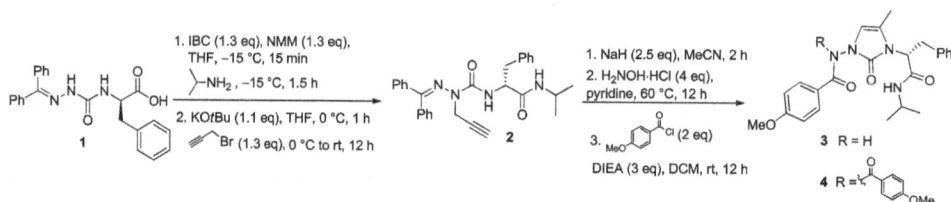

Scheme 1. Synthesis of N-(amido)imidazolin-2-ones 3 and 4.

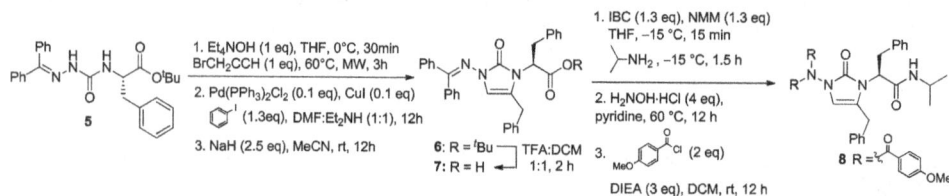

Scheme 2. Synthesis of N,N-bis-(p-methoxybenzamido)-4-benzyl-imidazolin-2-one 8.

Employing mixed anhydride conditions, protected aza-Gly-D-Phe **1** was converted to its *iso*-propylamide counterpart, which was selectively alkylated with propargyl bromide to furnish aza-propargylglycinamide **2** (Scheme 1). On treatment with sodium hydride, alkyne **2** underwent a 5-*exo-dig* cyclization to furnish the Nai residue. Removal of the hydrazone with hydroxyl amine hydrochloride in pyridine, and acylation with 2 or 5 equivalents of 4-methoxybenzoyl chloride gave respectively amides **3** and **4** in 56% and 70% yields over two steps [3,4].

4-Benzyl *N*-amino-imidazolin-2-one **8** was synthesized by alkylation of benzhydrylidene aza-glycinyl-L-phenylalanine *tert*-butyl ester **5**, using tetra-ethylammonium hydroxide and propargyl bromide [5], Sonogashira cross-coupling with iodobenzene, and 5-*exo-dig* cyclization (Scheme 2). The *tert*-butyl ester of **6** was removed with 50% TFA in DCM. The resulting acid **7** was coupled to *iso*-propyl amine under mixed anhydride conditions. Hydrazone removal with hydroxylamine hydrochloride in pyridine, followed by acylation with 2 equivalents of 4-methoxybenzoyl chloride gave 4-benzyl Nai model peptide **8** [4].

Table 1. Values from x-ray crystallography.

Type of turn	$\varphi i+1$	$\psi i+1$	$\varphi i+2$	$\psi i+2$	Phe χ^1
β-II'	60	−120	−80	0	n/a
Inverse γ	n/a	n/a	−70	60	n/a
3A	58.9	−153.3	−69.1	−4.6	54.6
3B	62.1	−166.1	−71.7	65.7	57.8
4	88.3	−177.3	−70.4	31.2	67.7
β-II	−60	120	80	0	n/a
8	−91.4	−174.9	79.2	−63.6	−166.7

Crystals of the Nai peptides **3**, **4** and **8** were examined by X-ray analysis (Table 1). In the crystal of **3**, two conformers were observed, featuring ten- and seven-membered hydrogen bonded type II' β- and inverse γ-turns (**3A** and **3B**). Although the 4-methyl Nai analogue **4** exhibited a ψ^{i+2}-dihedral angel geometry between ideal β-II' and γ-turn values, the 4-benzyl Nai analogue **8** exhibited an inverse γ-turn conformation. Moreover, the 4-position substitution influenced the neighbouring Phe-residue χ^1-dihedral angle, which adopted a *gauche* conformer for 4-methyl Nai analogues **3** and **4** and a *trans* orientation for 4-benzyl Nai analogue **8**. Currently inserting 4-methyl and 4-benzyl Nai residues into peptides with potential to serve as ligands for binding to the CD36 receptor, this crystallographic analysis is designed to facilitate understanding of their structure-activity relationships.

Acknowledgments

We thank the Natural Sciences and Engineering Research Council of Canada, the Canadian Institutes of Health Research (grant #TGC-114046), the Ministère du développement économique de l'innovation et de l'exportation du Québec (#878-2012, Traitement de la dégénerescence maculaire), Amorchem and Boehringer Ingelheim for financial support.

References

1. Proulx, C., Sabatino, D., Hopewell, R., Spiegel, J., Garcia-Ramos, Y., Lubell, W.D. *Future Medicinal Chemistry* **3**, 1139-1164 (2011).
2. Jamieson, A.G., Boutard, N., Beauregard, K., Bodas, M.S., Ong, H., Quiniou, C., Chemtob, S., Lubell, W.D. *J. Am. Chem. Soc.* **131**, 7917-7927 (2009).
3. Proulx, C., Lubell, W.D. *Org. Lett.* **14**, 4552-4555 (2012).
4. Proulx, C., Lubell, W.D. *Biopolymers, Peptide Science, in press* (2013), DOI: 10.1002/bip.22327.
5. Garcia-Ramos, Y., Proulx, C., Lubell, W.D. *Can. J. Chem.* **90**, 985-993 (2012).

Proceedings of the 23rd American Peptide Symposium
Michal Lebl (Editor)
American Peptide Society, 2013

Design and Synthesis of Small Molecule Peptidomimetics as Broad Spectrum Anticancer Drug Candidates

Lajos Gera[1], Daniel C. Chan[2], Paul A. Bunn, Jr.[2], and Robert S. Hodges[1]

[1]*Department of Biochemistry and Molecular Genetics, University of Colorado Denver, Aurora, CO, 80045, U.S.A.;* [2]*Cancer Center, University of Colorado Denver, Aurora, CO, 80045, U.S.A.*

Introduction

More than a decade ago we developed a highly potent bradykinin (BK) antagonist peptide dimer with excellent anticancer activity against small cell lung cancer (SCLC) (B9870; DArg-Arg-Pro-Hyp-Gly-Igl-Ser-DIgl-Oic-Arg, Hyp: *trans*-4-hydroxyproline; Igl: α-(2-indanyl)glycine; Oic: octahydroindole-2-carboxylic acid; GPI: 8.4 (pA$_2$ for BK antagonist activity on isolated guinea pig ileum); SHP-77: 65% (percent inhibition on growth of xenografts in nude mice, compound was injected i.p. at 5 mg/kg/day for 27 days) [1]. The transformation of our B9870 BK anticancer peptide dimer into peptidomimetics led us to the discovery of BKM570, a small molecule with highly potent anticancer activity (SHP-77: 91%), but moderate BK antagonist activity (GPI: 5.6) [1]. This first generation of our small molecules was found to be superior to the widely used but highly toxic cisplatin against SCLC *in vivo* and also showed strong cytotoxic effects in ovarian cancer cells [2]. This simple N-acyl-tyrosine-amide derivative BKM570 (Figure 1), which was chosen for further optimization studies, consists of subunits "A" acyl-group: pentafluorocinnamoyl, "B" amino acid residue O-(2,6-dichlorobenzyl)tyrosine and "C" sterically hindered cyclic amide derivative: 4-amino-2,2,6,6-tetramethylpiperidine.

Results and Discussion

Recently, we designed a second generation of our anti-cancer small molecules keeping the simple "A-B-C" structure. In this new generation [3], the "A" subunit contains an acyl-group derived from the non-steroidal anti-inflammatory drugs (NSAIDs) felbinac (4-biphenylacetic acid) in GH101, (S)-flurbiprofen ((S)-2-fluoro-α-methyl-4-biphenylacetic acid) in GH501, and (S)-ibuprofen ((2S)-2-(4-isobutylphenyl)propanoic acid) in GH503 (Figure 1). The NSAID felbinac, flurbiprofen and ibuprofen were preferred as an "A" subunit for the second generation because the NSAIDs and their amides were shown to exhibit anti-cancer activity [4]. Also masking the free carboxylic group as an amide can shift their enzyme selectivity from COX-1 to towards COX-2, thus lowering side effects [5]. (R)-flurbiprofen and ibuprofen also have anticancer effects [6]. We compared our GH501 flurbiprofen-conjugate to the flurbiprofen *in vitro* against eight lung cancer cells (H1299, H1650, H1975, H2935, HCC95, HCC193, LX-1 and SW1573) and our GH501 was 1200 times more potent than flurbiprofen itself.

A comparison study of BKM570 and the newly designed acyl-amino acid amide compounds GH101, GH501 and GH503 was completed by the US National Cancer Institute (NCI) *in vitro* against 60 human tumor cell lines of nine cancer types (breast, colon, central nervous system (CNS), leukemia, melanoma, non-small cell lung (NSCL), ovarian, prostate and renal cancer) at five doses. The IC$_{50}$ values (the drug concentration that inhibits the growth of cells by 50%) were in the sub-micromolar range (477 nM on average for the 9 cancer types and 60 cell lines). GH501, the second generation NSAID flurbiprofen modified anticancer small molecule, was superior to BKM570, against leukemia, colon, melanoma and NSCL cancer cell lines (Table 1). GH503 was superior to BKM570 with similar potency to GH501.

The newly designed second generation NSAID conjugate GH501 differs from GH101 in a methyl group on the α-carbon and a fluorine atom on the phenyl group in unit "A". GH503 differs from GH101 in a methyl group on the α-carbon and the phenyl group is replaced with a 4-carbon alkyl group. These subtle changes have dramatically increased the activity of GH501 and GH503. Concerning the GH501 flurbiprofen conjugate, it is important to note that its "B" component, containing a biphenyl moiety, has been reported in the literature to provide antitumor activity [7]; moreover, the "C" component, 4-amino-2,2,6,6-tetramethylpiperidine, too, has been stated to possess antitumor activity [8], by virtue of its

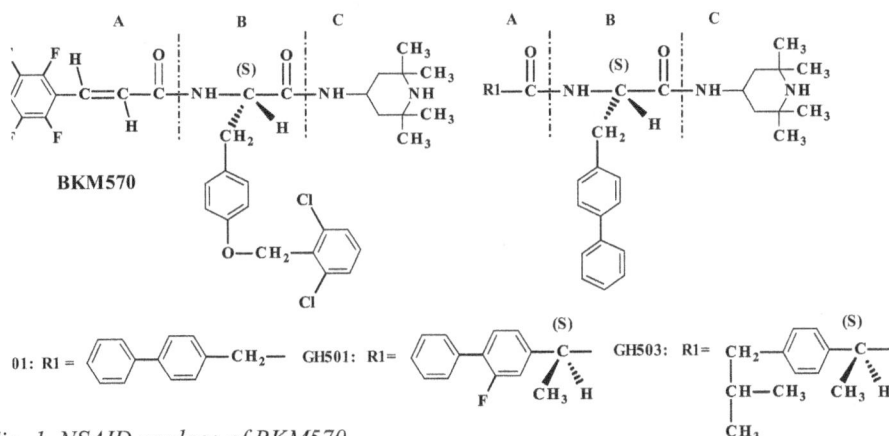

Fig. 1. NSAID analogs of BKM570.

antioxidant properties. These observations may explain our finding that the antitumor activity of the conjugate is more than three orders of magnitude greater than that of flurbiprofen itself.

These small molecules were synthesized in solution using Boc-chemistry and BOP coupling protocol. The compounds were purified by preparative HPLC and characterized by analytical HPLC and LC-MS.

Table 1. Average anticancer activity of 7-9 human tumor cell lines for each cancer type.

Cancer Type	Anticancer Small Molecules (GI_{50}, nanomolar)			
	GH101	GH501	GH503	BKM570
Leukemia	592	241	225	866
Colon	1006	446	362	846
Melanoma	1753	461	453	885
NSCLC	1251	373	369	845

NCI study results confirmed that GH501 and GH503 small molecules (500-700 Dalton range) with a simple "A-B-C" structure are highly potent anticancer compounds against human cancers. They are ready for further studies on their mechanism of action and anticancer drug development.

Acknowledgment

The authors are grateful to NCI for the NCI 60 human tumor cell line screening.

References

1. Whalley, E.T., Figueroa, C.D., Gera, L., Bhoola, K.D. *Expert Opin. Drug. Discov.* **7**, 1129-1148 (2012).
2. Jutras, S., Bachvarova, M., Keita, M., Bascands, J.L., Mes-Masson, A.M., Stewart, J.M., Gera, L., Bachvarov, D. *FEBS J.* **277**, 5146-5160 (2010); Erratum in: *FEBS J.* **278**, 999 (2011), Gera, L. [added].
3. Gera, L., Chan, D.C., Hodges, R.S., Bunn, P.A. Publication No. WO2011106721A1.
4. Marjanovich, M., et al. *Chem. Biol. Drug. Des.* **69**, 222-226 (2007).
5. Mehta, N., et al. *Int. J. Chem. Tech. Res.* **2**, 233-238 (2010).
6. Quann, E.J., et al. *Cancer Res.* **67**, 3524-3262 (2007).
7. Xiao, Q., et al. *J. Med. Chem.* **54**, 525-533 (2011).
8. Wasserman, V., et al. *Langmuir* **23**, 1937-1947 (2007).

Proceedings of the 23rd American Peptide Symposium
Michal Lebl (Editor)
American Peptide Society, 2013

NGF Loop 4 Dimeric Dipeptide Mimetic Active on Animal Models of Parkinson's, Alzheimer's Diseases and Stroke

T. Gudasheva, T.Antipova, P. Povarnina, and S. Seredenin

Zakusov Institute of Pharmacology RAMS, Moscow, 125315, Russia

Introduction

The ability of NGF to promote neuronal survival in pathological conditions makes this protein an attractive candidate for the treatment of neurodegenerative disorders. However, many problems are associated with its clinical use. The development of small molecule NGF mimetics is a promising approach to resolve the problems. To mimic NGF functions pharmacologically, we designed dimeric dipeptide named GK-2 (bis(N-succinyl-L-glutamyl-L-lysine)hexametylendiamide) on the basis of the most exposed to solvent beta-turn sequence Asp^{93}-Glu^{94}-Lys^{95} of loop4. The residue Asp^{93} was substituted by bioisostere, a succinic acid residue. Because NGF interacts with the TrkA receptor in the dimeric form, the agonistic activity was achieved by dimerizing the N-acyldipeptide by hexametylenediamine [1]. GK-2, like NGF, stimulated phosphorylation of TrkA, but not TrkB receptor and prevented glutamate- and H_2O_2-induced neuronal cell death at 10^{-9}M [2]. In the present study we investigated GK-2 ability to trigger phosphorylation of Akt and Erk proteins, which plays a key role in two major biochemical pathways mediated by TrkA receptor. In order to test the ability of GK-2 to mimic *in vivo* therapeutical properties of NGF we investigated the activity of the dipeptide on animal models of Parkinson's disease, Alzheimer's disease (AD) and brain ischemia.

Results and Discussion

Biochemical assays were done by Western blot analyses of the phosphorylation of Akt and Erk in different time points after exposing HT-22 cells to GK-2 (10^{-6}M) or NGF (100ng/ml). Phosphorylation of Akt was observed after 15, 30, 60 and 180 min after stimulations of the cells by GK-2 or 15, 60 and 180 min after stimulations of the cells by NGF (Figure 1). NGF induced phosphorylation of Erk after 60 and 180 min. At the same time there was no activation of Erk proteins by GK-2 at any time point (Figure 1). It is known that Akt activation is important for cell survival, whereas Erk activation is associated not only with cell survival but also with cell differentiation [3].

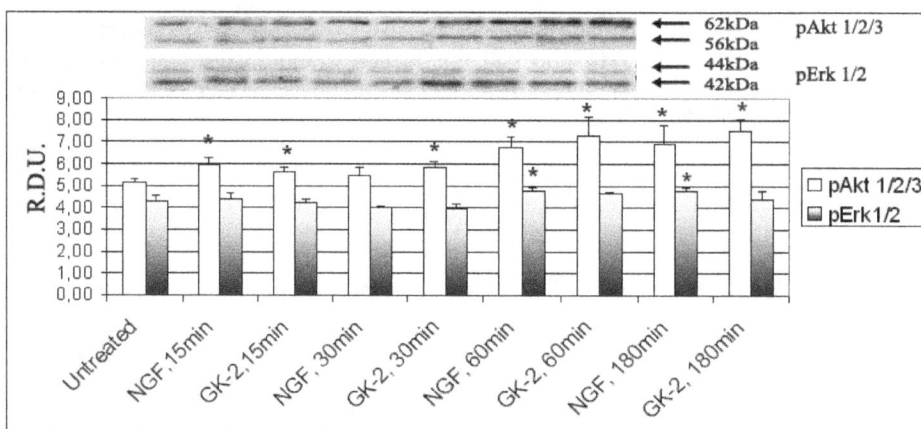

*Fig. 1. Kinases phosphorylation induced by GK-2. HT-22 cells were treated with GK-2 (10^{-6}M) or NGF(100 ng/ml). Changes in pAkt1/2/3 and pERK1/2 level were evaluated by Western blot analysis. Data presented as mean(SEM). * - p<0,05(Mann-Whitney U test).*

Table 1. Anti-cataleptogenic action of GK-2.

Experimental group	Dose of GK-2, i.p. (mg/kg)	Duration of catalepsy (sec), median (lower quartile-upper quartile)	Duration of catalepsy (% from haloperidol alone group)
Haloperidol alone		45 (21-120)	100
GK-2+haloperidol	0.01	6*(4-15)	14 *
	0.1	27*(11-120)	60 *
	0.5	3*(2-6)	7*
	1.0	4*(1-6)	9 *
	5.0	8*(5-120)	18 *

- p<0,05 compared with haloperidol alone group (Fisher exact test).

Indeed, dipeptide GK-2 does not induce differentiation of PC12 cells [2]. We can expect that GK-2 does not possess such an adverse effect of NGF as hyperalgesia, because this effect is associated with activation of MAP/Erk pathway [4].

We investigated anti-Parkinson properties of GK-2 on models of haloperidol-induced catalepsy, MPTP-induced parkinsonism and hemiparkinsonism, induced by 6-hydroxidopamine unilateral lesions of the nigrostriatal dopamine system. Peptide GK-2 decreased haloperidol-induced catalepsy in rats by 80-90%, being injected either i.p. (0.01-5 mg/kg) or per os (10 mg/kg) 24 h prior to administration of haloperidol (Table 1). In MPTP models acute injection of GK-2 (1 mg/kg, i.p.) 24 h prior to administration of MPTP (35 mg/kg, i.p.) reduced the manifestation of oligokinesia and rigidity in mice ($p<0.05$). Subchronical i.p. administration of GK-2 after 6-OHDA lesions totally abolished apomorphine-induced rotations in rats.

In models of AD, subchronical therapeutic i.p. administration of GK-2 significantly ($p<0.05$) attenuated habituation deficits in rats with fimbria-fornix lesions and spatial memory deficits in Morris water maze in rats subjected i.c.v. injection of streptozotocine.

GK-2 (subchronical therapeutic i.p. administration) showed high neuroprotective activity in rat models of brain ischemia. This dipeptide reduced cerebral infarct volume by 60% ($p<0.05$) and fully restored memory deficits in passive avoidance test in the model of photoinduced stroke of cerebral cortex. The same way administration GK-2 to rats with experimental focal ischemia provoked by unilateral occlusion of the middle cerebral artery significantly ($p<0.05$) improved neurological deficits and decreased the infarction area by 16%. GK-2 also fully prevented the deaths in rats subjected to incomplete global cerebral ischemia induced by permanent common carotid artery occlusion.

Overall, dimeric dipeptide mimetic of nerve growth factor GK-2 is neuroprotective compound that offers hope as potential treatment of Parkinson's disease, Alzheimer's disease and brain ischemia. Since GK-2 selectively activates Akt pathway, but not MAP/Erk pathway, we can expect that this compound has no side effects related to NGF treatment.

References

1. Seredenin, S.B., Gudasheva, T.A. US Patent Application US 2011/0312895 A1 (2011).
2. Gudasheva, T.A., Antipova, T.A., Seredenin, S.B. *Dokl. Biochem. Biophys.* **434**, 262-265 (2010).
3. Reichardt, L.F. *Philos. Trans. R. Soc. Lond. B Biol. Sci.* **361**, 1545-1564 (2006).
4. Obata, K., Noguchi, K. *Life Sci.* **74**, 2643-5263 (2004).

Proceedings of the 23rd American Peptide Symposium
Michal Lebl (Editor)
American Peptide Society, 2013

Dipeptide Analogue of Endogenous Cholecystokinin-4 with Anxiolytic and Analgesic Effects: Low Affinity and Multitarget Action

Larisa G. Kolik, Tatiyana A. Gudasheva, and Sergey B. Seredenin

State Zakusov Institute of Pharmacology RAMS, Moscow, 125315, Russia

Introduction

A series of dipeptides with CCK-positive and CCK-negative activity was designed on the basis of the structure of endogenous CCK_2-receptor agonist cholecystokinin-4 (CCK-4) [1]. It has been previously shown that the novel compound GB-115, ($Ph(CH_2)_5CO$-Gly-D-Trp-NH_2,), synthesized as dipeptide analogue of CCK-4, in behavioral models decreased the anxiety in rodents with "freezing" in an 'open field' test and the withdrawal-induced anxiety in long-term alcohol or diazepam experienced animals and attenuated reaction to thermal and chemical nociceptive stimulations [2]. The aim of the present work was to study the effects of GB-115 in various *in vitro* receptor binding and cellular functional assays and *in vivo* models to reveal the receptor targets of the novel compound.

Results and Discussion

GB-115 interaction with 109 different human and rat recombinant receptors selected on the basis of theoretical analysis of structure and pharmacological effects of GB-115 was studied in CHO and HEK-293 cells. In binding assays the results are expressed as a percent of control specific binding ((measured specific binding/control specific binding) x100) obtained in the presence of GB-115. This analysis was performed using software developed at Cerep (Hill software) and validated by comparison with data generated by SigmaPlot®4.0 for Windows®. In cellular functional assays the results are expressed as a percent of control specific agonist response ((measured specific response/control specific agonist response)x100) obtained in the presence of GB-115.

For the first time it was shown that GB-115 demonstrated low affinity to CCK_1 receptor (IC_{50}=8,1 · 10^{-5} M, K_i=6,1 · 10^{-5} M), kappa-opioid (KOP) receptors (IC_{50}=1,6 · 10^{-5} M, K_i=1,0 · 10^{-5} M) and bombesin receptor subtype-3 (BRS_3) (IC_{50}=1,6 · 10^{-5} M, K_i=9,7 · 10^{-6} M). In cellular functional assays GB-115 agonistic properties were revealed for KOP receptors (EC_{50} = 1.6·10^{-4}), GB-115 blocked the effects of BB_3 receptor agonist (IC_{50} = 3.7·10^{-5}M) and CCK_1 receptor agonist (IC_{50} = 7.3·10^{-5}M) (Figure 1). Surprisingly, there was no affinity to CCK_2 receptors demonstrated as expected.

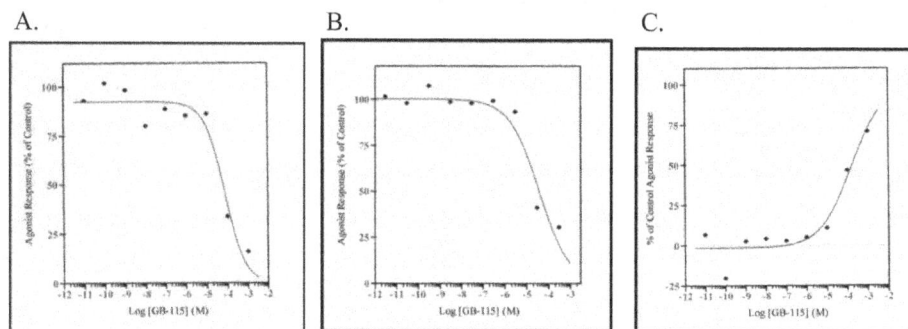

Fig. 1. GB-115 antagonist effects at the human CCK_1 receptor(A), at the human BRS_3 receptor(B) and agonist effect at the rat KOP receptor(C).

In behavioral tests GB-115 (0.0025-0.5 mg/kg, i.p.) dose-dependently reduced anxiety in BALB/c mice. Anxiolytic effect was antagonized by CCK-4 (Figure 2), but not with opiate antagonist naloxone and BRS_3 receptor agonist Bn(6-14). Moreover, GB-115 (0.1-20.0 mg/kg, p.o.) dose-dependently attenuated the nociceptive response in "tail flick" and "writing" tests in mice and this effect was prevented by pretreatment with nor-binaltorphimine (nor-BNI), a KOP receptor antagonist (Figure 3) and naltrexone iodide, peripheral non-selective opioid receptor antagonist.

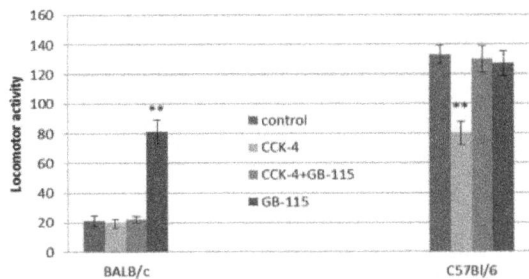

Fig. 2. Pretreatment with agonist of central CCK receptors CCK-4 (0,004 mg/kg, i.p.) prevents the development of anxiolysis induced by GB-115(0,05 mg/kg, i.p.) in "high anxious" BALB/c mice in the "open field" test. ** - $p<0.01$ as compared with control, ANOVA, post-hoc Dunnett's test; $n=10$-12 per group. Mean±SEM.

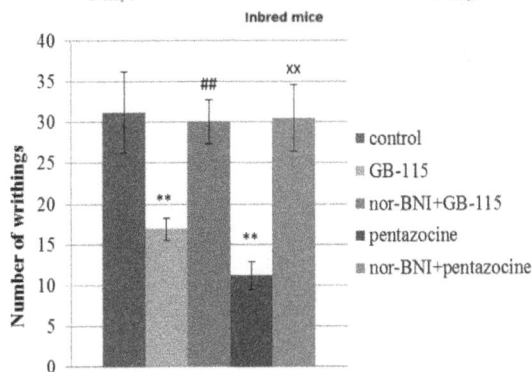

Fig. 3. Blockade of kappa-opioid receptors by nor-BNI (5.0 mg/kg, i.p.) prevented the antinociceptive effects of GB-115 (10.0 mg/kg, p.o.) in "writing test" in mice. **-$p<0.01$ as compared with control, ##-$p<0.01$ as compared with GB-115, xx-$p<0.01$ as compared with pentazocine, kappa-receptor agonist, (2.0 mg/kg, s.c.,positive control), ANOVA, post-hoc Dunnett's test; $n=9$-11 per group. Mean±SEM.

Three primary receptor targets (CCK_1, KOP and BRS_3 receptors) were determined for GB-115 which is in accordance with data obtained in behavioral studies demonstrated three dome-shaped curve "dose-effect". GB-115 antagonized novelty induced anxiety by blocking centrally located CCK_1-receptors at the low doses and reduced nociceptive response through activating peripheral KOP receptors at the high doses. This study indicated the novel target such as BRS_3 receptors, which may be indirectly involved in anxiolytic effects of GB-115 as BRS_3 antagonist by modulating orexinergic system hyperactivity [3] and preventing the decrease of GABA-ergic tone under emotional stress. Further, GB-115 may be better adapted than selective agents to the management of generalized anxiety and/or panic disorder with concomitant pain syndrome. On the other hand, GB-115 as a bi-functional agent may be of particular interest for visceral pain elimination with minimization of opioid side effects and reducing effective doses of opioid analgesics by antagonizing CCK-receptors.

Conclusion

It is important to continue the creative exploration of innovative strategies within a multitarget framework. GB-115 with complex mechanisms of action is well suited for the advancement of our understanding of the multiple roles that CCK plays in the brain and for the future development of CCK-based therapies.

References

1. Philippova, E., Gudasheva, T., Briling, V., et al. Abstract book from 27th Eur. Peptide Symp. Italy, Sorrento, 508-509 (2002).
2. Kolik, L.G., Gudasheva, T.A., Seredenin, S.B. Eur. Neuropsychopharm 15 (Suppl.2), S146 (2005).
3. Furunati, N., Hondo, M., Tsujino, N., Sakurai, T. J. Mol. Neurosci. 42, 106-111 (2010).

Proceedings of the 23rd American Peptide Symposium
Michal Lebl (Editor)
American Peptide Society, 2013

Synthetic Agonists for the CXCR4 Receptor: SAR, Signaling Pathways and Peptidomimetic Transition

Christine Mona[1], Marilou Lefrançois[1], Philip E. Boulais[1], Élie Besserer-Offroy[1], Richard Leduc[1], Nikolaus Heveker[2], Éric Marsault[1], and Emanuel Escher[1]

[1]Département de Pharmacologie, Université de Sherbrooke, Sherbrooke, J1H 5N4, Canada; [2]Centre de Recherche, Hôpital Sainte-Justine, H3T 1C5, Montréal, Canada

Introduction

CXCR4 a G-protein coupled receptor is an important pharmaceutical target in a variety of diseases including HIV and many forms of cancer. CXCR4 is therefore an important pharmaceutical target but besides the cognate agonist SDF-1 only synthetic antagonists/inverse agonists are targeting CXCR4 with pertinent affinities. CXCR4 has important household functions; those antagonists lead to the emergence of significant adverse drug effects. We recently describe nanomolar CXCR4 agonists [1] where the SDF-1 N-terminus was grafting to the inverse agonist T140 [2] (Figure 1). In order to translate these peptidic compounds into peptidomimetic agonists, permitting eventual pharmaceutical applications, the present contribution describes the SAR of the grafted SDF-N-terminus in order to pinpoint the agonist-antagonist transition and to determine the peptidomimetic transition strategies.

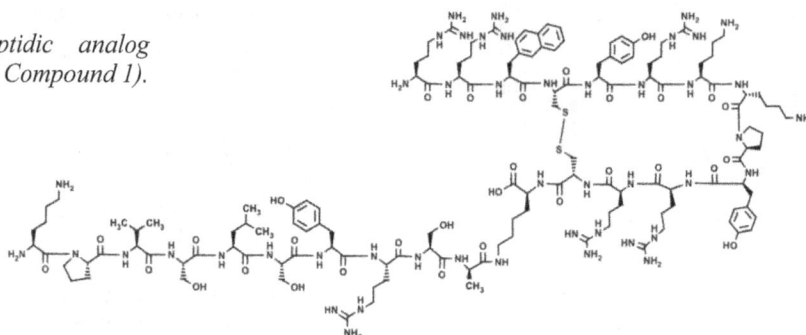

Fig. 1. Peptidic analog structure (i.e. Compound 1).

Results and Discussion

Peptides were synthesized using the Fmoc-based solid-phase strategy; as a first step the synthesis of the (Lys14 [ε-DDE]) T140 was completed and SDF-1 N-terminus grafted in position 14. Peptides were cleaved, purified, analyzed by HPLC indicating a purity greater than 97%, and molecular weights were confirmed by LC/MS. Peptide affinities were determined on CXCR4 binding assays (data not shown). Chemotactic activities were evaluated on Transwell migration assays on pre-B lymphocytes (REH cells). Signaling pathway Gi/o was evaluated with EPAC BRET assay on cAMP inhibition (Cotransfected HEK293 with HA-CXCR4 and EPAC Biosensor).

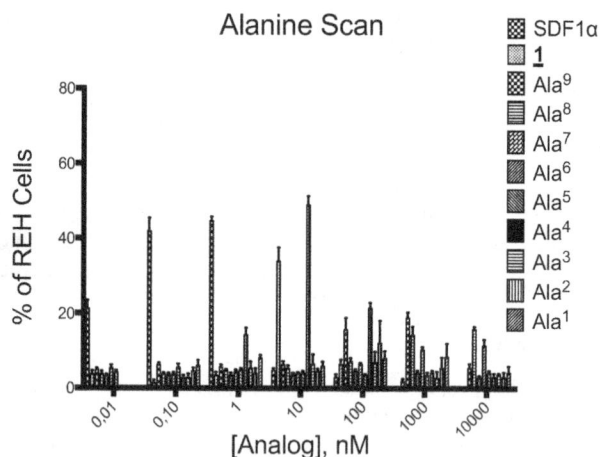

Fig. 2. Migration profiles of the Alanine Scan modified peptides.

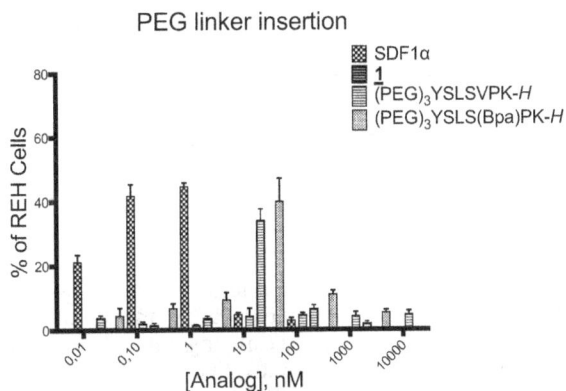

PEG linker insertion

Fig. 3. *Migration curves of PEG linker modified analogs.*

The relative importance of each *N*-terminal SDF-1 residues was evaluated by an Alanine scan and chemotactic activity screened by Transwell assays. Position 1, 2, 3 and 7 seems to be critical for the conservation of the chemotactic activity (Figure 2): all these alanine containing compounds were in fact competitive antagonists (data not shown). Position 1 and 2 are reputed to be essential for biological activity [3] therefore position 3 was studied. Screen in position 3 shown that hydrophobic amino acids with the ability to make π-stacking in position 3 like Trp, 2-Nal and Bpa (*p*-Benzoyl-phenylalanine) gave potent chemotactic chimeras with a maximum of migration at 10nM and efficacy comparable to SDF-1 (Max. of migration at 1nM: 53%). Tert-Leucine replacement gave partial chemotactic agonist.

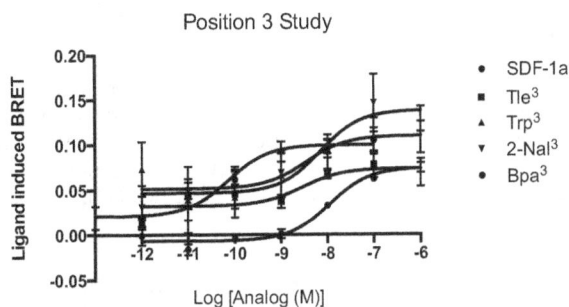

Fig. 4. *AMPc inhibition concentration-response curves of position 3 modified peptides.*

In order to progressively transform into peptidomimetic sequence in position 8 to 10 of SDF-1, *N*-terminus was replaced by a PEG linker. PEG insertion combined with hydrophobic modification in position 3 (i.e. Bpa) gave potent chemotactic agonist (Figure 3).

CXCR4 signals through Gi/o - therefore we evaluated the profile of our ligands using EPAC BRET assays [4]. All full chemotactic agonists were full agonists on the Gi/o pathway whereas partial agonists behaved like partial agonists (Figure 4).

Through this study, we have highlighted several important points for the conservation of the agonistic nature of our chimeras. The Alanine scan emphasizes the importance of residues like Lys^1, Pro^2, Val^3 and Tyr^7 and thus restricts the transition to peptidomimetic structures.

The position 3 of our chimeras seems to tolerate hydrophobic structural changes with the ability to make π-stacking. Hydrophobic amino acid in position 3 seems to positively modulate the agonistic behavior. Excessive flexibility appears to place the side chain into a less favorable conformation than the one required to induce the activation of the CXCR4 receptor. Nevertheless, some structural modifications (i.e. PEG) in position 8 to 10 combined with highly hydrophobic modification (i.e. BenzoylPhenylAlanine) in position 3 give potent CXCR4 agonist. Chemotaxis seems to correlate with the Gi/o pathway.

Acknowledgments

We would like to thank the *Conseil Régional de la Martinique* for studentship to Christine Mona. Christine Mona as a recipient of a APS Travel Award. Supported by Canadian Institutes of Health Research.

References

1. Lefrançois, M., et al. *ACS Med Chem Lett.* **2(8)**, 597-602 (2011).
2. Tamamura, H., et al. *FEBS Lett.* **569**(1-3), 99-104 (2004).
3. Chevigné, A., et al. *Biochemical Pharmacology* 82, 1438-1456 (2011).
4. Salahpour, A., et al. *Frontiers in Endocrinology* 3, (2012).

Proceedings of the 23rd American Peptide Symposium
Michal Lebl (Editor)
American Peptide Society, 2013

Chemical Synthesis and Analysis of Short Peptides Containing Modified Arginine Residues

I. Małuch, M. Lewandowska, T. Łepek, B. Lammek, and A. Prahl

Institute of Organic Synthesis, Department of Organic Chemistry, Faculty of Chemistry,
University of Gdańsk, Gdańsk, 80-952, Poland

Introduction

When designing peptide inhibitors of enzymes, including serine proteases, it is necessary to consider their resistance to enzymatic degradation. Very popular nowadays is modification of the peptide chain consisting of substitution of P1 arginine residue (binding with S1 pocket of the enzyme [1]) with compounds that are its mimetics. Owing to its specific structure, they have the potential to be more stable and less susceptible to enzymatic degradation. There is a wide range of known arginine mimetics, that incorporated into the peptide chain create strong inhibitors of serine proteases [2-5]. One of these compounds is 4-amidinobenzylamine (Amba). Many research groups were focused on synthesizing inhibitors of serine proteases (*e.g.* thrombin and urokinase) containing Amba in their structures [6-8].

Here we present results of the study verifying whether during coupling of 4-amidinobenzylamine to the *C*-terminus of the peptide chain, it is necessary to protect its amidino group, which hypothetically can be competitive for the free amino group. A set of analogues was synthesized using two various forms of Amba (Figure 1).

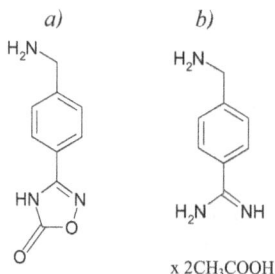

Fig. 1. Structures of
a) protected; b) unprotected
4-amidinobenzylamine.

Results and Discussion

Peptides with the following structure: Ac-Xaa-Xaa-Xaa-**His**-Xaa-Xaa-Xaa-**Amba** were prepared on 2-chloro(2'-chloro)trityl chloride resin by standard Fmoc solid phase peptide synthesis. The fully protected peptides were cleaved from the resin in mild acidic conditions. Coupling of an appropriate form of 4-amidinobenzylamine was carried out in the solution (overnight reaction, 0°C). Deprotected Ac-[peptide]-Amba was obtained by treatment with TFA. If protected Amba was used in the synthesis, hydrogenation was required to remove the oxycarbonyl moiety. Finally peptides were purified by reversed-phase HPLC and identified by LCMS ESI-IT-TOF.

Based on preliminary studies, we combined the two heptapeptides synthesized by different protocols and checked their HPLC profile and mass spectrum (Figure 2).

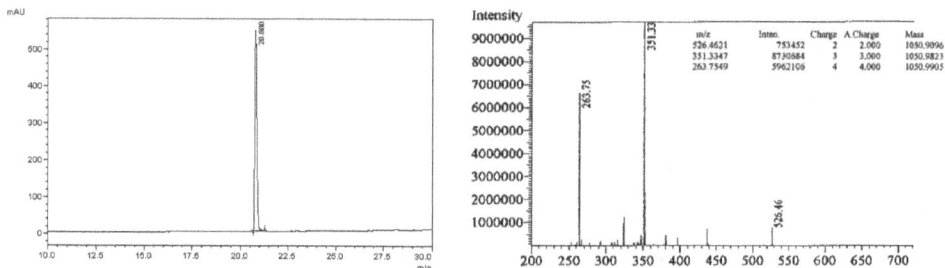

Fig. 2. HPLC profile (on the left) and mass spectrum with deconvolution (on the right) registered for combined samples of two heptapeptides. The elution gradient of solvents from 1% to 80% of B in A was applied for 30 min at the flow rate of 1ml/min. A-0.1% TFA/water; B-0.1% TFA in 80% acetonitrile/water.

Table 1. Inhibitory activities of synthesized compounds against hPACE4 and h-furin.

Analog (synthesized using appropriate form of Amba)	$K_i [\mu M]^a$	
	hPACE4	h-furin
Amba protected by oxycarbonyl group	0.21±0.02	3.53±0.22
Amba x 2CH₃COOH	0.17±0.01	3.55±0.15

^{a}*Inhibition constants are the average of four values obtained in two independent experiments.*

Moreover, both of synthesized analogues were separately tested for their inhibitory potency toward chosen proprotein convertases (hPACE4 and h-furin). Kinetic assays were prepared using spectrofluorimeter *GEMINI XS* provided with the program SoftMaxPro 3.3.1 (Molecular Devices). Program GraphPad Prism was used to calculate K_i values. All results are presented in Table 1.

Both of synthesized compounds demonstrate highly similar characteristics: retention times (20.8 min) and registered *m/z* values. Furthermore, determined inhibition constants closely reflect their similarity as well (0.21µM and 0.17µM for hPACE4, and 3.53µM and 3.55µM for h-furin).

In summary we can conclude, that both of analyzed methods of Amba coupling probably led us to the same final product. However, further studies need to be performed to confirm this thesis. Moreover, we designed and synthesized an additional three short peptides of the same sequence using three different forms of 4-amidinobenzylamine: protected Amba, Amba x 2CH₃COOH and Amba x 2HCl. All of them have the same HPLC profiles and mass spectrums. In the future, they will be precisely identified by analyzing their structures (conformational studies) and properties (capillary electrophoresis). Due to slight differences in the procedures of synthesis using these three forms of Amba, we will also verify their alkalinity.

Acknowledgments

The authors wish to thank Robert Day from the Faculty of Medicine, University of Sherbrooke (Québec, Canada) for the access to perform kinetic assays. We also thank Witold Neugebauer and Anna Kwiatkowska. This work was supported by the National Science Center in Poland with the grant no. 2012/05/N/ST5/01080.

References

1. Henrich, S., Cameron, A., Bourenkov, G.P., Kiefersauer, R., Huber, R., Lindberg, I., Bode, W., Than, M.E. *Nat. Struct. Biol.* **10**, 520-526 (2003).
2. Balakrishnan, S., Scheuermann, M.J., Zondlo, N.J. *ChemBioChem: J. Chem. Biol.* **13**, 259-270 (2012).
3. Hammamy, M.Z., Haase, C., Hammami, M., Steinmetzer, T. *ChemMedChem: Chemistry Enabling Drug Discovery* **8**, 231-241 (2013).
4. Dosa, S., Stimberg, M., Lülsdorff, V., Häußler, D., Maurer, E., Gütschow, M. *Bioorg. Med. Chem.* **20**, 6489-6505 (2012).
5. Ilas, J., Tomasić, T., Kikelj, D. *J. Med. Chem.* **51**, 2863-2867 (2008).
6. Gustafsson, D., Antonsson, T., Bylund, R., Eriksson, U., Gyzander, E., Nilsson, I., Elg, M., Mattsson, C., Deinum, J., Pehrsson, S., Karlsson, O., Nilsson, A., Sörensen, H. *Thromb. Haemost.* **79**, 110-118 (1998).
7. Künzel, S., Scheinitz, A., Reißmann, S., Stürzebecher, J., Steinmetzer, T. *Bioorg. Med. Chem. Lett.* **12**, 645-648 (2002).
8. Becker, G.L., Hardes, K., Steinmetzer, T. *Bioorg. Med. Chem. Lett.* **21**, 4695-4697 (2011).

Proceedings of the 23rd American Peptide Symposium
Michal Lebl (Editor)
American Peptide Society, 2013

The E3 Ubiquitin Ligase MARCH5: Integral Membrane Domain Links Mitochondria to Neurodegeneration in Alzheimer's Disease

Leah S. Cohen[1], Nicole LaMassa[2], and Alejandra Alonso[2]

[1]Department of Chemistry, The College of Staten Island (CSI), City University of New York (CUNY), Staten Island, NY, 10314, U.S.A.; [2]Center for Developmental Neuroscience, CSI, CUNY, Staten Island, NY, 10314, U.S.A.

Introduction

Mitochondrial dynamics is the process during which mitochondria undergo fission or fusion in order to regulate their health and number. When mitochondria do not function properly this affects many processes such as the generation of ATP and Ca^{2+} regulation [1] and proper localization of mitochondria, especially in neurons [2]. The number of neurodegenerative diseases, including Alzheimer's Disease (AD), that have been linked to changes in mitochondrial dynamics have been increasing [1,3-8]. The proteins that are known to be involved in both fusion and fission of the mitochondria are nuclear encoded and then targeted to the organelle. The proteins involved in both fusion [9-11] (Mfn1, Mfn2 and OPA1) and fission [12,13] (Drp1 and hFis1) appear to be regulated by the mitochondrial membrane-bound E3 ubiquitin ligase, MARCH5. It was found to interact with and ubiquitinate both Mfn1 and Mfn2 [14-16] and both Drp1 and hFis1 have been shown to coimmunoprecipitate with it [16]. Furthermore, the functions of MARCH5 and Drp1 appear to be codependent as stable fission complexes are dependent on the presence of both proteins [16]. Mutations of the domain responsible for the ubiquitination activity cause changes in the appearance of the mitochondria in mammalian cell lines [16,17]. The ubiquitination helps to maintain a functional balance of the mitochondrial dynamics that results in proper regulation of the mitochondria [14-16,18]. The reliance of both fission and fusion on proper function of MARCH5 indicates that it may be a central processing point for proteins involved in mitochondrial dynamics.

Results and Discussion

Information outside of the RING domain may be important for MARCH5 function. The cytoplasmic loop of MARCH5 appears to be important for the proper localization and/or insertion into the mitochondrial membrane. Removal of this loop in deficient mitochondrial localization as it was found only in the cytoplasm of the cell [19]. We began to analyze the non-RING portion of MARCH5 by generating two mutations - one in the first TM domain (TM1), MARCH5(C97A), and one in the cytoplasmic loop, MARCH5(C188A). We also generated a mutation in the RING domain that had been previously characterized MARCH5(H43W) [16]. Vectors encoding the mutated MARCH5 genes were transfected into CHO cells and analyzed by confocal microscopy (Figure 1). The MARCH5(H43W) mutant resulted in similar clustering as previously described as does the MARCH5(C97A) mutant protein. Interestingly, the protein MARCH5(C188A) appears to be found in the cytoplasm and the nucleus of the cell. This preliminary data indicates that a single residue, not necessarily the entire loop, can play a role in localization.

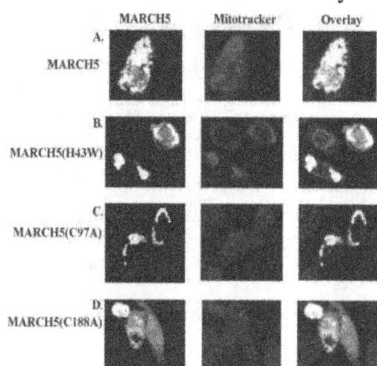

Fig. 1. CHO cells were transfected with MARCH5-EYFP, WT and mutant, and were analyzed by confocal microscopy. In the overlay, the nucleus is stained with DAPI.

To investigate the role of MARCH5 in AD we cotransfected the MARCH5 constructs and Tau or pseudophosphorylated Tau, which mimics PH-Tau, Tau(S199E/T212E/T231E/S262E/R406W) (the patho-

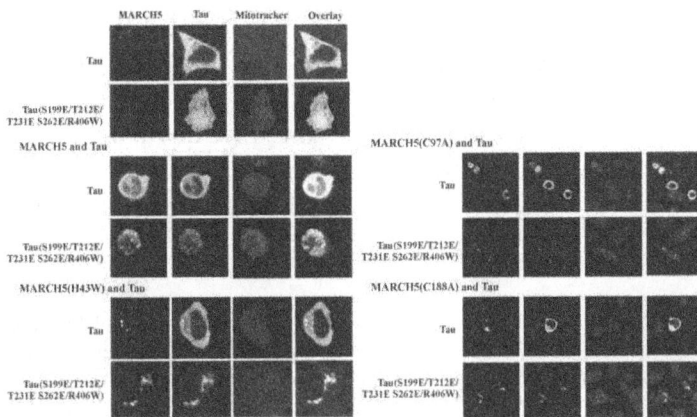

Fig. 2. MARCH5-EYFP and GFP-Tau were co-transfected into CHO cells. Mitotracker is a mitochondrial marker. The overlay contains DAPI to stain the nucleus. The proteins appear to colocalize in the cell and result in changes of localization when expressed by themselves.

logical human form of Tau) (Figure 2). The MARCH5 and Tau proteins appear to colocalize in the cells. Co-expression of mutants of each protein altered the subcellular localization of the other proteins when expressed alone. For example, mutated MARCH5 co-expressed with PH-Tau does not always colocalize with the mitochondria. Conversely, PH-Tau, found to translocate to the nucleus [20], when co-expressed with mutant MARCH5 remained in the cytoplasm. These results suggest an interaction between these two proteins that may be the cause and/or result of mitochondrial dysfunction in AD patients. Based on these data we have looked at the localization of MARCH5 in the CA3 region of the hippocampus in PH-Tau transgenic mice and found that there is an increase in the number of MARCH5 immunostained particles that are smaller is size than control mice suggesting that the expression of PH-Tau potentially disrupts endogenous MARCH5 localization to the mitochondria or disrupts mitochondrial localization and size in the cells. Furthermore, we are focusing on trying to express the protein for use in structural studies by NMR to better determine the relationship between the structure and function of MARCH5.

Acknowledgments

This work was supported by NIH Grant R15AG034524-01; Alzheimer's Association (Chicago, IL) Grant IIRG-09-133206, Brooklyn Home for the Aged Man grant support for developing Alzheimer Research at CSI and Leah S. Cohen as a recipient of an APS Travel Award

References

1. Su, B., Wang, X., et al. *Biochim. Biophys. Acta* **1802**, 135-142 (2010).
2. Verstreken, P., Ly, C.V., et al. *Neuron* **47**, 365-378 (2005).
3. Chen, H., Chan, D.C. *Curr. Opin. Cell Biol.* **18**, 453-459 (2006).
4. Wang, X., Su, B., et al. *J. Neurosci.* **29**, 9090-9103 (2009).
5. Winklhofer, K.F., Haass, C. *Biochim. Biophys. Acta* **1802**, 29-44 (2010).
6. Morais, V.A., De Strooper, B. *J. Alzheimers Dis.* **20 Suppl 2**, S255-263 (2010).
7. Shi, P., Wei, Y., Zhang, J., et al. *J. Alzheimers Dis.* **20 Suppl 2**, S311-324 (2010).
8. Santos, R.X., Correia, S.C., et al. *J. Alzheimers Dis.* **20 Suppl 2**, S401-412 (2010).
9. Santel, A., Fuller, M.T. *J. Cell Sci.* **114**, 867-874 (2001).
10. Ishihara, N., Eura, Y., Mihara, K. *J. Cell Sci.* **117**, 6535-6546 (2004).
11. Cipolat, S., Martins de Brito, O., et al. *Proc. Nat. Acad. Sci. USA* **101**, 15927-15932 (2004).
12. Smirnova, E., Griparic, L., et al. *Mol. Biol. Cell* **12**, 2245-2256 (2001).
13. Yoon, Y., Krueger, E.W., et al. *Mol. Cell. Biol.* **23**, 5409-5420 (2003).
14. Yonashiro, R., Ishido, S., et al. *EMBO J.* **25**, 3618-3626 (2006).
15. Nakamura, N., Kimura, Y., et al. *EMBO Rep.* **7**, 1019-1022 (2006).
16. Karbowski, M., Neutzner, A.; Youle, R.J. *J. Cell Biol.* **178**, 71-84 (2007).
17. Yonashiro, R., Sugiura, A., et al. *Mol. Biol. Cell* **20**, 4524-4530 (2009).
18. Park, Y.Y., Lee, S., et al. *J. Cell Sci.* **123**, 619-626 (2010).
19. Yonashiro, R., Kimijima, Y., et al. *Proc. Nat. Acad. Sci. USA* **109**, 2382-2387 (2012).
20. Alonso, A.D., Di Clerico, J., et al. *J. Biol. Chem.* **285**, 30851-30860 (2010).

Proceedings of the 23rd American Peptide Symposium
Michal Lebl (Editor)
American Peptide Society, 2013

Recognition of Tumor-Associated MUC1 Glycoprotein by Galectin-3

Svetlana Yegorova[1], Maria C. Rodriguez[1,2], Anais E. Chavaroche[1], D. Minond[1], and Mare Cudic[1]

[1]Torrey Pines Institute for Molecular Studies, Port St. Lucie, FL, U.S.A.;
[2]Florida Atlantic University, Boca Raton, FL, U.S.A.

Introduction

The functional significance of glycosylation changes during the onset of cancer [1]. MUC1, a high-molecular weight glycoprotein, secreted and expressed at the cellular surface of various epithelial tissues, is the major carrier of altered glycosylation in carcinomas [2]. Given the evidence that the human β-galactoside-specific lectin, galectin-3, can interact with mucins and also weakly with free Thomsen-Friedenreich (TF) antigen, we here initiate the study of the interaction of MUC1 (glyco)peptides with this endogenous lectin. TF antigen has been shown to be actively involved in tumor metastasis, promoting several key cell-cell interactions *via* association with galectin-3 [3,4]. In turn, galectin-3 promotes cancer cell invasion and metastasis.

Results and Discussion

The assembly of glycosylated MUC1 peptides was performed using standard Fmoc automated solid-phase synthetic chemistry, and the building block approach was used for the incorporation of the Fmoc-protected *O*-glycosylated Thr (Figure 1). We have evaluated two synthetic approaches towards the glycoamino acid building block that relies on the synthesis of a suitable protected 2-azido disaccharide glycosyl donor. The azido group was found to be a very efficient non-participating substituent at C-2, resulting in predominantly α-glycosidic bond formation, and can be easily converted into the desired acetamido group, reduced and subsequently acetylated, or reductive acetylation can be performed in a one-step procedure with thioacetic acid, before or after on-resin peptide assembly [5]. The first synthetic strategy was based on the protocol previously described for the synthesis of Fmoc-Ser/Thr-OPfp building blocks containing α-linked linear tri- and heptasaccharides [5]. The second approach explored the one-pot azidochlorination procedure recently described by Plattner et al. [6] since the synthesis of 2-azido glycosyl donors, in this case, 2-azido-2-deoxy-β-D-galacto-pyranoside, *via* the azidonitration procedure introduced by Lemieux [7] is rather laborious. The *O*-glycosylated Thr building block was obtained after purification by silica gel chromatography in 62% and 40% yields, respectively. The purity of the building block Fmoc-Thr-OPfp containing α-linked TF and glycosylated MUC1 peptides was confirmed by HPLC chromatography and MALDI-MS.

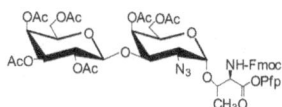

T* = Thr *O*-linked TF antigen

MUC1-Thr[4]:
HGVT*SAPDTRPAPGSTAPP

MUC1-Thr[9]:
HGVTSAPDT*RPAPGSTAPP

MUC1-Thr[16]:
HGVTSAPDTRPAPGST*APP

Fig. 1. Fmoc protected O-glycosylated Thr and glycosylated MUC1 fragments.

Binding affinities of glycosylated MUC1 fragments that carry TF antigen for galectin-3 were assessed by an AlphaScreen-based assay, and isothermal titration calorimetry (ITC). We have recently demonstrated that AlphaScreen assay in a competitive binding configuration can be efficiently utilized for discovery of new inhibitors of galectin-3 and offers several advantages over the existing methods for monitoring low affinity glycan-lectin interactions [8]. His-tagged galectin-3 was bound to nickel-chelate Acceptor beads (capacity: 600 nM) while streptavidin-coated Donor beads (capacity: 30 nM) were used to attach the galectin-3 binding partner. In the competitive binding mode, increasing the concentration of the inhibitor leads to disruption of the association between the beads, and therefore a decrease in AlphaScreen signal. The resulting data points were used to calculate values of the

Table 1. IC_{50} values determined using the AlphaScreen competitive binding assay for ASF/galectin-3 binding pair and the corresponding K_d values measured by ITC.

Compound	$IC_{50}/\mu M^a$	$K_d/\mu M^b$
TF-αThr	28.1	288
MUC1-Thr[4]	18.1	28
MUC1-Thr[9]	3.1	45
MUC1-Thr[16]	56.6	n. o.[c]
MUC1	110	no binding

[a]The IC_{50} values were obtained by non-linear regression analysis using the Graph Pad Prism 5.04.; [b]K_d for full length galectin-3 was obtained using the one-site fit model of the binding data with MicroCal analysis software (Origin 7.0); [c]not observed.

glycosylated MUC1 fragments that carry TF antigen for galectin-3 (Table 1). Interestingly, glycosylated MUC1-Thr[9] peptide was a better inhibitor than TF glycan alone suggesting the binding relevance of the peptide scaffold presenting the carbohydrate ligand, and in agreement with our recent findings from NMR studies and molecular dynamics simulations of the binding of Thr O-linked TF antigen to chimera-type avian galectin-3 [9]. The binding profile of MUC1-Thr[16] and MUC1 indicated possible non-specific binding. Binding constants determined from ITC experiments further confirmed the relevance of the peptide scaffold presenting the carbohydrate ligand in binding. The measured K_d for galectin-3 and glycosylated MUC1-Thr[4] ($K_d = 28$ μM) and MUC1-Thr[9] ($K_d = 45$ μM) were 10 x lower in comparison to TF glycan alone ($K_d = 288$ μM). No binding was observed for the glycopeptide MUC1-Thr[16], as well as for the non-glycosylated control version of MUC1 peptide (Table 1). In order to explore the effect of the galectin-3 N-terminal domain in binding, we have also performed binding studies with the galectin-3 carbohydrate recognition domain (CRD). The affinity of the glycosylated MUC1-Thr[9] ($K_d = 103$ μM) was slightly lower compared to the binding affinities of the full-length galectin-3.

In summary, presentation of the carbohydrate ligand by the natural peptide scaffolds can have a major impact on recognition. Better understanding of this process, that combines the contributions of the protein and glycan components, is essential for studying the functional relevance of tumor-associated carbohydrate antigens (TACAs) on MUC1.

Acknowledgments

This study was supported by startup funds provided by the State of Florida, Executive Officer of the Governor's Office of Tourism, Trade and Economic Development and by an American Cancer Society Institutional Research Grant, Junior Faculty Development Award IRG-08-063-01 to M. Cudic.

References

1. Adamczyk, B., Tharmalingam, T., Rudd, P.M. *Biochim. Biophys. Acta Gen. Subj.* **1820**, 1347-1353 (2012).
2. Hollingsworth, M.A., Swanson, B.J. *Nat. Rev. Cancer* **4**, 45-60 (2004).
3. Khaldoyanidi, S.K., Glinsky, V.V., Sikora, L., Glinskii, A.B., Mossine, V.V., Quinn, T.P., Glinsky, G.V., Sriramarao, P. *J. Biol. Chem.* **278**, 4127-4134 (2003).
4. Yu, L.-G., Andrews, N., Zhao, Q., McKean, D., Williams, J.F., Connor, L.J., Gerasimenko, O.V., Hilkens, J., Hirabayashi, J., Kasai, K., Rhodes, J.M. *J. Biol. Chem.* **282**, 773-781 (2007).
5. Cudic, M., Ertl, H.C.J., Otvos, Jr., L. *Bioorg. Med. Chem.* **10**, 3859-3870 (2002).
6. Plattner, C., Hofener, M., Sewald, N. *Org. Lett.* **13**, 545-557 (2011).
7. Lemieux, R.U., Ratcliffe, R.M. *Can. J. Chem.* **57**, 1244-1251 (1979).
8. Yegorova, S., Chavaroche, A.E., Rodriguez, M.C., Minond, D., Cudic, M. *Anal. Biochem.* **439**, 123-131 (2013).
9. Yongye, A., Calle, L., Arda, A., Jimenez-Barbero, J., André, S., Gabius, H.-J., Martinez-Mayorga, K. Cudic M. *Biochemistry* **51**, 7278-7289 (2012).

Proceedings of the 23rd American Peptide Symposium
Michal Lebl (Editor)
American Peptide Society, 2013

Synthesis and Bioactivity Analysis of a-Factor Analogs Containing Different Alkyl Esters at the C-Terminal Cysteine Using Trityl Side-Chain Anchoring

Veronica Diaz-Rodriguez[1], Elena Ganusova[2], Jeffrey M. Becker[2], and Mark D. Distefano[1]

[1]Department of Chemistry, University of Minnesota, Minneapolis, MN, 55414, U.S.A.; [2]Department of Microbiology, University of Tennessee, Knoxville, TN, 37996, U.S.A.

Introduction

a-Factor is a mating pheromone secreted by a haploid cell of *S. cerevisiae* that is involved in mating. This dodecapeptide has attracted considerable attention in the field of protein prenylation due to its similarity with the C-terminal portion of larger, farnesylated proteins.

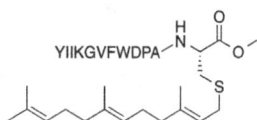

Fig. 1. Structure of the yeast mating pheromone a-factor.

Moreover, in common with these proteins, the farnesyl moiety and the methyl ester group incorporated at the C-terminal cysteine of the **a**-factor peptide have been shown to be critical for its bioactivity [1,2]. Here we describe the use of our recently developed trityl side-anchoring method [3] for the synthesis of **a**-factor and **a**-factor analogs containing different alkyl esters at the C-terminus. The bioactivity of the peptides was analyzed using a growth arrest assay [4].

Results and Discussion

Farnesylated proteins are post-translationally modified proteins that are involved in signal transduction pathways and have been found to be critical for the progression of diseases such as cancer. Due to the importance of the C-terminal methyl ester group in the structure of many mature farnesylated proteins, the development of methods that produce peptides incorporating this modification in a more effective way are needed. A new cysteine anchoring method was recently developed for the synthesis of peptides containing C-terminal cysteine methyl esters [3]. With this method we achieved the preparation of **a**-factor; a dodecapeptide found in yeast cells that contains the modifications present in mature farnesylated proteins (C-terminal methyl ester and farnesyl group at C-terminal cysteine thiol). Here, we investigate the scope of this methodology and demonstrate that this method can be used for the synthesis of peptides containing different alkyl esters at the C-terminus other than methyl esters. The trityl side-chain anchoring method consists of the attachment of an N-protected cysteine alkyl ester to a trityl-containing resin (*via* the side-chain thiol) followed by preparation of the desired peptides using Fmoc-based SPPS (Figure 2a). The obtained peptides, containing a C-terminal methyl, ethyl, isopropyl or benzyl ester (Figure 2b), were obtained in high yields (>80%) and with high purity (>70% prior to purification) as shown by RP-HPLC (Figure 2c). Farnesylation of peptides b1-b4 was performed, using conditions developed by Naider [5,6], to yield the desired **a**-factor analogs (Figure 3).

To verify that the synthesized **a**-factor analogs are biologically active, we tested their ability to cause growth arrest of yeast cells expressing the Ste3p **a**-factor receptor. In that assay, previously characterized, **a**-factor with wild-type potency stimulated growth arrest with an end point of 0.12 ng [4], identical to the end point obtained after analysis of the synthetic **a**-factor produced in this study. Analysis of the **a**-factor analogs (data not shown) containing ethyl ester and isopropyl ester showed a decrease in biological activity compared to the wild-type **a**-factor. The data showed that the ethyl ester containing **a**-factor have more potency than isopropyl ester containing **a**-factor. Bioactivity assays for the benzyl ester containing **a**-factor are in progress. This study indicates that while these **a**-factor analogs have somewhat lower activity, they still possess significant bioactivity and hence should be useful for future studies.

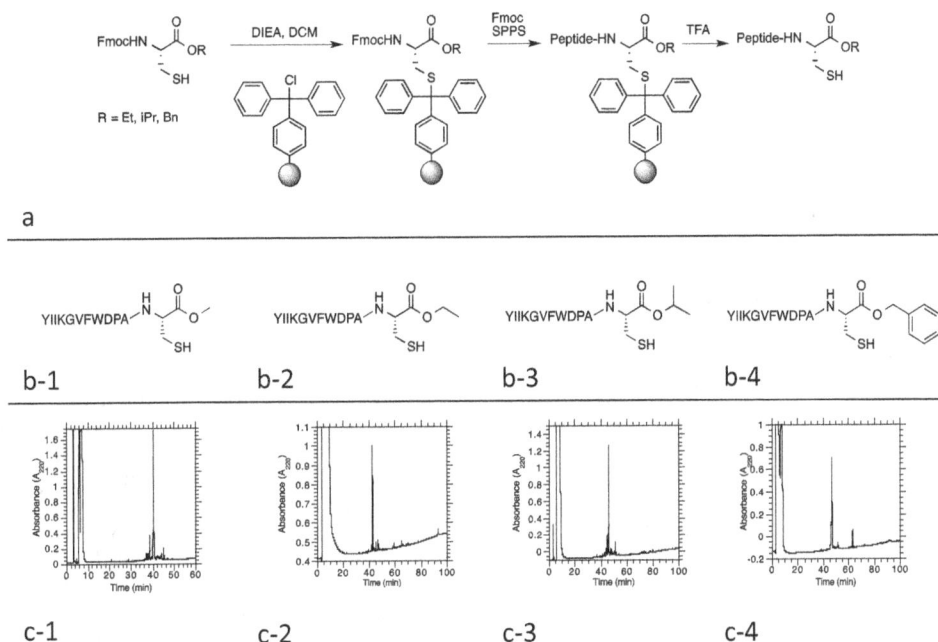

Fig. 2. a) Trityl-based side-chain anchoring strategy; b) Chemical structure of synthesized peptides containing C-terminal methyl (b-1), ethyl (b-2), isopropyl (b-3) and benzyl (b-4) esters; c) HPLC chromatogram of crude peptides in b after cleavage from the trityl resin.

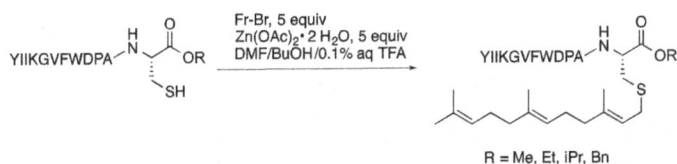

Fig. 3. Farnesylation reaction used for the synthesis of the desired **a**-factor and **a**-factor peptide analogs.

In summary, the scope of the trityl side-chain anchoring method was investigated for the synthesis of peptides with different alkyl ester moieties at the C-terminus. We successfully synthesized **a**-factor and **a**-factor analogs containing ethyl, isopropyl and benzyl esters and analyzed their ability to arrest growth of yeast cells. The success of this trityl side-anchoring method for the synthesis of C-terminal ester modified peptides opens the door to new ways to study protein prenylation and other structurally related biological processes.

Acknowledgments

This research was supported by the National Institute of Health Grants GM008700 (L.Q.), GM084152 and GM058842 (M.D.D.), 5T32GM008347-22 (W.H.), and GM22087 (J.M.B.)

References

1. Dawe, A.L., Becker, J.M., Jiang, Y., Naider, F., Eummer, J.T., Mu, Y.Q., Gibbs, R.A. *Biochemistry* **36**, 12036 (1997).
2. Sherrill, C., Khouri, O., Zeman, S., Roise, D. *Biochemistry* **34**, 3553 (1995).
3. Diaz-Rodriguez, V., Mullen, D.G.; Ganusova E., Becker, J.M., Distefano, M.D. *Org. Lett.* **14**, 5648 (2012).
4. Xue, C.B., Caldwell, G.A., Becker, J.M., Naider, F. *Biochem. Biophys. Res. Commun.* **162**, 253 (1989).
5. Xue, C.B., Becker, J.M., Naider, F. *Tetrahedron Lett.* **33**, 1435 (1992).
6. Mullen, D.G., Kyro, K., Hauser, M., Gustavsson, M., Veglia, G., Becker, J.M., Naider, F., Distefano, M.D. *Bioorg. Med. Chem.* **19**, 490 (2011).

Proceedings of the 23rd American Peptide Symposium
Michal Lebl (Editor)
American Peptide Society, 2013

Photoactive Prenylated Peptides for Studying Protein Isoprenylation

Jeffrey Vervacke[1], Amy Funk[2], Nathan Schuld[3], Carol Williams[3], Christine Hrycyna[2], and Mark D. Distefano[1]

[1]Departments of Chemistry and Medicinal Chemistry, University of Minnesota,
Minneapolis, MN, 55455, U.S.A.; [2]Department of Chemistry, University of Purdue,
West Lafayette, IN, 47907, U.S.A.; [3]Department of Pharmacology and Toxicology,
Medical College of Wisconsin, Milwaukee, WI, 53226, U.S.A.

Introduction

Prenylated proteins contain a farnesylated or geranylgeranylated cysteine residue near their C-termini. These proteins are involved in a variety of signal transduction pathways that regulate diverse cellular behaviors including cell growth, differentiation and survival. Given their critical cellular functions, they are important targets for the development of new therapeutic agents. Synthetic prenylated peptides that incorporate photoactive groups are useful tools for probing the enzymology and function of protein prenylation and we are currently using such probes to study two aspects of prenylation. In the first, peptides based on the sequence of a-factor, a farnesylated dodecapeptide, are being used to probe the interaction between ICMT (isoprenylcysteine methyltransferase) and its protein substrate. In the second, farnesylated and geranylgeranylated peptides derived from the C-terminus of Rap1B are being used to study the binding of Rap1B to SmgGDS, a regulator of small GTPases that possess C-terminal polybasic domains. Results from photolabeling and pull-down experiments are described.

Results and Discussion

Photoaffinity labeling has been a useful tool for identifying interaction sites between peptide-protein interacting partners [1]. Here, we describe an additional photolabeling approach for mapping out the isoprenoid binding sites with ICMT and SmgDGS. Utilizing the 12-aa a-factor-derived peptide sequence, four peptides were developed incorporating a C-terminal prenylated cysteine to study ICMT (Figure 1, **1-4**). These photoactive peptides were designed to crosslink to and label residues in or near the isoprenoid binding sites of ICMT by appending either a benzophenone (BP) or a diazirine (Diaz) moiety to the end of an isoprenoid chain as a mimic for the naturally recognized farnesyl cysteine moiety. Benzophenone and diazirine are useful in photoaffinity labeling experiments due to their chemical stability and ability to form a covalent adduct with the target protein upon UV irradiation [2,3]. By adding a Dde protected Lys to the N-terminus of **a**-factor, orthogonal Fmoc/Dde removal conditions were used to incorporated a biotin on the N-terminus for streptavidin enrichment and a fluorophore onto the Lys side chain for fluorescent detection. Peptides **2-4** were studied for substrate recognition, and a comparison of photolabeling efficiency with ICMT between the benzophenone and diazirine versions.

Yeast membranes over-expressing ICMT (Ste14p) were photolyzed in the presence of the a-factor analogues peptides for 45 minutes prior to freeze quenching and

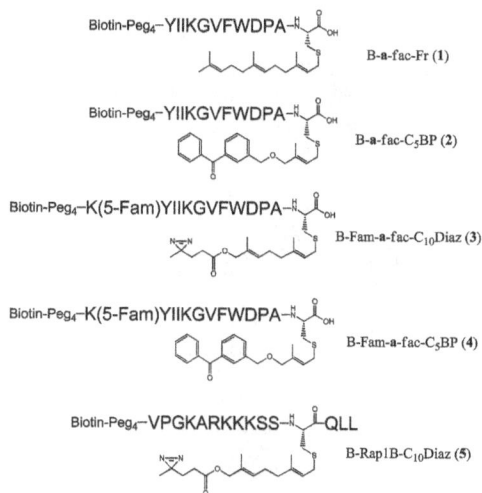

Fig. 1. A-factor/Rap1B peptides for photolabeling.

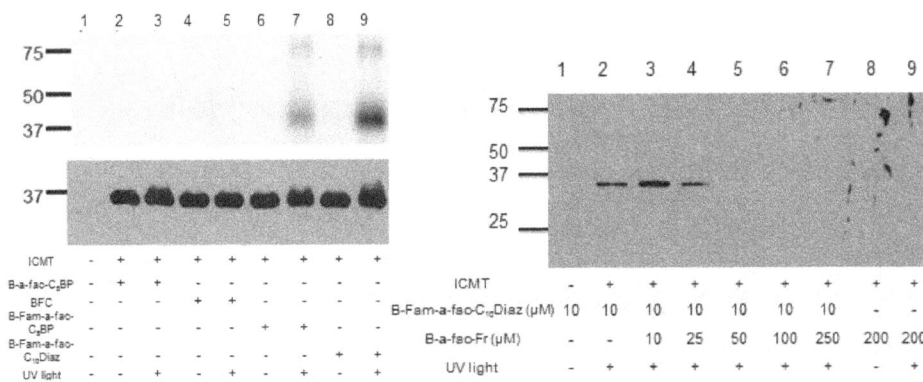

Fig. 2. Photolabeling experiment with ICMT.   Fig. 3. ICMT photolabeling competition experiment.

enrichment with a streptavidin pulldown. Following SDS-PAGE electrophoresis, the protein was transferred to a PVDF membrane and cross-linking was visualized using antibodies against HA-tagged Ste14p. The labeling efficiencies of **2** and **4** were approximately the same, with **4** labeling the enzyme slightly better than **1** (Figure 2, lanes 3 and 7). Peptide **3** labeled Ste14p 100% more efficiently than peptide **4** based on fluorescent densitometry measurements (Figure 2, top image, lanes 7 and 9). The two-fold increase suggests that the diazirine-containing probe is a better farnesyl mimic than the benzophenone-functionalized one. The increased labeling efficiency of **4** along with the ease of fluorescent detection should facilitate future structural studies with ICMT. Evaluation of the substrate specificity of **4** was also examined through a competition experiment with a biotinylated, farnesylated **a**-factor peptide **1**. As seen in the western blot, (Figure 3), the extent of photolabeling with peptide **4** was diminished with increasing concentrations of the competitor peptide **1**. The decrease in the formation of the photoadduct indicates that the two peptides were competing for the same binding site in Ste14p.

Additional photoaffinity labeling work was performed with the GTPase regulator SmgGDS. Based on Rap-1B, a 15 residue peptide was synthesized bearing a *C*-terminal "CAAX" box, a biotinylated *N*-terminus, and a diazirine-containing isoprenoid (Figure 1, **5**). Peptide **5** was incubated with cellular lysate overexpressing SmgGDS and irradiated with UV

light for varying time periods. The western blot for HA-tagged SmgGDS, (Figure 4), shows that with increasing UV exposure, the extent of labeling of SmgGDS by peptide **5** increasees up to 30 minutes. Optimized exposure time will be used in future labeling experiments with SmgGDS to maximize photoadduct formation and assist in determining the site of crosslinking.

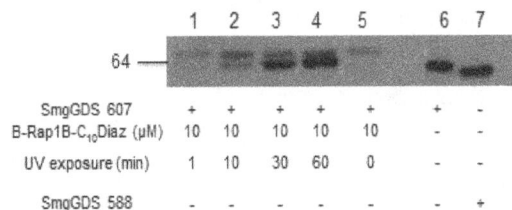

Fig. 4. Photolabeling experiment with SmgGDS.

Acknowledgments

The University of MN Center for Mass Spectrometry and Proteomics for technical assistance regarding MS/MS sequencing. Supported by the National Institutes of Health (GM058842 and GM084152).

References

1. Vervacke, J., Yen-Chih, W., Distefano, D.M. *Current Medicinal Chemistry* **20**, 1585-1594 (2013).
2. Dorman, G., Prestwich, G.D. *Biochemistry* **33**, 5661-5673 (1994).
3. Korshunova, G., Sumbatyan, N., Topin, A., Mtchedlidze, M. *Molecular Biology* **34**, 823-839 (2000).

Proceedings of the 23rd American Peptide Symposium
Michal Lebl (Editor)
American Peptide Society, 2013

Preparation of Glycosylated 5-Hydroxylysine Suitable for SPPS and Evaluation of its Influence on Melanoma Interactions with Type IV Collagen Peptides

Maciej J. Stawikowski, Roma Stawikowska, and Gregg B. Fields

Torrey Pines for Molecular Studies, Port St. Lucie, FL, 34957, U.S.A.

Introduction

Invasion of the basement membrane is believed to be a critical step in the metastatic process. Melanoma cells have been shown previously to bind distinct triple-helical regions within basement membrane (type IV) collagen [1]. Additionally, tumor cell binding sites within type IV collagen contain glycosylated 5-hydroxylysine (Hyl) residues. It has been previously reported that the type IV collagen $\alpha 1(IV)382-393$ sequence interacts with the $\alpha_2\beta_1$ integrin [2], and the $\alpha 1(IV)531-543$ sequence binds to the $\alpha_3\beta_1$ integrin [3]. In the present study, we have utilized triple-helical peptide models of the type IV collagen $\alpha 1(IV)382-393$ and $\alpha 1(IV)531-543$ sequences to investigate the influence of glycosylation on adhesion and spreading of four different melanoma cell lines.

Results and Discussion

Fmoc-D,L-Hyl[O-β-(Ac₄Gal)Nε-(Cbz)]-OPF (Figure 1) has been successfully synthesized in 6 steps, starting from commercially available unprotected amino acid (Figure 2). 9-BBN was employed for simultaneous α-amino and carboxyl group protection [4]. After selective protection of the ε-amino group of 5-hydroxylysine with the Cbz group, the amino acid was O-galactosylated under Koenigs-Knorr conditions using silver silicate as promoter. Following BBN deprotection, the α-amino group was protected with Fmoc and the carboxyl group was treated with pentafluorophenol to yield the fully protected amino acid. The protected building block was then incorporated into collagen type IV peptide sequences (Table 1). Along with glycopeptides **1** and **3**, their non-glycosylated counterparts (peptides **2** and **4**) were synthesized possessing Lys instead of Hyl. Following synthesis and purification, glycopeptides/peptides were characterized by MALDI-TOF MS, HPLC and CD. All peptides were tested *in vitro* for cell adhesion activity as described previously [5]. Four different melanoma cell lines were used

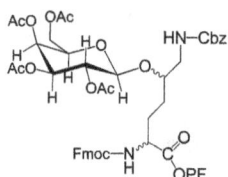

Fmoc-DL-Hyl[(O-β-Gal(Ac₄))(Nε-Cbz)]-OPF

Fig. 1. Fully protected amino acid for suitable for SPPS.

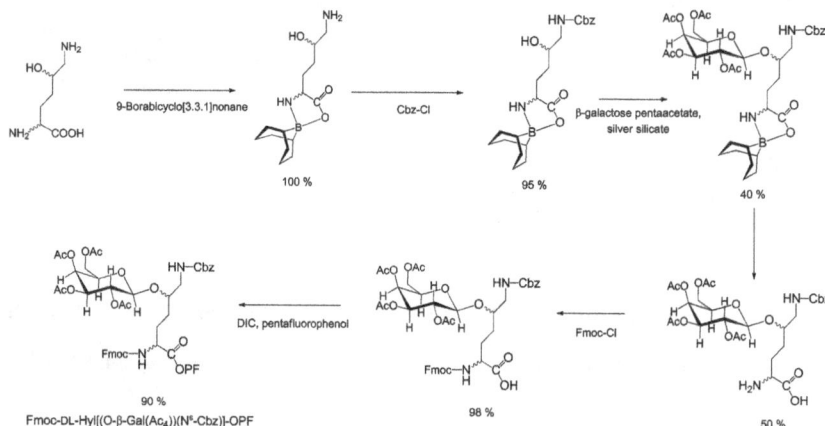

Fmoc-DL-Hyl[(O-β-Gal(Ac₄))(Nε-Cbz)]-OPF

Fig. 2. Synthesis of glycosylated 5-hydroxylysine building block suitable for SPPS.

Table 1. Type IV collagen peptides containing glycosylated 5-hydroxylysine and their lysine counterparts.

#	Collagen sequence	Triple-helical peptide sequence	Integrin binding
1	α1(IV)382-393(Gal)	C_{10}-(Gly-Pro-Hyp)$_4$-Gly-Ala-Hyp-Gly-Phe-Hyp-Gly-Glu-Arg-Gly-Glu-**Hyl(Gal)**-(Gly-Pro-Hyp)$_4$-Tyr-NH$_2$	$α_2β_1$
2	α1(IV)382-393	C_{10}-(Gly-Pro-Hyp)$_4$-Gly-Ala-Hyp-Gly-Phe-Hyp-Gly-Glu-Arg-Gly-Glu-**Lys**-(Gly-Pro-Hyp)$_4$-Tyr-NH$_2$	$α_2β_1$
3	α1(IV)531-543(Gal, Gal)	C_{10}-(Gly-Pro-Hyp)$_5$-Gly-Glu-Phe-Tyr-Phe-Asp-Leu-Arg-Leu-**Hyl(Gal)**-Gly-Asp-**Hyl(Gal)**-(Gly-Pro-Hyp)$_4$-NH$_2$	$α_3β_1$
4	α1(IV)531-543	C_{10}-(Gly-Pro-Hyp)$_5$-Gly-Glu-Phe-Tyr-Phe-Asp-Leu-Arg-Leu-**Lys**-Gly-Asp-**Lys**-(Gly-Pro-Hyp)$_4$-NH$_2$	$α_3β_1$

for cell adhesion assays, SK-MEL-2, WM-115, WM-266-4, and M14#5. Two of these lines (WM-115 and WM-266-4) came from the same patient and represent primary and metastatic stages, respectively. All peptides were assayed in 96-well format plates, incubated initially for 16 h at 4°C. Glycopeptide/peptide concentrations were 50, 10, 5, 1, and 0.1 μM. Appropriate controls were also utilized. Cells were allowed to adhere for 60 min at 37°C. Quantification of adhesion was measured based on the CellTiter-Glo® luminescence cell viability assay.

The adhesion assay experiment showed very interesting results. All melanoma cell lines were adherent to non-glycosylated α1(IV)382-393 (peptide **2**), whereas α1(IV)531-543 (peptide **4**) was active only at higher concentrations (5-50 μM). $α_2β_1$ integrin mediated adhesion was slightly inhibited by glycosylation of the α1(IV)382-393 sequence (glycopeptide **1**). Interestingly, $α_3β_1$ integrin mediated adhesion was significantly inhibited by glycosylation of the α1(IV)531-543 sequence (glycopeptide **3**) at each concentration except 50 μM, where non-specific adhesion would be expected. To determine the influence of collagen glycosylation on melanoma cell spreading, we are currently performing cell spreading assays using the aforementioned cell lines.

Integrin-mediated melanoma adhesion can be modulated by collagen glycosylation, in similar fashion to our prior conclusions for CD44-mediated adhesion [5]. We are currently exploring the possibility that tumor cells may modify the extent of extracellular matrix glycosylation to promote metastasis.

Acknowledgments

This work was supported by the State of Florida.

References

1. McCarthy, J.B., Vachhani, B., Iida J. *Peptide Science* **40**, 371-381 (1996).
2. Knight, C.G., Morton, L.F., Peachey, A.R., Tuckwell, D.S., Farndale, R.W., Barnes, M.J, *J. Biol. Chem.* **275**, 35-40 (2000).
3. Miles, A.J., Skubitz, A.P., Furcht, L.T., Fields, G.B. *J. Biol. Chem.* **269**, 30939-30945 (1994).
4. Syed, B.M., Gustafsson, T., Kihlberg, J. *Tetrahedron* **60**, 5571-5575 (2004).
5. Lauer-Fields, J.L., Malkar, N.B., Richet, G., Drauz, K., Fields, G.B. *J. Biol. Chem.* **278**, 14321-14330 (2003).

Proceedings of the 23rd American Peptide Symposium
Michal Lebl (Editor)
American Peptide Society, 2013

The Role of Protein Deamidation in Cardiovascular Disease

Esther Cheow[1*], Piliang Hao[1*], Vitaly Sorokin[2], Chuen Neng Lee[2],
Dominique PV de Kleijn[2], and Siu Kwan Sze[1*]

[1]*School of Biological Sciences, Nanyang Technological University, Singapore;* [2]*Cardiovascular Research Institute and Surgery, National University Health System, Singapore; *equal contribution*

Introduction

Cardiovascular disease (CVD) is a major cause of morbidity and mortality worldwide. Its underlying cause is the development of arteriosclerosis with atherosclerotic plaques resulting in stroke, myocardial infarction, heart failure, and other diseases. The underlying molecular disease mechanisms, however, remain poorly understood. Degenerative protein post-translational modifications (DPTMs), e.g. deamidation, racemization, oxidation, and glycation of long-lived proteins are attributed to many critical events in degenerative disease initiation, progression and ageing [1]. Despite the obvious biological importance of the DPTMs, the molecular basis of the DPTMs in relation to diseases and ageing has not been extensively explored thus far due to technical challenges in studying DPTMs. Deamidation of asparagine (Asn) and glutamine (Gln) residues in a protein can occur spontaneously under physiological conditions *via* a non-enzymatic process [2-4], as illustrated in Figure 1. This causes time-dependent amino acid damage and a change in protein's structure, functions and stability. For example, Asn deamidation produces a mixture of asparagine, *n*-aspartic acid (Asp), and isoaspartic acid (*iso*Asp) [4]. IsoAsp residue lengthens the protein backbone by one methylene unit ($-CH_2-$) and concurrently shortens the side-chain. The deamidated protein may subsequently misfold and aggregate, leading to the loss in protein functions [5]. Deamidation usually accumulates over time and have been observed in many proteins associated with ageing and diseases including cancer and neurodegenerative diseases (Alzheimers, Huntington and Parkinsons). Whilst it is clear that deamidated proteins in the heart and blood vessels affect human longevity adversely, and have potential application as early diagnostic and prognostic biomarkers for cardiovascular diseases, their exact roles in pathogenesis of atherosclerosis, hypertension and other cardiovascular systemic diseases remain unexplored.

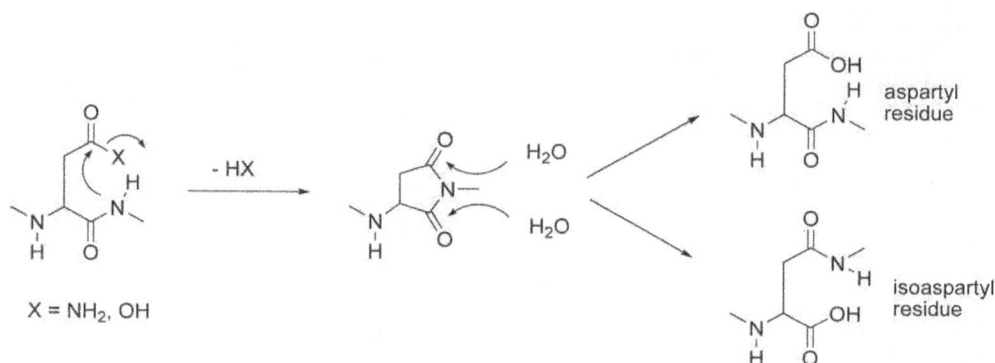

Fig. 1. Spontaneous deamidation of asparagine and isomerization of aspartic acid through a succinimide intermediate. An isoAsp residue in which the protein backbone is lengthened and the sidechain is shortened by one methylene unit ($-CH_2-$) can be produced in the process.

Results and Discussion

We applied proteomic methods to profile aortic tissues, atherosclerotic plaques, and blood plasma of clinical samples. 1955 unique N-deamidated peptides were identified and quantified from 443 unique proteins. This first proteome-wide identification of a large number of deamidated peptides in the cardiovascular system enabled us to identify functional deamidated peptide motifs that possibly encode biological functions. By using motif-x (motif extractor) v1.2 [6] coupled with the human protein database, we have extracted numerous peptide sequence motifs from our deamidated peptide dataset, one being the NGR motif as showed in Figure 2. A literature search disclosed that deamidation of NGR to isoDGR produced a gain-of-function integrin-binding motif that mimics the RGD motif [7-9] that promotes endothelial cell adhesion and proliferation. Paradoxically, degenerative deamidation in proteins possibly impart novel functions to the proteins with isoDGR sequences, i.e. the isoDGR motif is a gain-of-function deamidation. Many proteins in the cardiovascular system and plasma were detected with the deamidated isoDGR motif, implicating it in the pathogenesis of arterial dysfunctions. This may open up new directions for research in cardiovascular disease. For example, our analysis identified the isoDGR integrin recognition motif in many proteins involved in cell adhesion functions that may advocate the initiation and progression of atherosclerosis in CVD. We speculate that protein deamidation in the cardiovascular system has two consequences: 1) it causes protein damage that result in compromise arterial structure integrity, accelerating arteriosclerosis and hypertension; and 2) significant accumulation of ECM proteins with isoDGR activates novel ligand-integrin-mediated cell adhesion that promotes recruitment and sequestration of mononuclear cells into the artery and atherosclerotic plaque, leading to arterial dysfunctions, atherosclerotic plaque instability and increased risk for cardiovascular events like stroke and myocardial infarction.

Fig. 2. Some of the NGR containing motifs identified in the deamidated peptides in the cardiovascular system by motif-x program.

Acknowledgments

This work is supported by the Singapore Ministry of Health's National Medical Research Council (NMRC/CBRG/0004/2012).

References

1. Truscott, R.J.W. *Rejuvenation Research* **13**, 83-89 (2010).
2. Robinson, N.E., Robinson, A.B., Molecular clocks: Deamidation of Asparaginyl and Glutaminyl Residues in Peptides and Proteins. *Althouse Press: Cave Junction, OR (*2004).
3. Hao, P., Ren, Y., Alpert, A.J., Sze, S.K. *Mol. Cell. Proteomics* **10**, (10), O111 009381 (2011).
4. Hao, P., Qian, J., Dutta, B., Cheow, E.S., Sim, K.H., Meng, W., Adav, S.S., Alpert, A., Sze, S.K. *J. Proteome Res.* **11**, 1804-1811 (2012).
5. Moreau, K.L., King, J.A. *Trends Mol. Med.* **18**, 273-282 (2012).
6. Schwartz, D., Gygi, S.P. *Nature biotechnology* **23**, 1391-1398 (2005).
7. Corti, A., Curnis, F. *J. Cell Sci.* **124**, (Pt 4), 515-522 (2011).
8. Spitaleri, A., Mari, S., Curnis, F., Traversari, C., Longhi, R., Bordignon, C., Corti, A., Rizzardi, G.P., Musco, G. *J. Biol. Chem.* **283**, 19757-19768 (2008).
9. Zou, M., Zhang, L., Xie, Y., Xu, W. *Anticancer Agents Med. Chem.* **12**, 239-246 (2012).

Proceedings of the 23rd American Peptide Symposium
Michal Lebl (Editor)
American Peptide Society, 2013

Strategy for the Synthesis of Isotope-Labeled Branched Protein Mimics

Sabine Abel[1], Bernhard Geltinger[1], Dirk Schwarzer[2], and Michael Beyermann[2]

[1]*Leibniz-Institut fuer Molekulare Pharmakologie (FMP), Robert-Roessle-Strasse 10, 13125 Berlin, Germany;* [2]*IFIB Interfaculty Institute of Biochemistry, Eberhard Karls University Tuebingen, 72076 Tuebingen, Germany*

Introduction

Structural investigation of membrane-spanning proteins such as G-protein coupled receptors (GPCRs) is hampered by enormous experimental difficulties. Thus, partial domains or domain mimics of such proteins are examined. Recently, we reported the synthesis of a 23 kDa (CRF$_1$), a GPCR of the B1 subfamily, and demonstrated that the mimic exhibits CRF$_1$-like binding to natural peptide ligands [1]. The mimic consists of the extracellular receptor domains (*N*-terminus (ECD1) and 3 loops (ECD 2-4)), that are assumed to represent the main binding sites of peptide ligands, bound to a linear peptide template (see Figure 1). To accomplish the investigation of its 3D structure and protein dynamics by nuclear magnetic resonance (NMR) measurements a strategy for the synthesis of an isotope-labeled branched protein mimic has been developed.

Results and Discussion

In our former approach chemically prepared cyclic mimics of ECD2-4 were introduced to the template *via* thiol-maleimide reaction. The ECD1 expressed in *E. coli* was coupled enzymatically to the template exploring sortase A. Therefore, the *N*-terminus (ECD1) can be isotope-labeled by expression with ^{15}N ammonia and ^{13}C glucose. The chemical synthesis of isotope-labeled extracellular loops using ^{15}N-^{13}C-labeled amino acids is unaffordable and causes, therefore, a change in the strategy. To gain access to isotope-labeled loop mimics we show here exemplarily that the cyclic extracellular receptor domain 3 (ECD3) has been

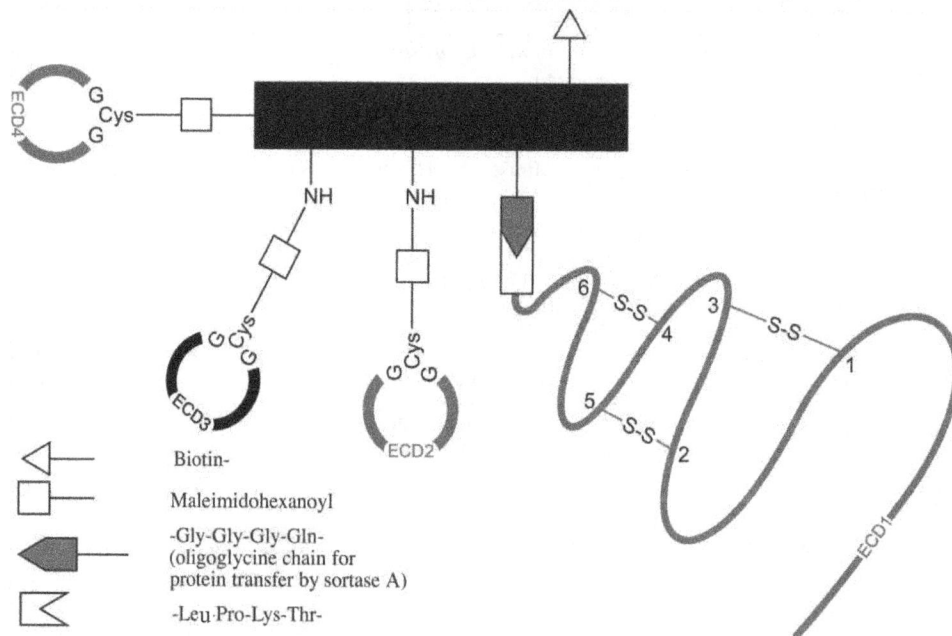

Fig. 1. Protein mimic consisting of the ectodomains of the GPCR-receptor CRF1 (expressed loop black).

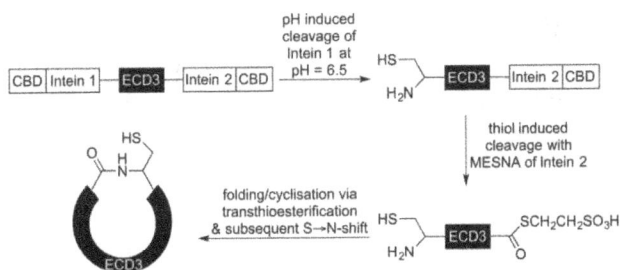

Fig. 2. Scheme of the EPL of ECD3.

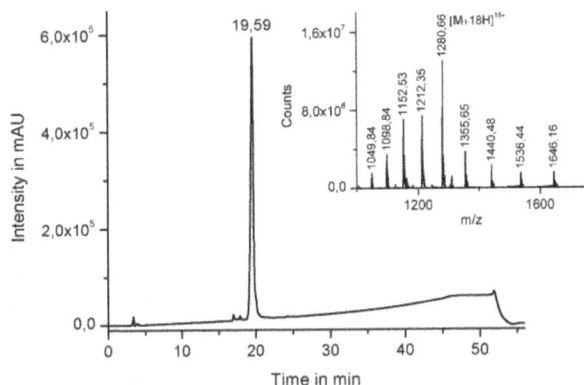

Fig. 3. HPLC profile (uv, 220 nm) and ESI-ToF-MS-spectrum of pure CRF$_1$ mimic with expressed ECD3.

produced efficiently by expressed protein ligation (EPL, Figure 2). For this, the pTWIN2 vector with *Synechocystis DnaB* as the *N*-terminal intein (Intein 1) and *Methanobacterium thermo-autotrophicum rir 1* as the *C*-terminal intein (Intein 2) has been used in origami cells to generate a peptide bearing an *N*-terminal cysteine and at the *C*-terminus a reactive thioester for the subsequent head-to-end cyclisation *via* trans-thioesterification initiated by sodium 2-mercaptoethansulfonate [2]. After purification via reversed-phase high performance liquid chromatography (RP-HPLC) the ECD3 was again chemically coupled to the receptor template. By sortase A-mediated ligation of ECD1 to the loop-template construct we obtained the pure CRF$_1$ mimic with the expressed as well as with the chemically synthesized loop (Figure 3).

Here, we show the preparation of a cyclic peptide representing an extracellular receptor loop via expressed protein ligation and its incorporation into a CRF$_1$ mimic. This opens an avenue for an alternative labeling approach compared to the expensive chemical synthesis. The result illustrates the power of combined recombinant, enzymatic, chemical, and expressed protein ligation semisynthesis strategies which make isotope-labeled, branched protein mimics available.

Acknowledgments

We would like to thank Kirill Piotukh from the Leibniz-Institut fuer Molekulare Pharmakologie (FMP; Berlin, Germany) for the preparation of the sortase A ligation enzyme.

References

1. Pritz, S., Kraetke, O., Klose, A., Klose, J., Rothemund, S., Fechner, K., Bienert, M., Beyermann, M. *Angew. Chem. Int. Ed. 47*, 3642-3645 (2008).
2. Xu, M.-Q., Evans, T.C., Jr. *Methods* **24**, 257-277 (2001).

Proceedings of the 23rd American Peptide Symposium
Michal Lebl (Editor)
American Peptide Society, 2013

Nanoparticle Peptide Synthesis NPPS

Gerardo Byk[1], Victoria Machtey[1], Aryeh Weiss[2], and Raz Khandadash[1]

[1]Deptartment of Chemistry, Laboratory of Nano-Biotechnology; [2]School of Engineering;
Bar Ilan University, 52900-Ramat-Gan, Israel

Introduction

Nanoparticles (NP) modification techniques are mostly based on the conjugation of a single molecule (protein etc.) by covalent bonding of complementary functional groups found on both components. Because of the difficulty in isolating NPs by filtration/precipitation or dialysis, conjugation is mostly restricted to a single or few coupling reactions due to the difficult removal of excess reagents after every reaction. Our NPs possess both bio and chemical compatibility; these qualities rend their modification a difficult task due to non-aggregative behavior that prevents their precipitation. This physical difficulty was overcome by a new approach that exploits magnetic susceptibility of materials for their easy and fast separation. The presented method might be applied to other NPs, thus extending the scope of Merrifield synthesis to the nanometric range.

Results and Discussion

The overall nanoparticle peptide synthesis (NPPS) process is shown in Figure 1. The method consists of embedding appropriate NPs into a magnetic matrix (step 1) followed by the multistep synthesis of the desired peptides facilitated by the magnetic susceptibility of the magnetic matrix necessary for washings/reactions along the intermediate steps (step 2). Once the synthesis is completed the matrix is treated with HCl for 6h and after appropriate neutralization, the obtained NPs are dialysed once for removal of iron salts (step 3).

As proof of feasibility, we have synthesized on different NPs sizes, a nuclear localization sequence (NLS) peptide from the sv40 large T antigen (PKKKRKV) [1] and a HIV Tat-derived peptide (GRKKRRQRRRPPQ) [2]. These peptides have been shown to penetrate into cells and/or localize in the nucleus. Additionally, we have synthesized a tumor homing pro-apoptotic peptide [KLAKLAK]2 fused to GGKRK (KLA). This peptide has

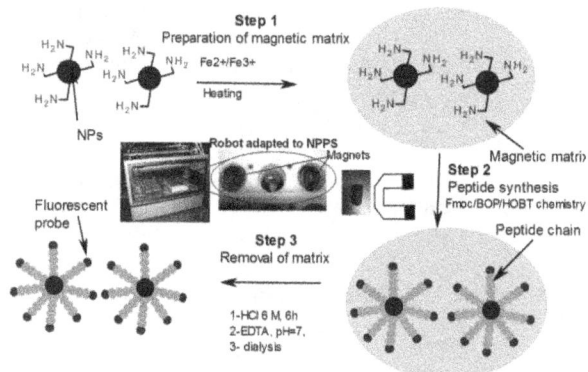

Fig. 1. Nanoparticle peptide synthesis (NPPS) process.

been shown to be highly toxic when specifically targeted to tumors [3]. Cell penetration processes mediated by these peptides, can be observed if fluorescent probes are linked to the peptide-NPs. One of the possible applications for NLS/TAT NPs is the facilitated cell penetration of drugs loaded into the NPs, but can also be exploited for observing transport to and from the nucleus. The peptides were synthesized both manually or using an Advanced ChemTech 496MOS robot that was especially adapted to the NPPS by placing *cylindrical* magnets (N50) on both flanks of each reaction well so that the magnetic matrix adheres to the walls of the well permitting the contact of solvents and reagents with the matrix and appropriate evacuation of solvents/reagents through the lower commercially available PTFE filter (see Figure 1). To allow a direct structural chemical analysis of the obtained peptides, the magnetic matrix embedded NPs were first reacted with a base-labile HMBA linker. This permits partial cleavage of the product peptides by hydrolysis of the ester bond for direct high resolution mass spectrometric (HRMS) and HPLC analyses in solution. Peptides were synthesized using Fmoc chemistry with coupling reagent BOP/HOBT. An amino-hexanoyl (Ahx) spacer was placed at the *N*-terminal of all the peptides that was reacted with a

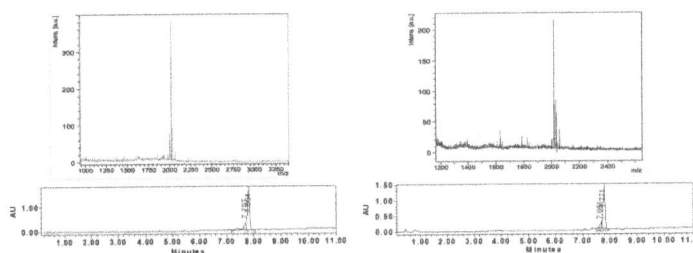

Fig. 2. HPLC (220 nm) and HRMS analysis of: NLS-FITC cleaved from NPs 190 nm (left panel) as compared to NLS-FITC cleaved from commercial Rink linker (right panel).

Table 1. Synthesis of bio-relevant peptide sequences on NPs. a - Free carboxylate, b - Fully unprotected peptide. FITC = fluorescein isothiocyanate, FAM = 5(6)-carboxyfluorescein, SRB = sulforhodamine B.

#	Support	Peptide	Calc. MS	Obt. MS
1	Rink Amide Resin	FITC-NLS	2019.9822	2019.989[a]
2	Magnetic embedded NPs (190 nm)	FITC-NLS	2019.9822	2019.982[a]
3	Magnetic embedded NPs (190 nm)	SRB-TAT	2372.2815	2372.262[b]
4	Magnetic embedded NPs (190 nm)	FAM-KLA	2634.6222	2634.617[b]
5	Magnetic embedded NPs (100 nm)	FAM-TAT	2190.1904	2190.178[b]
6	Magnetic embedded NPs (100 nm)	SRB-KLA	2816.7133	2816.721[b]

molecular probe to give labeled peptide-NPs conjugates. The products were of good quality as compared to crudes obtained using the same chemistry but synthesized on conventional solid supports (see Figure 2).

Peptide-NPs were recovered from the matrix after treatment with HCl for 6h followed by addition of EDTA, neutralization to pH=6 and single dialysis process to remove iron-EDTA soluble complex species. The HPLC analysis of the peptides cleaved directly from the magnetic matrix embedded NPs as well as the HR-MS analysis confirmed that the syntheses were successful (see Table 1).

Overall, we have demonstrated for the first time that embedding nanoparticles into a magnetic matrix allows multistep synthesis on NPs using the well-known Merrifield synthetic methodology. The NPs can be recovered and eventually used "as is" for cell signaling *via* the molecules synthesized on their surface and appropriate molecular probes. A crucial advantage of the current approach is that magnetic susceptibility for the synthesis on the NPs is mediated by an external, permeable and disposable magnetic matrix that embeds the NPs and can be removed at the end of the multistep process leaving intact the biocompatible polymeric NPs with no magnetic material suspected to be toxic in biological applications. This method might be useful for different type of NPs we are currently testing and for a variety of applications such those we have recently shown in the field of cell screening [4,5].

Acknowledgments

This work was financed by the ISF grant number 830/11 and "The Marcus Centre for Medicinal Chemistry of Bar Ilan University". RK and VM are indebted to the BIU President Scholarships and RK is also indebted to Israel Council for Higher Education for the Converging Technologies Fellowship.

References

1. Kalderon, D., Roberts, B.L., Richardson, W.D., et al. *Cell* **39**, 499-509 (1984).
2. Vives, E., Brodin, P., Lebleu, B. *J. Biol. Chem.* **272**, 16010-16017 (1997).
3. Agemy, L., Friedmann, D., et al. *Proc. Nat. Acad. Sci. USA* **108**, 17450-17455 (2011).
4. Byk, G., Partouche, S., Weiss, A., et al. *J. Comb. Chem.* **12**, 332-345 (2010).
5. Khandadash, R., Partouche, S., Weiss, A., et al. *Open. Opt. J.* **5**, 17-27 (2011).

Proceedings of the 23rd American Peptide Symposium
Michal Lebl (Editor)
American Peptide Society, 2013

High Purity Synthesis of PTHrP (1-34) with Infrared Heating on the Tribute®-UV-IR Peptide Synthesizer

James P. Cain[1], Christina A. Chantell[1], Michael A. Onaiyekan[1],
Mahendra Menakuru[1], Fabio Rizzolo[2,3], Giuseppina Sabatino[2,4],
Olivier Monasson[2,3], Paolo Rovero[2,5], and Anna-Maria Papini[2,3,4]

[1]Protein Technologies, Inc. Tucson, AZ, 85714, U.S.A. (www.ptipep.com) Email:
info@ptipep.com; [2]Laboratory of Peptide & Protein Chemistry & Biology (www.peptlab.eu);
[3]PeptLab@UCP c/o Laboratoire SOSCO, Université de Cergy-Pontoise, 5 Mail Gay Lussac,
F-95031, Cergy-Pontoise, France; [4]Department of Chemistry "Ugo Schiff", University of
Florence, Via della Lastruccia 13, I-50019, Sesto Fiorentino, Italy; [5]Department NEUROFARBA,
University of Florence, Via U. Schiff 6, I-50019, Sesto Fiorentino, Italy

Introduction

A variety of parameters may be optimized in the course of a solid-phase synthesis of a peptide, including the reaction time, activator type, reagent excess and concentration, solvent, and resin loading. The application of heat, typically using oil baths, heating elements, or microwaves, has emerged as an additional tool for peptide synthesis [1].

Recently, infrared (IR) heating has been introduced on the Tribute® UV-IR peptide synthesizer. Heating with IR is as rapid as that with microwaves. In contrast to microwave instruments, however, multiple reaction vessels can be heated with IR simultaneously at precisely controlled temperatures, allowing parallel synthesis.

The difficult peptide PTHrP(1-34) (Figure 1) was synthesized in order to illustrate the efficiency of IR heating. As previously reported [2], this synthesis usually leads to deletion sequences lacking Arg, Leu, and His residues. This challenging model peptide contains a cluster of Arg prone to gamma-lactam formation, owing to the sterically hindered side-chain protecting groups, thus reducing the coupling yield. Moreover, His tends to undergo racemization. Previously the peptide was synthesized in 77% purity using a microwave peptide synthesizer.

Ala-Val-Ser-Glu-His-Gln-Leu-Leu-His-Asp-Lys-Gly-Lys-Ser-Ile-Gln-Asp-Leu-Arg-Arg-
Arg-Phe-Phe-Leu-His-His-Leu-Ile-Ala-Glu-Ile-His-Thr-Ala-NH$_2$

Fig. 1. Sequence of PTHrP(1-34) peptide synthesized with heat on the Tribute® UV-IR platform.

This sequence is the biologically active N-terminal region of PTHrP, an important regulator of bone formation and glandular development. PTHrP has been found to play a role in the development of lung [3], breast [4], and a number of other cancers, and elevated levels of the peptide are used as a diagnostic for malignant tumors [5].

Results and Discussion

A single major peak was observed in the HPLC trace of the crude peptide (Figure 2), along with very low levels of impurities of similar retention times. Peak integration indicates a crude purity of 87%. The correct mass was also observed by LC-MS.

The 87% purity observed is about 10% higher than previously found for the microwave synthesis. As both systems rely on rapid, non-contact heating using the transmission of electromagnetic energy, it is possible that this difference in the results can be attributed to differences in the underlying platform. The microwave synthesis was performed on an automated single-vessel instrument that uses only nitrogen bubbling for agitation, while the Tribute® uses vortex mixing (or vortex mixing with nitrogen bubbling) to ensure that a homogeneous temperature distribution is maintained. PTI instruments also use a matrix valve block system for fluid deliveries that minimizes dead volume and reagent carryover.

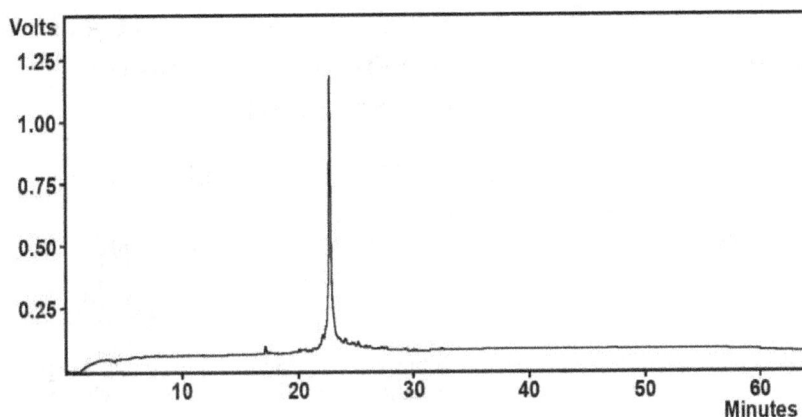

Fig. 2. HPLC trace of PTHrP(1-34) peptide synthesized with heat on the Tribute® UV-IR platform.

References

1. Pedersen, S.L, Tofteng, A.P., Malik, L., Jensen, K.J. *Chem. Soc. Rev.* **41**, 1826-1844 (2012).
2. Rizzolo, F., Testa, C., Lambardi, D., Chorev, M., Chelli, M., Rovero, P., Papini, A.M. *J. Pept. Sci.* **17**, 708-714 (2011).
3. Hastings, R.H. *Respir. Physiol. Neurobiol.* **142**, 95-113 (2004).
4. Li, J., Karaplis, A.C., Huang, D.C., Siegel, P.M., Camirand, A., Yang, X.F., Muller, W.J., Kremer, R. *J. Clin. Invest.* **121**, 4655-4669 (2011).
5. (a) Dumon, J.C., Jensen, T., Lueddecke, B., Spring, J., Barlél, J., Body, J. *J. Clin. Chem.* **46**, 416-418 (2000); (b) Grill, V., et al. *J. Clin. Endocrinol. Metab.* **73**, 1309-1315 (1991).

Proceedings of the 23rd American Peptide Symposium
Michal Lebl (Editor)
American Peptide Society, 2013

Fast Solid-Phase Peptide Synthesis of β-Amyloid (1-42) and the 68-mer Chemokine SDF-1α on the Symphony X™ Multiplex Peptide Synthesizer

James P. Cain, Christina A. Chantell, Michael A. Onaiyekan, and Mahendra Menakuru

Protein Technologies Inc., Tucson, AZ, 85714, U.S.A.; Tel: +1-520-629-9626,
Website: www.ptipep.com, Email: info@ptipep.com;

Introduction

Human β-amyloid (1-42) peptide (Figure 1) is a major component of the plaque deposits found in the brains of Alzheimer's disease (AD) patients [1]. Continued research into the pathology and potential therapeutics for AD has sustained demand for this peptide. Synthesis of the β-amyloid (1-42) peptide by conventional solid phase peptide synthesis has been reported to be difficult due to the high hydrophobicity of the C-terminal segment and on-resin aggregation [2].

DAEFRHDSGYEVHHQKLVFFAEDVGSNKGAIIGLMVGGVVIA

Fig. 1. Sequence of β-amyloid (1-42).

Previously we have demonstrated that the total synthesis time for β-amyloid (1-42) peptide can be reduced from approximately 54 hours using a traditional conservative approach to less than 12 hours on the Symphony® and Prelude™ synthesizers with fast reaction and wash times, while actually improving the purity of the peptides produced [3]. The choice of resin as well as coupling reagent was crucial for optimizing this synthesis.

Another peptide to have attracted interest is human stromal cell-derived factor 1α (SDF-1α), or CXC chemokine ligand 12α (CXCL12α), a member of the chemokine family of peptides involved in basal leukocyte trafficking and homing, as well as in development [4]. The interaction of this 68-residue peptide (Figure 2) with its receptor, CXCR4, is involved in HIV pathogenesis [5] and tumor metastasis such as in breast cancer [6] and lung cancer [7].

KPVSLSYRCPCRFFESHVARANVKHLKILNTPNCALQIVARLKNNNRQVCIDPKLKW
IQEYLEKALNK

Fig. 2. Sequence of SDF-1α.

In earlier studies SDF-1α was synthesized in a total time of 22 hours using fast deprotection and coupling times on low-loaded resin with HCTU as coupling reagent [8].

Our optimized synthetic protocols have been applied to the synthesis of β-amyloid (1-42) and SDF-1α on the recently introduced Symphony X® synthesizer. This instrument can accommodate the synthesis of up to 24 peptides simultaneously using different scales and protocols.

Results and Discussion

The speed of automated peptide synthesis will depend both on the optimization of reaction parameters and the throughput of the instrument used. The choice of coupling and deprotection reagents, reagent excesses and concentrations, solvents, resin type, and resin loading can all be varied in order to obtain high purity products with relatively short total reaction times. Variations in these factors can have particularly pronounced effects in the synthesis of long or difficult peptides. Yet the full benefits of optimized chemistry may not be realized when applied to slow or inappropriate automated platforms. Very active coupling reagents such as HATU and HCTU, for instance, are not well-suited for use on many of the robotic multiple peptide synthesizers available on the market, owing to the long times required for them to dispense reagents. The PTI Symphony X™, the latest addition to the

PTI product line, was designed with these considerations in mind. This instrument features up to 24 independent reaction vessels, or twelve with pre-activation, along with UV monitoring of deprotection reactions and IR heating.

The synthesis of 24 β-amyloid (1-42) peptides was complete in about 14.5 hours, and 24 SDF-1α peptides were synthesized in about 31 hours. This level of throughput in an automated synthesizer is unprecedented. The speed of the Symphony X multiplex platform stems from the ability to simultaneously add reagents to multiple vessels, while also conducting washes of valve blocks and other components at the same time, as described in U.S. patent 5203368.

References

1. Burdick, D., Soreghan, B., Kwon, M., Kosmoski, J., Knauer, M., Henschen, A., Yates, J., Cotman, C., Glabe, C. *J. Biol. Chem.* **267**, 546-554 (1992).
2. (a) Quibell, M., Turnell, W.G., Johnson, T. *J. Org. Chem.* **59**, 1745-1750 (1994); (b) Tickler, A., Clippingdale, A.B., Wade, J.D. *Prot. Pept. Lett.* **11**, 377-384 (2004).
3. (a) Hood, C.A., Fuentes, G., Patel, H., Page, K., Menakuru, M., Park, J. *J. Pept. Sci.* **14**, 97-101 (2008); (b) Fuentes, G., Hood, C., Park, J.H., Patel, H., Page, K., Menakuru, M. *Adv. Exp. Med. Biol.* **611**, 173-174 (2009).
4. Kunkel, S.L., Godessart, N. *Autoimmunity Rev.* **1**, 313-320 (2002).
5. Berson, J.F., Long, D., Doranz, B.J., Rucker, J., Jirik, F.R., Doms, R.W. *J. Virol.* **70**, 6288-6295 (1996).
6. Muller, A., Homey, B., Soto, H., Ge, N., Catron, D., Buchanan, M.E., McClanahan, T., Murphy, E., Yuan, W., Wagner, S.N., Barrera, J.L., Mohar, A., Verastegui, E., Zlotnik, A. *Nature* **410**, 50-56 (2001).
7. Phillips, R.J., Burdick, M.D., Lutz, M., Belperio, J.A., Keane, M.P., Strieter, R.M. *Am. J. Respir. Crit. Care Med.* **167**, 1676-1686 (2003).
8. Patel, H., Chantell, C.A., Fuentes, G., Menakuru, M., Park, J.H. *J. Pept. Sci.* **14**, 1240-1243 (2008).

Proceedings of the 23rd American Peptide Symposium
Michal Lebl (Editor)
American Peptide Society, 2013

SpheriTide™ Resin Comparison Study for the Synthesis of Difficult Peptides on the Symphony X™ Multiplex Peptide Synthesizer

James P. Cain, Christina A. Chantell, Michael A. Onaiyekan, and Mahendra Menakuru

Protein Technologies Inc., Tucson, AZ, 85714, U.S.A.;
Website: www.ptipep.com, Email: info@ptipep.com;

Introduction

The most commonly used resins for SPPS are composed of polystyrene crosslinked with divinylbenzene (PS-DVB), functionalized with the appropriate handles and linkers. The polymer backbone in this case is relatively hydrophobic. A number of alternative solid-phase resins have been developed that incorporate polar functionalities, such polyethylene glycol (PEG) chains. These include the Tentagel resins, in which PEG spacers are grafted to the PS-DVB backbone, and ChemMatrix resin, which replaces the PS-DVB backbone altogether with cross-linked PEG chains.

Recently SperiTech Inc. has introduced the SpheriTide™ resins to the market. These are comprised of a poly-lysine backbone, crosslinked either with sebacic or nitrilotriacetic acid, functionalized with an amine handle and available with standard linkers attached. To the best of our knowledge, no synthesis results have yet been reported using this resin.

We have synthesized the difficult peptide $Ala_{10}Lys$ (Figure 1) using the Symphony X™ peptide synthesizer on SpheriTide™ resins as well as other common solid supports and analyzed the purity and yields of the crude peptides produced.

AAAAAAAAAK

Fig. 1. Sequence of the poly-alanine (Ala)$_{10}$Lys peptide synthesized in this study.

Peptides containing poly-alanine tracts have been associated with several human diseases [1] and have been used to form model beta sheet systems for studying Alzheimer's disease [2,3]. Due to their high propensity to aggregate after the fifth residue, these sequences are difficult to synthesize by conventional Fmoc solid-phase peptide synthesis. In earlier work the optimization of the synthetic protocol, and particularly the addition of 2% diazabicyclo [5.4.0] undec-7-ene (DBU) to the deprotection mixture, allowed this difficult sequence to be produced in comparatively high purity in just 5.5 hours [4].

Results and Discussion

A similar profile was observed for the peptide produced with each resin, but with somewhat different levels of impurities. It should be noted that the best and worst results differ by only 15%. Low-loaded Rink MBHA resin produced the most pure peptide, followed by SpheriTide™ KS, SpheriTide™ KN, TentaGel™, and ChemMatrix™ resins (Table 1). The functionality of the linker in the two varieties of SpheriTide™ resin apparently has no effect on the purity in this case.

Table 1. Purity and yield of crude products using SpheriTide™ and common resins.

Resin	Rink MBHA	SpheriTide KS	SpheriTide KN	TentaGel RAM	ChemMatrix
% Purity	73	66	66	63	58
% Yield	106	65	83	71	48

Differences were observed in the yields of crude products. Rink MBHA gave the highest crude yields, followed by SpheriTide^{TM} KN, TentaGel^{TM}, SpheriTide^{TM} KS, and ChemMatrix^{TM}.

References

1. Tickler, A., Clippingdale, A.B., Wade, J.D. *Prot. Pept. Lett.* **11**, 377-384 (2004).
2. Brown, L.Y., Brown, S.A. *TRENDS Gen.* **20**, 51-58 (2004).
3. Forood, B., Prez-Pay, E., Houghten, R.A., Blondelle, S.E. *Biochem. Biophys. Res. Comm.* **211**, 7-13 (1995).
4. Blondelle, S.E., Forood, B., Houghten, R.A., Prez-Pay, E. *Biochemistry* **36**, 8393-8400 (1997).
5. Fuentes, G.E., Hood, C.A., Patel, H., Menakuru, M. *PharmaChem* 12-14 (2010).

Proceedings of the 23rd American Peptide Symposium
Michal Lebl (Editor)
American Peptide Society, 2013

Chemical Synthesis of Human Relaxin-2

Lin Chen, Aksana V. Dalhitski, Michael P. Fleming, R. Andrew Hamilton, Songyu Liu, Jill R. Moore, Gillian M. Nicholas, Ileana I. Nuiry, A. James Vieth, and Gregory P. Withers

Corden Pharma Colorado, Inc., 2075 North 55th Street, Boulder, CO, 80301, U.S.A.

Introduction

Human Relaxin-2 (HR-2) is the peptide hormone that plays a vital regulatory role in mammalian pregnancy optimizing the many physiological changes taking place during pregnancy. It also acts as a pleiotropic hormone which protects heart, lungs, and kidneys due to its strong anti-fibrotic and vasodilator activities [1]. Currently, HR-2 and similar peptide hormones are mainly produced by biosynthesis. There are some published examples of small scale chemical syntheses [2,3]. We have identified a scalable chemical process for synthesis of HR-2. Taking advantage of our expertise in large scale peptide manufacture (including enfuvirtide, GLP-1 analogs and linaclotide), the process development focused on a convergent peptide synthesis to generate HR-2 at a multi-gram scale and to enable us to scale the process to multi-kilogram production.

Results and Discussion

As shown in the Figure 1, Chain A (1-24) of HR-2 was assembled linearly on 2-CTC resin using Fmoc chemistry, cleaved/globally deprotected and purified by HPLC while the Chain B (1-29) was synthesized on 2-CTC resin in three fragments: Chain B1 (1-12), Chain B2 (13-24) and Chain B3 (25-28) due to the severe aggregation issues if synthesized linearly. Each fragment was cleaved from the resin, isolated and then coupled together in solution to produce Chain B: (1) the Chain B3 (25-28) was first coupled with *C*-terminal Serine methyl ester to afford Chain B3' (25-29); (2) Chain B3' then coupled with Chain B2 (13-24) to yield Chain B2+3' (13-29); finally Chain B2+3' (13-29) coupled with Chain B1 (1-12) to give Chain B (1-29).

The fully assembled Chain B (1-29) was globally deprotected and purified by HPLC. Human Relaxin-2 was obtained by folding using purified Chain A (1-24) and Chain B (1-29) in a redox solution and followed by two pass final HPLC purification.

There are several challenges to overcome in order to make this chemical process scalable. Here are some we have encountered:

(1) The Chain A [24]Cys racemization could be minimized by selecting proper coupling conditions.
(2) The Chain B2 [13]Arg deletion was caused by the difficult Fmoc removal. Several conditions were evaluated such as a) DBU/piperidine: b) elevated temperature and c) repeated treatements. We found that although DBU/piperidine was effective to remove Fmoc, the overall peptide purity went down.

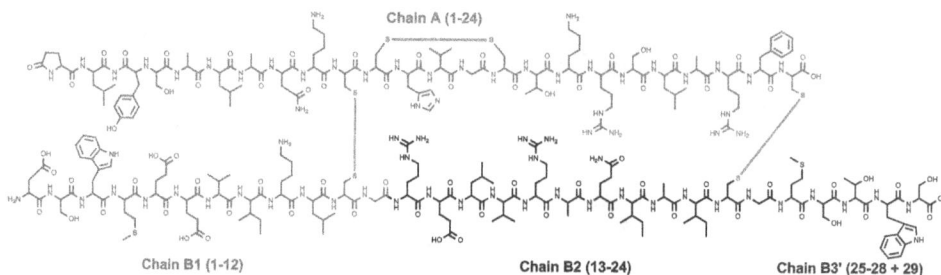

Fig. 1. Human Relaxin-2 (HR-2) sequence.

Table 1. HPLC purities of each process steps.

Process Step	Chain A Build	Chain B1 Build	Chain B2 Build	Chain B3 Build	
Purity (% AN)	70	80	92	95	
Process Step	Chain A Crude	Chain B Crude	Chain B1+2+3'	Chain B2+3'	Chain B3'
Purity (% AN)	52	39	57	73	89
Process Step	Purified Chain A	Purified Chain B	Final Purification 1st pass	Final Purification 2nd pass	
Purity (% AN)	85	75	97	98	

(3) The Chain B fragment isolations after cleavage were complicated due to gelling issues of the fragments, especially with Fragments 1 and 2. By optimizing the cleavage conditons and isolation solvent system, the fragments were isolated.

(4) Gelling in fragment condensation steps caused incomplete couplings as well as protected peptide intermediate isolation problems. The usage of LiBr reduced the impact of the gelling.

(5) During the purification, finding dissolution conditions for crude Chain A and Chain B presented multiple challenges because the crude peptides were either not completely soluble or caused gel formation. Multiple solvent systems were tried and proper dissolution conditions were identified.

(6) Due to high aggregation propensity of Chain B, it failed to elute from the HPLC column. The peptide tends to be retained on the column. This problem caused both analytical and purification issues. The HPLC conditions we developed greatly minimized the issues.

(7) We also found that the purification fractions turned to gel if they were held for long periods.

(8) At the folding step, the reaction foams. While organic solvents help reduce foaming, it is not scalable due to flammability concerns. We developed the process to scale up the folding step. The folding reaction was monitored using Mettler Toledo InPro Sensors.

In conclusion, a scalable chemical process for production of HR-2 has been developed (HPLC purities of the process steps are shown in Table 1). Five grams of human relaxin-2 of >98.0% AN purity were produced.

References

1. Teichman, S.L., Unemori, E., Teerlink, J.R., Cotter, G., Metra, M. *Curr. Heart Fail. Rep.* **7**, 75-82 (2010).
2. Wade, J.D., Lin, F., Hossain, M.A., Shabanpoor, F., Zhang, S., Tregear, G.W. *Relaxin and Related Peptides: Fifth International Conference: Ann. N.Y. Acad. Sci.* **1160**, 11-15 (2009).
3. Barlos, K.K., Gatos, D., Vasileiou, Z., Barlos, K. *J. Pept. Sci.* **16**, 200-211 (2010).

Proceedings of the 23rd American Peptide Symposium
Michal Lebl (Editor)
American Peptide Society, 2013

Fmoc-Sec(Xan)-OH: A Selenocysteine SPPS Derivative with a TFA-Labile Sidechain Protecting Group

Stevenson Flemer Jr.

Department of Biochemistry, University of Vermont College of Medicine, B415 Given Building, 89 Beaumont Ave; Burlington, VT, 05405, U.S.A; sflemer@uvm.edu

Introduction

Advancement in the field of protein engineering is promoted only when the synthetic tools available to the peptide chemist are sufficient to allow this progress. In the synthetic arena of disulfide-bonded proteins in particular there exist distinctive challenges which rely on continuous improvements in architectural strategies and stepwise oxidative closure of the disulfide connectivities. The substitution of selenocysteine (Sec) for cysteine (Cys) in protein models allows for the installation of the more robust diselenide bridges in place of native disulfide structure to afford peptide chimeras with interesting physical properties, increased stabilities, and elevated therapeutic potential. Construction of Sec-containing peptides and small proteins can be difficult to carry out using biological vectors and is therefore typically carried *via* chemical means using appropriately protected Sec derivatives.

Fmoc SPPS is frequently the method of choice for many peptide chemists, as it employs more gentle piperidine-mediated N^α deprotection while avoiding the harsh and more toxic HF cleavage conditions typically used. Currently, all known Sec SPPS derivatives (commercially-available or otherwise) bear selenol sidechain protection which is orthogonal to standard TFA cleavage conditions for Fmoc/tBu SPPS, requiring additional deprotective steps or inclusion of harsh additives to the TFA cocktail. Presented here is the first example of a TFA-labile sidechain protectant for Sec in the form of the SPPS derivative Fmoc-Sec(Xan)-OH. Xanthenyl sidechain protection, known for the corresponding Cys derivative [1], offers a new and more convenient avenue toward the chemical synthesis of native Sec-containing peptides and proteins. TFA lability of a Sec sidechain protectant adds a new practical vector of orthogonality to the repertoire of avenues available to the peptide chemist.

Results and Discussion

Fmoc-Sec(Xan)-OH **3** was prepared in two high-yielding steps from commercially-available L-selenocystine **1** *via* the versatile *bis* Fmoc diselenide intermediate **2** [2] (Figure 1). Interestingly, it was found that methods involving the more traditional methods of initial diselenide scission (ie: borohydride reduction) were inefficient due to the basic conditions typically employed for this process. Instead, the acidic biphasic Zn-mediated method of Santi [3] with some modification was most amenable to maintaining the intermediate free selenol long enough for complete reaction with the xanthenyl cation electrophile. This synthetic scheme was remarkably amenable to scaleup, allowing the production of multigram quantities of the derivative in high yield. **3** was found to be bench-stable for long periods, with no requirements for special handling or usage.

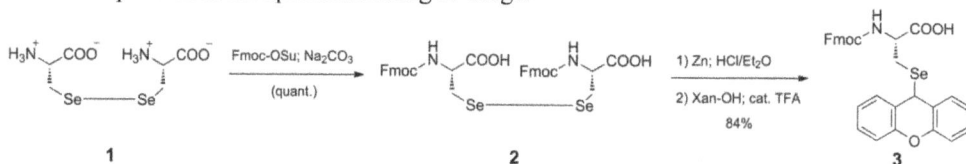

Fig. 1. Synthesis of Fmoc-Sec(Xan)-OH 3 from L-selenocystine.

In order to ascertain its suitability for use in SPPS, derivative **3** was incorporated into two test peptide sequences, both of which were chosen to emphasize the behavior of this Sec derivative in different environments following TFA deprotection. As shown in Figure 2, single Sec-containing MMP3-I [4] model peptide **4** and dual Sec-containing Lys-16 Glutaredoxin 10-17 analog **5** [5] were synthesized on resin and cleaved using a standard TFA cocktail. As expected, peptide **4** was isolated as its corresponding diselenide dimer **6** in good crude purity. Most gratifyingly, peptide model **5** was isolated exclusively as the intramolecular diselenide **7** in exceptional crude purity, without any trace of oligomeric structure.

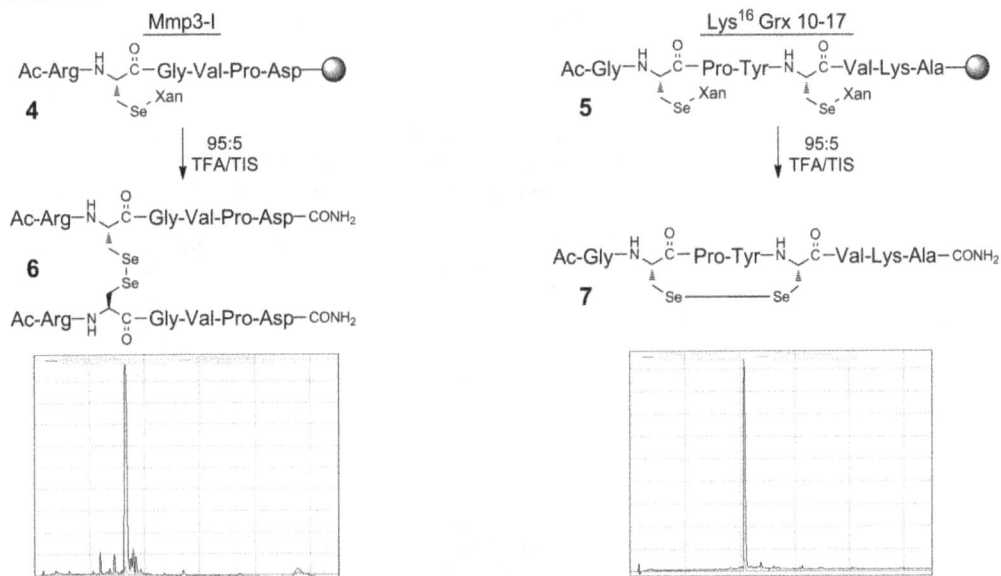

Fig. 2. Deprotection profile of model peptides showing HPLC spectra of crude isolates.

The scarcity of existing Sec sidechain protection protocol is striking when compared with the enormous number of protecting groups for its chalcogen analog cysteine. The advent of new Sec protecting groups allow for new vectors of orthogonality to be applied to differentiated Sec pairs in engineered peptide and protein systems. The perpendicular orthogonality profile between the TFA-labile Sec(Xan) group and all other currently existing Sec sidechain protectants holds great promise in the design and synthesis of multiple selenylsulfide- and diselenide-containing peptide systems constructed via the stepwise approach. Indeed, our lab is presently carrying out the production of new and novel Sec protecting groups orthogonal to presently-existing ones in order to hopefully inspire a renaissance within the field and, once having reached a critical mass, we hope it will nucleate a renewed interest in the syntheses of multiple diselenide-containing peptide and protein analogs.

References

1. Han, Y., Barany, G. *J. Org. Chem.* **62**, 3841-3848 (1997).
2. Agan, M., Schroll, A. *Abstracts of Papers, 241st ACS National Meeting*, Anaheim, CA, March 27-31 (2011).
3. Santi, C., Santoro, S., Testaferri, L., Tiecco, M. *Synlett* **39**, 1471-1474 (2008).
4. Hanglow, A.C., Lugo, A., Walsky, R., Finch-Arietta, M., Lusch, L., Visnick, M., Fotouhi, N. *Agents & Actions* **39**, C148-150 (1993).
5. Besse, D., Moroder, L. *J. Pept. Sci.* **3**, 442-453 (1997).

Proceedings of the 23rd American Peptide Symposium
Michal Lebl (Editor)
American Peptide Society, 2013

PAET Resin for the Synthesis of Peptide Thioesters

Robert O. Fox and Kris F. Tesh

Department of Biology & Biochemistry, The University of Houston, Houston, TX, 77004-5001, U.S.A.

Introduction

The synthesis of proteins by native chemical ligation (NCL) provides the opportunity to produce pure proteins with novel chemistries. Thioester peptides can be directly synthesized using Boc/benzyl solid phase peptide synthesis (SPPS) methods. For Fmoc SPPS, the use of piperidine base to remove the Fmoc protecting group is incompatible with the direct synthesis of peptide thioesters. A number of methods have been reported that introduce the thioester after SPPS. Some of these methods involve the synthesis as peptide C-terminal amide adducts followed by an N,S-acyl shift reaction to produce a thioester. Recently, bis(2-sulfanylethyl)amino-trityl-polystyrene (SEA) resin has been reported [1,2], where the peptide is synthesized as the secondary amide, and the N,S-acyl shift occurs by reaction of one of the two 2-sulfanylethyl moieties. The 2-(propylamino)ethanethiol 2-chlorotrityl (PAET) resin reported herein is easily synthesized, and was used to demonstrate the efficacy of a single ethanethiol group to support thioester formation and subsequent NCL with peptide-*paet* C-terminal secondary amides.

Experimental Methods

PAET Resin Synthesis (Figure 1): Cystamine 2-chlorotrityl resin **1** (524 mg, 0.25 mmol) (Sigma Aldrich) was reacted at 40°C overnight in methanol (10 mL) with propionaldehyde (359 µL, 20 eq) to form a Schiff base **2**. The resin was washed for 5 min 4x with 5 mL methanol. The resin **2** was incubated with 2-picoline borane complex (535 mg, 20 equ) at 40°C overnight to reduce the Schiff base to the secondary amine PAET resin **3**. The resin was washed with 4 x 5 mL methanol and dried under reduced pressure.

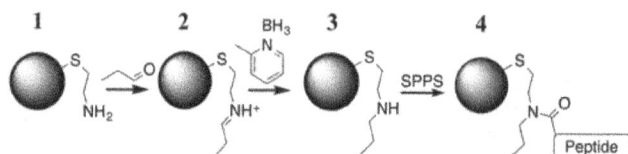

Fig. 1. Synthesis of PAET resin and use in SPPS.

Peptide Synthesis: SPPS was carried out using Fmoc chemistry on a CEM Liberty 12 instrument with microwave assisted coupling following the CEM standard protocols, with HBTU/DIEA for coupling to yield a resin bound protected peptide **4**. The first residue was added to the PAET resin using double coupling with the standard protocol. The peptide GGLLAYG-*paet* **5** was synthesized as described above, cleaved with 94%TFA/2.5%H$_2$O/1.5%TIS/2.0%EDT, and purified by HPLC.

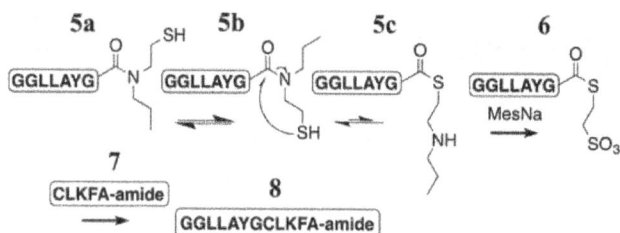

Fig. 2. Native chemical ligation.

Native Chemical Ligation (Figure 2): Peptide **5** was incubated with a ~10-fold excess of CLKFA-*NH$_2$* **7** in 0.5 M phosphate buffer containing 100 mM MesNa and 50 mM TCEP at 40°C. Samples were analyzed by HPLC and mass spectrometry at 0, 2, 4 and 6 hrs.

Results and Discussion

PAET Resin Synthesis: The PAET resin was easily synthesized from commercially available cystamine 2-chlorotrityl resin. The conversion of the cystamine groups to form 2-(propylamino)ethanethiol 2-chlorotrityl (PAET) resin **3** was not complete as judged by the formation of some peptide cystamine product. The sequential Schiff base formation followed by reduction was adopted to eliminate a possible second alkylation that could occur if the steps were carried out simultaneously, which would result in a tertiary amine that would not support SPPS.

Native Chemical Ligation: The PAET resin was used to synthesize peptide **5** (GGLLAYG-*paet*). Deprotection yields the secondary amide that can isomerize between two configurations (**5a** & **5b**), and **5b** can undergo an N,S-acyl shift to form the peptide thioester **5c** (Figure 2). Reaction of **5c** with 2-mercaptoethane *s*ulfonate Na (MesNa) yields a stable

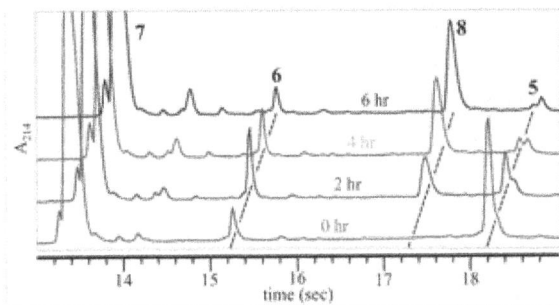

Fig. 3. Native chemical ligation.

GGLLAYG-*mesna* thioester **6**. An NCL reaction with peptides **5** and **7** was carried out at 40°C and monitored by HPLC (Figure 3). The thioester peptide **5** (18.2 min) was depleted by 6 hr. The MesNa adduct **6** (15.2 min) formed initially during sample preparation (0 hr), peaked in concentration at 2 hr, and was largely consumed by 6 hr. The NCL product **8** (17.2 min) formed readily, presumably by reaction of peptide **7** with both peptides **5c** and **6**. All peptides were characterized by mass spectrometry: **5** (GGLLAYG-*paet*, obs. 751.5, calc. 751.5), **6** (GGLLAYG-*mesna*, obs. 774.6, calc. 774.5), **7** (CLKFA-NH_2, obs. 580.6, calc. 580.7), **8** (GGLLAYGCLKFA-NH_2, obs. 606.7, calc. 606.5, 2+ charge state).

The NCL reactivity of peptide-*paet* products, (with a *C*-terminal secondary amide containing a single ethanethiol) compares favorably with published results [1,2] for peptide-*sea* peptides, (which contain two ethanethiol groups), indicating that the isomerization of the secondary amide is not an impediment to thioester formation. Similar results have been obtained with N-sulfanylethylanilide (SEAlide) peptides [3]. The PAET resin is easily prepared and supports the synthesis of peptide-*paet* products that react readily in NCL reactions.

Acknowledgments

We thank Deqian Liu for assistance with the mass spectrometry. Supported by funding from The University of Houston.

References

1. Dheur, J., Ollivier, N., Vallin, A., Melnyk, O. *J. Org. Chem.* **76**, 3194-3202 (2011).
2. Hou, W., Zhang, X., Li, F., Liu, C.F. *Org. Lett.* **13**, 386-389 (2011).
3. Tsuda, S., Shigenaga, A., Bando, K., Otaka, A. *Org. Lett.* **11**, 823-826 (2009).

Proceedings of the 23rd American Peptide Symposium
Michal Lebl (Editor)
American Peptide Society, 2013

Exploiting Furan's Versatile Reactivity in Reversible and Irreversible Orthogonal Peptide Labeling

Kurt Hoogewijs[1], Dieter Buyst[2], Johan M. Winne[1], José C. Martins[2], and Annemieke Madder[1]

[1]Ghent University, Organic and Biomimetic Chemistry Research Group, Krijgslaan 281 S4, 9000, Ghent, Belgium; annemieke.madder@ugent.be; [2]Ghent University, NMR and Structure Analysis, Krijgslaan 281 S4, 9000, Ghent, Belgium

Introduction

Modern chemical biology oriented research frequently relies on the ability to site-selectively label macromolecules such as peptides, carbohydrates and oligonucleotides with fluorophores. Most strategies exploit the nucleophilic functionalities naturally present in biomolecules, providing efficient but less selective labeling. The need for more orthogonal strategies can be met by the introduction of unnatural amino acids. Recently our group developed a method for the labeling of peptides by generation of a reactive aldehyde in the sequence. The non-natural 2-furylalanine was incorporated into a peptide and subsequently converted into an α,β-unsaturated aldehyde by selective oxidation [1]. Reductive amination then introduces the desired fluorophore [2,3]. However, the furan diene has also shown to be a useful partner in Diels-Alder reactions with commercially available maleimides as dienophiles. Due to the necessity of a large excess of maleimide, the long reaction times as well as the formation of several diastereoisomers in this reaction, this reaction was shown to be less suited for quantitative labeling of peptides. The more recently described 1,2,4-triazole-3,5-diones (TAD) present an excellent alternative, requiring less equivalents as well as shorter reaction times [4,5].

Results and Discussion

Peptide **1** was synthesized by automated peptide synthesis using the Fmoc/tBu strategy with HBTU as coupling reagent (Figure 1) [6]. 20 equiv. of N-phenylmaleimide were added in a minimal amount of toluene and the reaction mixture was left at room temperature for 48h. The Diels-Alder product was cleaved of the resin with TFA/TIS/H_2O (95:2.5:2.5), 2h at RT for HPLC analysis (Figure 1b). A mixture of Diels-Alder adducts along with starting material was observed, due to formation of several diastereoisomers during the cycloaddition reaction. By performing the Diels-Alder reaction at 70°C, a more selective reaction occurred; yet starting material was still observed (Figure 1c). These undesirable results could be explained by the reversible nature of this reaction mixture. Indeed, heating of the resin-bound labeled peptide **3** to 70°C for 24h in toluene, results in a complete retro-Diels-Alder conversion (Figure 1d).

When making use of 4-phenyl-1,2,4-triazole-3,5-dione (PTAD) as dienophile, the reaction is driven to completion in only 15 minutes, using just 3 equivalents of PTAD in dichloromethane. Subsequent cleavage from the resin and deprotection of the peptide for HPLC analysis

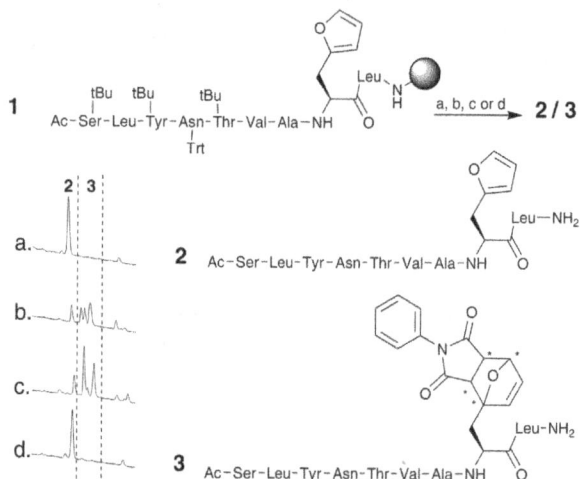

Fig. 1. Relevant peptide structures and corresponding HPLC traces of reaction mixtures after cleavage of the solid support and deprotection. With: a) starting material; b) and c) Diels-Alder products; and d) retro-Diels-Alder products. (Figure adapted from reference 4).

shows that all starting material is indeed converted (Figure 2). This reaction thus provides advantages compared to the Diels-Alder reaction, both in terms of reaction rate as well as number of equivalents used. However, by detailed NMR analysis of the formed adduct, we provided evidence that it was in fact not the Diels-Alder type adduct which was formed. Protons at C-7 and C-8 are observed as mutually coupled doublets (Figure 3 a,b), indicating there are no other neighbouring protons. In addition, a signal broadened through exchange with residual water in DMF-d7, can be seen at 11,61 ppm (Figure 3c),

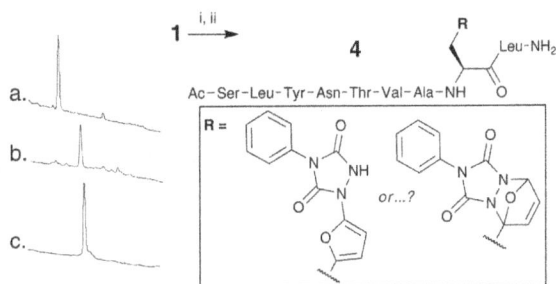

Fig. 2. Reaction conditions: (i) PTAD in DCM, 15 min. at RT; (ii) TFA/TIS/H$_2$O (95:2,5:2,5), 2h at RT. HPLC traces of reaction mixtures for: a) Peptide 1 after cleavage and deprotection with TFA/TIS/H$_2$O (95:2,5:2,5); b) Crude HPLC revealed complete conversion; c) Labeled peptide 4 after HPLC purification. (Figure adapted from reference 4).

favoring structure 5 over the Diels-Alder adduct 6, since only the former features an exchangeable hydrogen in the side chain. This hypothesis was confirmed by measurements at -50°C, where the exchange broadening is sufficiently reduced (Figure 3d).

In conclusion, as the incorporation of furan moieties in peptides and proteins has recently been firmly established in various contexts, the current methodology ideally complements the toolbox of bio-orthogonal labeling reactions. In conjunction with our previously developed furan-oxidation based peptide labeling and nucleic acid crosslinking methodologies [7-10] the current work again testifies of

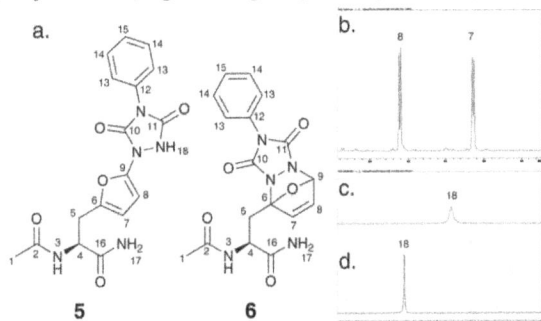

Fig. 3. a) Proposed structures for the furylalanine-PTAD adduct; b) 2 doublets between 6.2 and 6.6 ppm; c) broadened signal at 11,6 ppm assigned to exchangeable proton 18 in 5; d) measurements at -50°C sharpen and move the signal downfield by slowing down exchange. (Figure adapted from ref. 4).

the usefulness and versatile application of a simple and small aromatic furan moiety for the decoration and conjugation of different biomacromolecules.

Acknowledgments

We thank Prof. Dr. Alain Krief for valuable suggestions and BOF-UGent (01J06111) for financial support. The 700 MHz equipment is part of the Interuniversitary NMR Facility funded in part by the FFEU-ZWAP initiative of the Flemish Government.

References

1. Hoogewijs, K., Deceuninck, A., Madder, A. Org. Biomol. Chem. 10, 3999-4002 (2012).
2. Deceuninck, A., Madder, A. Chem. Commun. 340-342 (2009).
3. Hoogewijs, K., Buyst, D., Martins, J., Madder, A. J. Pept. Sci. 18, S73-S73 (2012).
4. Hoogewijs, K., Buyst, D., et al. Chem. Commun. 49, 2927-2929 (2013).
5. Ban, H., Gavrilyuk, J., Barbas, C.F. J. Am. Chem. Soc. 132, 1523 (2010).
6. Chersi, A., Ferracuti, S., Falasca, G., Butler, R.H., Fruci, D. Anal. Biochem. 357, 194-199 (2006).
7. Stevens, K., Madder, A. Nucleic Acids Res. 37(5), 1555-1565 (2009).
8. Op de Beeck, M., Madder, A. J. Am. Chem. Soc. 133, 796-(2011).
9. Jawalekar, A.M., de Beeck, M.O., et al. Chem. Commun. 47(10), 2796-2798 (2011).
10. Op de Beeck, M., Madder, A. J. Am. Chem. Soc. 134 (26), 10737-10740 (2012).

Proceedings of the 23rd American Peptide Symposium
Michal Lebl (Editor)
American Peptide Society, 2013

The Synthesis of ADP-Ribosylated Peptides

Hans A.V. Kistemaker, Gerbrand J. van der Heden van Noort, Herman S. Overkleeft, Gijsbert A. van der Marel, and Dmitri V. Filippov

Leiden Institute of Chemistry, Leiden University, P.O. Box 9502, 2300 RA Leiden, The Netherlands

Introduction

Adenosine diphosphate ribosylation (ADPr) is a peculiar type of protein glycosylation that occurs in both *mono-* and *poly*meric form and is considered to play an important role in a wide range of biological processes (i.e. DNA damage repair). The construction of well-defined ADP-ribosylated peptides and analogues thereof would be of significant help in gaining a better understanding of the role and function of ADP-ribosylation. One of the crucial steps in the synthesis of ADP-ribosylated peptides is the construction of the α-glycosidic linkage between the ribofuranosyl moiety and the amino acid side chain. We demonstrate a new and versatile method for the α-selective ribosylation of various amino acids and the incorporation in a peptide via Solid Phase Peptide Synthesis (SPPS).

Results and Discussion

We reasoned that the preparation of side-chain ribosylated amino acids might proceed with better α-selectivity when using acid catalysed glycosylation. Therefore, different ribofurano-

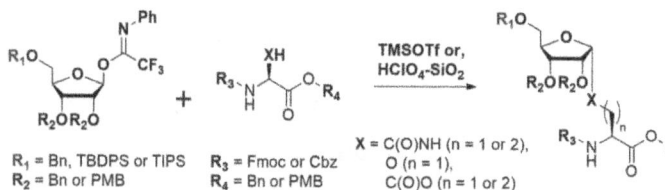

syl N-phenyltrifluoro-acetimidates were synthesized and tested in the ribosylation of various amino acids (Figure 1). The newly synthesized imidate donors were first tested in the ribosylation of amino acids with carboxyben-zyl (Cbz) and benzyl (Bn) protection. The

Fig. 1. Schematic for the ribosylation of amino acids.

reactions were performed in DCM and the donors were activated by the addition of a catalytic amount of TMSOTf.

Table 1. Ribosylation of various amino acids.

Donor		Acceptor	Temp (°C)	Yield (%)	Ratio (α/β)
R_1	R_2	(Cbz-X-OBn)			
Bn	Bn	Gln	-20 to rt	56	69/31
TBDPS	Bn	Gln	-20 to rt	57	95/5
TIPS	Bn	Gln	-20 to rt	66	95/5
TIPS	PMB	Gln	-20 to rt	52	85/15
Bn	Bn	Asn	-20 to rt	68	96/4
TIPS	Bn	Asn	-20 to rt	68	98/2
Bn	Bn	Ser	-50	78	75/25
TBDPS	Bn	Ser	-50	60	100/0
TIPS	Bn	Ser	-50	56	100/0
TIPS	Bn	Glu	-50	84	77/23
TIPS	Bn	Asp	-70	63	50/50

Table 2. Ribosylation of Fmoc-glutamine and asparagine.

Donor		Acceptor	Temp (°C)	Yield (%)	Ratio (α/β)
R_1	R_2	(Fmoc-X-OBn)			
TBDPS	Bn	Asn	0	44	97/3
TBDPS	Bn	Asn	rt	63	93/7
TBDPS	Bn	Gln	0	49	72/28
TIPS	Bn	Gln	rt	59	96/4*
TIPS	PMB	Gln	-20 to rt	60	78/22
TIPS	PMB	Gln	rt	69	93/7*

*$HClO_4$-SiO_2 was used as activator.

The stereochemical outcome was determined by H-NMR spectroscopy [1] and the results clearly show the influence of the 5'-modification on the stereochemical outcome of the reaction. Introducing the more bulky TIPS or TBDPS protecting group results in a tremendous increase in the α-selectivity (Table 1). For the ribosylation of glutamic- and aspartic acid only moderate α-selectivities could be achieved. However, this is the first time that these compounds could be synthesized and the anomers could be separated by silica gel chromatography. For the ribosylated amino acids to be useful in SPPS, multiple protective group manipulations were needed. Therefore, the more straightforward ribosylation of Fmoc-glutamine and asparagine was explored. The solvent was changed to Dioxane/DCM (1/1) for solubility reasons and the ribosylation of asparagine remained to be highly α-selective even at room temperature (Table 2). Ribosylation of glutamine proved to be more troublesome with loss in α-selectivity but this could be overcome by using $HClO_4$-SiO_2 as activator. The

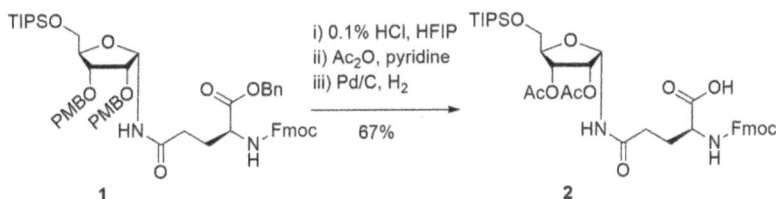

Fig. 2. Synthetic route to a useful Fmoc building block for SPPS.

ribosyl imidate donor was further optimized by replacing the benzyl ethers with para-methoxybenzyl ethers (PMB). This allowed to selectively replace them with acetyls. Final hydrogenolysis step yields the desired Fmoc building block **2** which can be used in SPPS (Figure 2). Ribosylated Fmoc building block **2** was incorporated in a peptide via SPPS as described by van der Heden van Noort, et al. [2].

The optimized methodology to ribosylate amino acids in a highly α-selective manner will be extended to other amino acids which will be used in SPPS. Furthermore, the phosphorylation and pyrophosphate formation of these compounds is currently investigated in order to obtain higher yields of mono-ADP-ribosylated peptides.

Acknowledgments

This work was funded by the Netherlands Organization of Scientific Research (NWO). The authors would like to thank the American Peptide Society and Leiden University Fund for financial support for H.A.V.K to participate in the International Peptide Symposium.

References

1. Kistemaker, H.A.V., van der Heden van Noort, G.J., Overkleeft, H.S., van der Marel, G.A., Filippov, D.V. *Org. Lett.* **15**, 2306-2309 (2013).
2. van der Heden van Noort, G.J., van der Horst, M.G., Overkleeft, H.S., van der Marel, G.A., Filippov, D.V. *J. Am. Chem. Soc.* **132**, 5236-5240 (2010).

Proceedings of the 23rd American Peptide Symposium
Michal Lebl (Editor)
American Peptide Society, 2013

Exploration of the Scope of Suzuki–Miyaura Cross-Coupling in Peptide Ligation

Tae-Kyung Lee, Bikash Manandhar, and Jung-Mo Ahn*

Department of Chemistry, University of Texas at Dallas, Richardson, TX, 75080, U.S.A.

Introduction

Suzuki–Miyaura cross-coupling reaction is one of the most powerful methods for the construction of carbon-carbon bonds. In a typical Suzuki–Miyaura reaction, an aryl boronic acid is cross-coupled with an aryl or vinyl halide or pseudo-halide in the presence of a palladium catalyst and a base in organic solvent at elevated temperature [1,2]. Whereas the Suzuki–Miyaura reaction has a high potential in peptide ligation and conjugation due to advantages like bio-orthogonality and tolerance toward a broad range of functional groups, it has been rarely used in peptide chemistry presumably because of relatively incompatible reaction conditions to peptides, such as high temperature and use of organic solvents [3-6].

In this study, we examined the scope and limitations of the Suzuki–Miyaura cross-coupling as a tool for peptide ligation. The reaction conditions were optimized by surveying various catalysts, bases, temperature, solvents, and additives. We also determined the effects of amino acid side chains on the ligation reactions of peptides. Following the optimized conditions, we then demonstrated the feasibility of ligations between two long peptides.

Results and Discussion

As reactants of Suzuki–Miyaura cross-coupling, iodo-peptide and peptide boronic acid were synthesized on Rink amide resin by using standard Fmoc-tBu solid-phase peptide synthesis protocol. These peptides were cleaved from the resin with TFA and purified by RP-HPLC. We first studied the impact of bases and solvents at a range of temperature on the cross-coupling reaction of the iodo-peptide and the peptide boronic acid (Figure 1). As a base, K_2CO_3 was found to be quite efficient compared to KF and TEA. Screening of solvents showed that aqueous DMF, 2,2,2-trifluoroethanol (TFE) and 10% sodium dodecyl sulfate (SDS) solutions afforded the cross-coupled peptides in good yields, whereas 6M urea solution was not effective for the ligation. In particular, the coupling reaction at 40°C in 10% aqueous SDS solution achieved a remarkably high yield (95%, Figure 1a). The concentration of SDS can be lowered to 0.3% without compromising the high yield. On the other hand, Triton and glycerol were less efficient for the cross-coupling reactions (Figure 1b). With the optimized reaction conditions in hand (K_2CO_3, 10% SDS), we then evaluated various Pd catalysts for the efficiency of the ligation reaction. Whereas all of the tested catalysts gave the cross-coupled peptide in moderate to high yield, $PdCl_2(dppf)$ turned out to be the most suitable one (Figure 1c).

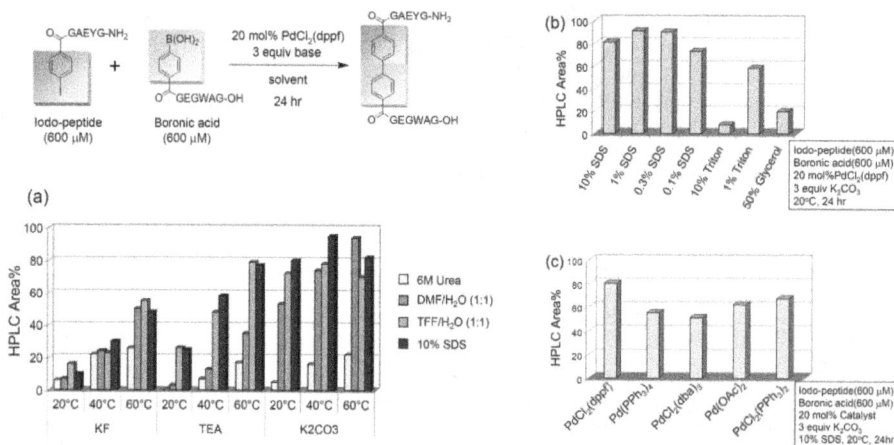

Fig. 1. Optimization of Suzuki–Miyaura cross-coupling reactions of peptides. Choice of (a) base, solvent, and temperature; (b) additive and concentration; and (c) catalyst.

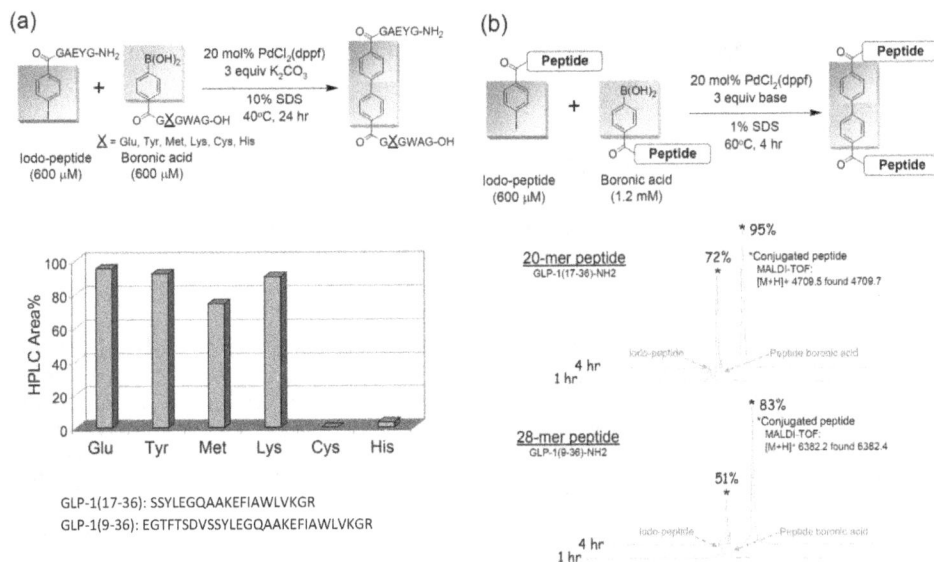

Fig. 2. Suzuki–Myaura ligation of peptides. (a) Effect of amino acid side chains on the cross-coupling reaction. (b) Ligation of long peptides.

In order to study the effect of amino acids on the cross-coupling reaction, we prepared a series of peptide boronic acids bearing various side chain functional groups (Figure 2a). Many amino acids were found to be compatible to the cross-coupling reaction giving 70-90% yield. However, Cys and His were not tolerated for the ligation presumably due to chelation of palladium with thiols and imidazoles. This suggests that protecting groups may be required for ligating peptides containing His or Cys. We then carried out ligation reactions between long peptides. Two 20-mer peptides were ligated at 40°C for 24 hr giving a 40-mer peptide product in 59% yield (data not shown). The yield of the ligation was increased to 95% by doing the reaction at 60°C for 4 hr. Remarkably, 28-mer peptides were successfully ligated generating a 56-mer peptide in 83% yield (Figure 2b). It is notable since a few methods were reported for efficient ligation of two long peptides in high yields.

In summary, we have optimized reaction conditions for Suzuki–Myaura cross-coupling of peptides. Denaturing solvents were found to be important for high-yielding ligation of long peptides and the composition of peptides may affect the ligation yield. Peptides containing Cys or His were unable to produce ligated products albeit many other amino acids were well tolerated. Under the optimized reaction conditions, ligations of long peptides were efficiently carried out showing a high potential of the reaction.

Acknowledgments

This work was supported by the Welch Foundation (AT-1595), Juvenile Diabetes Research Foundation (37-2011-20), and Cancer Prevention and Research Institute of Texas (RP100718).

References

1. Miyaura, N., Suzuki, A. *Chem. Rev.* **95**, 2457-2483 (1995).
2. Alonso, F., Beletskaya, I.P., Yus, M. *Tetrahedron* **64**, 3047-3101 (2008).
3. Ahn, J.-M., Wentworth, P., Janda, K.D. *Chem. Commun.* 480-481 (2003).
4. Ojida, A., Tsutsumi, H., Kasagi, N., Hamachi, I. *Tetrahedron Lett.* **46**, 3301-3305 (2005).
5. Doan, N.-D., Bourgault, S., Létourneau, M., Fournier, A. *J. Comb. Chem.* **10**, 44-51 (2008).
6. Chalker, J.M., Wood, C.S.C., Davis, B.G. *J. Am. Chem. Soc.* **131**, 16346-16347 (2009).

Proceedings of the 23rd American Peptide Symposium
Michal Lebl (Editor)
American Peptide Society, 2013

Synthesis of a Double Marker Synthon (NMR and Fluorescent) for Peptide Labeling

Marc-Andre Bonin[1,2], Pierre Baillargeon[1,*], Martin Lepage[2], and Witold A. Neugebauer[1]

[1]Department of Pharmacology, [2]Department of Nuclear Medicine and Radiobiology, Université de Sherbrooke, Sherbrooke (Québec), J1H 5N4, Canada; [*]currently teaching at CEGEP de Sherbrooke, 475 rue du Cégep, Sherbrooke (Québec), J1E 4K1, Canada

Introduction

Multimodality imaging can non-invasively monitor the distribution of peptide-based imaging probes *in vivo*. Fluorescent dyes are major tools for biomolecules labeling. They are applied in fluorescence microscopy [1], confocal microscopy [2], flow cytometry [3] and other fluorescent techniques. Paramagnetic Gd[III] chelates can achieve high proton relaxivity [4] as contrast agents in magnetic resonance imaging (MRI). Our intention was to combine those two markers (fluorescence and MRI) into one unit ready to couple to biomolecules such as peptides. We propose the synthesis of a double marker building block with *bis*-amino acids as lysine, ornithine, *2, 4*-diaminobutric acid or *2, 3*-diaminopropionic acids in the core (Figure 1). A *tris-t*-butyl ester DOTA chelator (in free acid form) for [68]Ga (positron emission tomography, PET) or Gd (MRI) is coupled on N^α-amine site and other β, γ, δ and ε-amino group of selected *bis* amino acids were coupled with a fluorescent agent (fluorescein) *via*

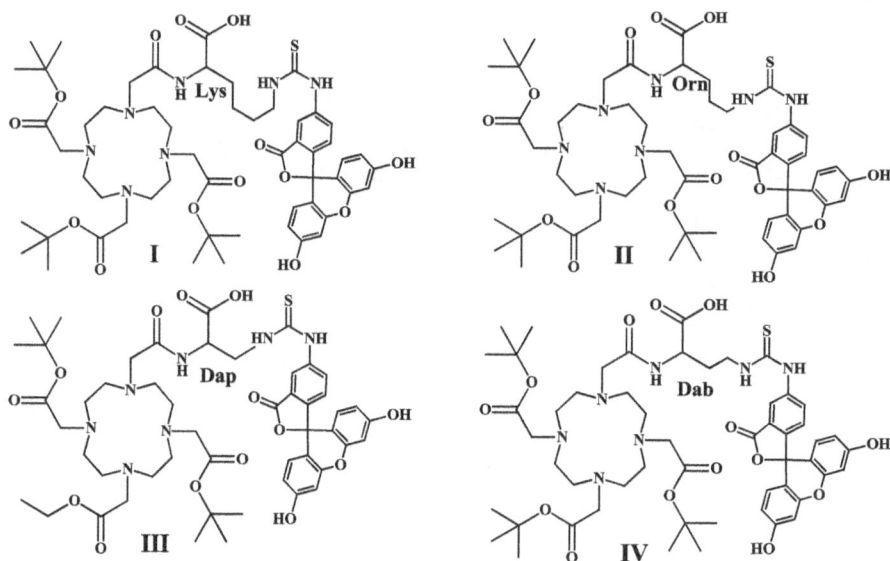

Fig. 1. Structures of fluorescein-DOTA double markers.

thiourea link. Use of β, γ, δ and ε-amino site for isothiocyanate coupling fits perfectly with requirements for stable thiourea bond formation [5]. As fluorescent agents we used FITC, but other fluorescent agents could easy be used instead if they are resistant to peptide cleavage conditions. Synthesis on chlorotrityl resin results in building blocks ready to couple to a selected peptide. Naturally occurring protein amino acid lysine (**I**) in peptide sequence at *N*-terminus of labeled peptide would be the most appropriate as synthon. In some cases other more enzymatically resistant *bis*-amino acids [ornithine (**II**), *2,3*-diaminopropionic acid (**III**) or *2,4*-diaminobutric acid (**IV**)] were used.

Results and Discussion

We synthesized four dual markers based on *di*-amino acid core as lysine, ornithine, *2,4*-diaminobutric acid and *2,3*-diaminopropionic acid (**I, II, II,** and **IV**). Synthesis scheme of those synthons is shown in Figure 2 as lysine derivative. The next step (not performed yet) in use of those synthons is coupling them to biomolecule of choice on solid phase with HCTU (3 eq.) and DIPEA (5eq.) as used in similar type of coupling [6]. Then *t*-Bu esters are cleaved and Gd^{III} chelate formed in solution or on solid phase if resin-peptide nature allows.

In case that steric hindrance may interfere, coupling of the synthon to peptide on solid phase could be achieved *via* a spacer (β-alanine, ε-aminocaproic acid or *11*-aminoundecanoic acid) at the *C*-site of core amino acid (lysine, ornithine, *2,4*-diaminobutric acid and *2,3*-diaminopropionic acid).

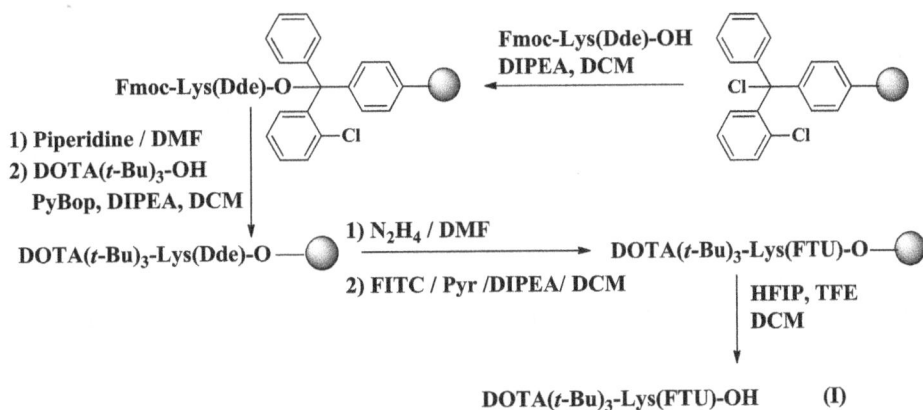

Fig. 2. Scheme of double marker synthesis.

Abbreviations: **HFIP**-*1,1,1,3,3,3-hexafluoro-2-propanol;* **TFE**-*2,2,2-trifluoroethanol;* **FITC**-*fluorescein isothiocyanate;* **FTU**-*fluorescein thiourea linked;* **DIPEA**-*N,N'-diisopropylethylamine;* **Dde**-*1-(4,4-dimethyl-2,6-dioxocyclohex-1-ylidene)ethyl;* **Pyr**-*pyridine;* **DOTA**-*1,4,7,10-tetraazacyclododecane-1,4,7,10-tetraacetic acid.*

An alternative solution to obtain double markers on biomolecule would be to couple DOTA(*tri*-ester) to the molecule, which usually is difficult, and then couple a fluorescent component (preferably at the last step). Fluorescent moieties are generally in form of isothiocyanate, such that coupling must occur to a site other than the N^α amino acid to avoid removal of the amino group of the last amino acid.

Acknowledgments

Supported by grant # 89722 from the Canadian Institutes of Health Research.

References

1. Lichtman, J.W., Conchello, J.A. *Nat. Methods* **2**(12), 910-919 (2005).
2. Minsky, M. *Scanning* **10**, 128-138 (1988); USA patent filed in 1957 and granted 1961. US 3013467.
3. Fulwyler, M.J. *Science* **150**, (698), 910-911 (1965).
4. Tóth, É. Bolskar, R.D., Borel, A., González, G., Helm, L., Merbach, A.E., Sitharaman, B., Wilson, L.J. *J. Am. Chem. Soc.* **127**, 799-805 (2005).
5. Jullian, M., Hernandez, A., Maurras, A., Puget, K., Amblard, M., Martinez, J., Subra, G. *Tetrahedron Letters* **50**, 260-263 (2009).
6. Bryson, D.I., Zhang, W., Ray, W.K., Santos, W.L. *Mol. BioSyst.* **5**, 1070-1073 (2009), Supporting Information.

Proceedings of the 23rd American Peptide Symposium
Michal Lebl (Editor)
American Peptide Society, 2013

Comparison of Heating Platforms for Fast, High Purity Peptide Synthesis: Infrared Heating on the Tribute®-IR vs. Microwave Peptide Synthesizers

James P. Cain, Christina A. Chantell, Michael A. Onaiyekan, and Mahendra Menakuru

Protein Technologies, Inc. Tucson, AZ, 85714, U.S.A.;
Website: www.ptipep.com, Email: info@ptipep.com;

Introduction

Like the choice of reagents, reaction times, resin type and loading, temperature is a variable that can be optimized in peptide syntheses. Adding heat may accelerate deprotection and coupling reactions for certain difficult sequences. In order to illustrate the advantages of using heat as part of a synthesis strategy, many manufacturers of microwave synthesizers have reported that peptides synthesized on their instrument at room temperature exhibit low purities and yields, but the results improve when heat is added. Unfortunately, what this may show in some cases is that the underlying platform itself does not perform well at room temperature, or is not compatible with all useful chemistries.

In contrast, the Tribute® UV-IR has been designed to accommodate a wide range of desired chemistry, including the use of highly efficient activators like HATU. This IR heating platform offers the advantages of parallel synthesis on more than one reaction vessel, vortex mixing for homogeneous temperature profiles, and the potential for scalability.

We have synthesized a difficult peptide with and without heat on the Tribute® and compared the results with previously reported syntheses on a microwave instrument. The sequence is a modified form of the 65-74 fragment of the acyl carrier peptide, a well-known difficult sequence commonly used to test the efficacy of various synthesis protocols. In the modified form, the original alanines are replaced with two adjacent sterically hindered aminoisobutyric acid (Aib) residues (Figure 1), resulting in a very challenging sequence (Aib-ACP).

Val-Gln-Aib-Aib-Ile-Asp-Tyr-Ile-Asn-Gly

Fig. 1. Structure of Aib-ACP peptide synthesized with and without heat on the Tribute® UV-IR and a microwave platform.

Results and Discussion

Aib-ACP was synthesized on the Tribute® UV-IR and compared to results from a microwave robotic peptide synthesizer [1] at room temperature and at 75°C. In each case, the Tribute® UV-IR produced the highest purity results. The heated results are quite similar, with the Tribute® UV-IR producing a slightly higher purity peptide (65% vs. 63% purity). On the other hand, the product of the longer room temperature synthesis is much more pure (84%) than the previously reported value for the room temperature synthesis on the microwave instrument (67%).

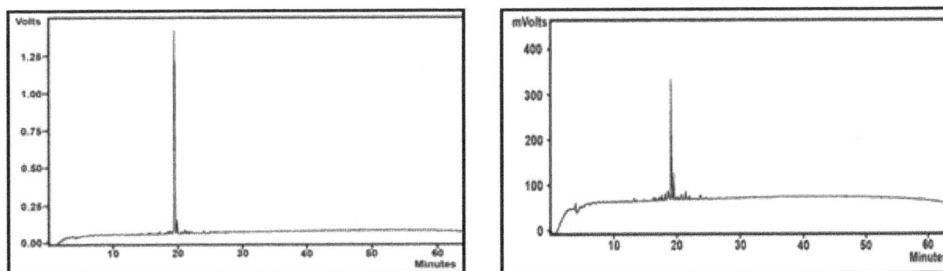

Fig. 2. HPLC traces of crude Aib-ACP peptides synthesized on the Tribute® UV-IR peptide synthesizer at a) room temperature and b) 75°C.

In both cases, DIC/HOAt chemistry produced crude percent purities in the 60s. On the microwave robotic platform, HATU at room temperature produced a similar crude percent purity to DIC/HOAt - also in the 60s. However, HATU at room temperature on the Tribute® UV-IR produced 84% crude purity - 20% higher than the other platform. It is well known that HATU is the most efficient activator available in the market today. Unfortunately, on the microwave robotic platform, its full potential may not be realized. This has been seen on other robotic synthesizers as well. It has been noted that with these instruments "the distribution of protected amino acids and coupling reagents takes a significant part of the synthesis procedure..." and therefore "...the faster coupling reagents are not providing a significant advantage" [2]. On these units, the time to dispense the coupling reagents is quite long. For easy couplings, long periods of time in contact with highly efficient activators like HATU may result in increased side products, lowering the overall purity of the peptide. The comparatively slow activator DIC is more forgiving, which is why it is often the activator of choice on such platforms.

In contrast, the Tribute® UV-IR is a solid, reliable platform regardless of your chemistry. Whether you are adding heat or using highly efficient activators such as HATU, the Tribute® UV-IR demonstrably gives you the best results.

References

1. Pedersen, S.L. and Jensen, K.J. P-071, *22nd American Peptide Symposium,* San Diego, CA, June 2011.
2. Hachman, J., Lebl, M. *Biopolymers* **84**, 340-347 (2006).

Proceedings of the 23rd American Peptide Symposium
Michal Lebl (Editor)
American Peptide Society, 2013

An Improved Synthetic Approach to Head-to-Tail Cyclic Tetrapeptides

P. Anantha Reddy, Sean T. Jones, Anita H. Lewin, Hernán A. Navarro, and F. Ivy Carroll

Center for Organic and Medicinal Chemistry, Discovery Sciences, Research Triangle Institute, Research Triangle Park, NC, 27709-2194, U.S.A.

Introduction

A number of cyclic tetrapeptides are bioactive natural products (fungal metabolites). For example, trapoxin B [1], chlamydocin [2], HC-toxin [3], WF-3161 [4], Cyl-2 [5] and more recently CJ-15,208 [6]. Their broad range of biological activity has spurred scientists to investigate both their *in vitro* and *in vivo* biological effects and use these compounds as templates in the design of newer drug candidates to explore more potent drugs. However, formation of 12 member constrained cyclic structure from a linear tetrapeptide can be difficult since more favored dimers and trimers could occur due to intermolecular condensation. Herein we report a simple, straightforward and general procedure for the synthesis of proline containing cyclic tetrapeptides. Peptides synthesized by this procedure were found to be stable both in solution and as solids over extended period of time. The individual linear tetrapeptide precursors were synthesized using either [1+(2+1)] or (2+2) solution phase fragment condensations. Thus, the following are cyclic tetrapeptides synthesized using this protocol:

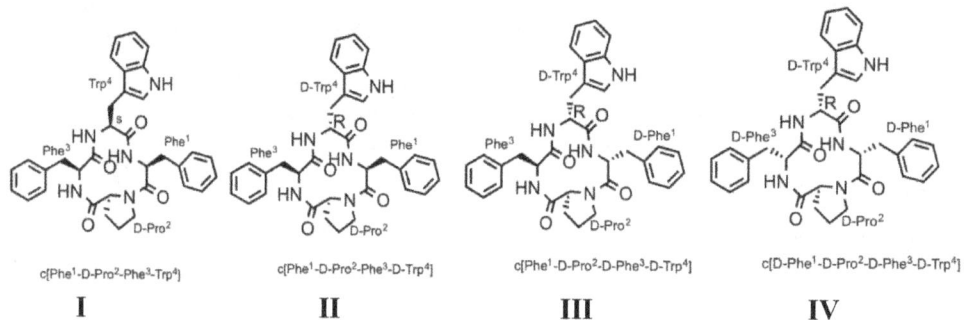

I	II	III	IV
c[Phe¹-D-Pro²-Phe³-Trp⁴]	c[Phe¹-D-Pro²-Phe³-D-Trp⁴]	c[Phe¹-D-Pro²-D-Phe³-D-Trp⁴]	c[D-Phe¹-D-Pro²-D-Phe³-D-Trp⁴]

Results and Discussion

The broad range of biological activities of several naturally occurring cyclic tetrapeptides; e.g., the antitumor activity of WF-3161; the ability of trapoxins A and B to flatten the *sis* oncogene-transformed NIH3T3 cells; and the kappa opioid receptor inhibitor activity of novel cyclotetrapeptide, CJ-15,208, make these peptides important model templates for use in the design of therapeutically useful drug candidates. In order to investigate their structure-activity relationships (SAR) there is an acute need for an easy access to synthetic analogs. In addition, the synthetic approach to obtain multiple analogs for use in high throughput screening needs to be as simple and straightforward as possible. Herein we report a general approach to synthesize cyclic tetrapeptides using solution phase peptide synthesis. As outlined in Figure 1, the synthesis of cyclic peptide **I** was accomplished using a (1+2+1) condensation approach. The key step is the head-to-tail cyclization using a reverse addition of the tetrapeptide to a condensation agent containing HATU/HOAt/DIEA mixture. Experimental and purification protocols are described below.

Synthesis of TFA•c[Phe-D-Pro-Phe-Trp] (CJ-15,208). A solution of the linear peptide, TFA•Phe-D-Pro-Phe-Trp-OH (50 mg, 0.000072 mol) in DMF (200 mL) was added slowly dropwise over a 10 h period to a solution of HATU (46 mg, 0.00012 mol), HOAt (16 mg, 0.00012 mol), and DIEA (0.10 mL, 0.00058 mol) in DMF (300 mL). The mixture was then allowed to stir at room temperature for an additional 24 h. After this time, the solvent was evaporated under reduced pressure. The residue was dissolved in EtOAc (200 mL), and the solution was washed successively with 2 N citric acid (2×10 mL), brine (2×20 mL), 10% aq. $NaHCO_3$ (2×10 mL) and brine (2×20 mL) then dried over anhydrous $MgSO_4$. The solvent was evaporated to get the cyclic tetrapeptide, c[Phe-D-Pro-Phe-Trp] (CJ-15,208) (**I**) as a brown solid (28 mg). The synthesis was repeated several times with similar results. Similarly, peptides **II**, **III**, and **IV** were also obtained and characterized.

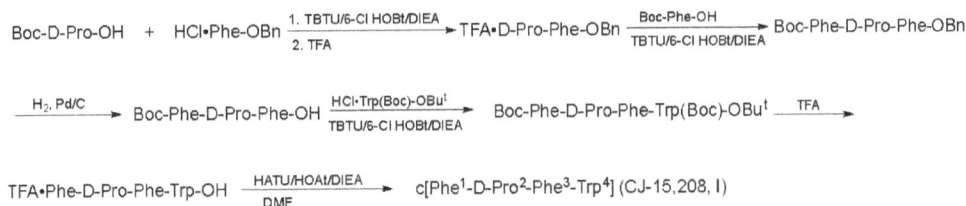

Boc-D-Pro-OH + HCl·Phe-OBn $\xrightarrow[\text{2. TFA}]{\text{1. TBTU/6-Cl HOBt/DIEA}}$ TFA·D-Pro-Phe-OBn $\xrightarrow[\text{TBTU/6-Cl HOBt/DIEA}]{\text{Boc-Phe-OH}}$ Boc-Phe-D-Pro-Phe-OBn

$\xrightarrow{\text{H}_2, \text{Pd/C}}$ Boc-Phe-D-Pro-Phe-OH $\xrightarrow[\text{TBTU/6-Cl HOBt/DIEA}]{\text{HCl·Trp(Boc)-OBu}^t}$ Boc-Phe-D-Pro-Phe-Trp(Boc)-OBu^t $\xrightarrow{\text{TFA}}$

TFA·Phe-D-Pro-Phe-Trp-OH $\xrightarrow[\text{DMF}]{\text{HATU/HOAt/DIEA}}$ c[Phe^1-D-Pro^2-Phe^3-Trp^4] (CJ-15,208, I)

Fig. 1. Synthesis of CJ-15,208.

Purification and characterization. General procedure: A sample of the above crude peptide was purified to homogeneity using preparative RP HPLC on Vydac C_{18} column (218TP1022) utilizing a gradient 40%B→50%B over 50 min, at 15 mL/min flow rate with UV detection at 254 nm and solvent A being 0.08%TFA/H_2O and solvent B being 0.08%TFA/CH_3CN. Fractions eluting at R_t 40 min were collected, pooled and evaporated to yield I as a white solid (18 mg): TLC single spot, R_f 0.703, on silica gel using EtOAc/hexanes/MeOH (50:49.9:0.1); HPLC single peak (>99%), R_t 28.19 min, on Vydac C_{18} column (218TP54) using a gradient 30%B→95%B over 65 min at 1.0 mL/min flow rate and with UV detection at 220 nm and solvent A being H_2O and solvent B being MeOH; MS (ESI) *m/z* 578.7 (M + H), *m/z* 600.8 (M + Na); $[\alpha]_D^{22} = -65.4°$ (c 0.0520, DMSO).

In vitro **pharmacological evaluation.** These peptides were evaluated for opioid receptor affinity by radioligand inhibition assay using cloned opioid receptors. The results are shown in Table 1 and are consistent with the reported data [7,8] (for peptides **I** and **II**) and confirm the identity of the peptides synthesized. Peptide **III** and **IV**, however, were found to be inactive.

Table 1. In vitro binding data for peptides I-IV. Inhibition of agonist-stimulated $[^{35}S]GTP\gamma S$ binding in cloned human μ, δ, and κ opioid receptors.

Compound	μ, DAMGO K_e (nM)	δ, DPDPE K_e (nM)	κ, U69,593 K_e (nM)	μ/κ	δ/κ
I (CJ-15,208)	6.4 ± 1.0	72 ± 29	12.5 ± 2.6	0.51	5.8
II (D-Trp^4-CJ-15,208)	24.6 ± 5.0	1490 ± 196	8.1 ± 0.8	3	184
III (D-Phe^3, D- Trp^4-CJ-15,208)	817 ± 166	8848 ± 2875	1180 ± 806	1.44	7.49
IV (D-Phe^1, D-Phe^3, D-Trp^4-CJ-15,208)	3128 ± 939	inactive	inactive	-	-

Acknowledgments

We are indebted to the National Institute on Drug Abuse (NIDA) (Contract No. NO1DA-8-7763) for support of this work. We thank Mr. Keith Warner for technical assistance in obtaining the bioassay data.

References

1. Itazaki, H., Nagashima, K., Sugita, K., Yoshida, H., et al. *Antibiot.* **43**, 1524-1532 (1990).
2. Closse, A. and Huquenin, R. *Helv. Chim. Acta.* **57**, 533-545 (1974).
3. Pingle, R.B. *Plant Physiol.* **46**, 45-49 (1970).
4. Umehara, K., Nakahara, K., Kiyota, S., Iwami, M., et al. *J. Antibiot.* **36**, 478483 (1983).
5. Hirota, A., Suzuki, A., Suzuki, H., Tamura, S. *Agric. Biol. Chem.* **37**, 643-0647 (1973).
6. Saito, T., Hirai, H., Kim, Y. I., Kojima, Y., Matsunaga, Y., et al. *J. Antibiot.* **55**, 847-854 (2002).
7. Dolle, R.E., Michaut, M., et al. *Bioorganic & Med. Chem. Lett.* **19**, 3647-3650 (2009).
8. Aldrich, J.V., Kulkarni, S.S., Senadheera, S.N., et al. *ChemMedChem* **6**, 1739-1745 (2011).

Proceedings of the 23rd American Peptide Symposium
Michal Lebl (Editor)
American Peptide Society, 2013

Development of Efficient Synthetic Method for *N*-Amino Acyl *N*-Sulfanylethyl Anilide Linkers as Peptide Thioester Equivalent

Ken Sakamoto, Kohei Sato, Akira Shigenaga, Daisuke Tsuji, Kohji Itoh, and Akira Otaka

Institute of Health Biosciences and Graduate School of Pharmaceutical Sciences, The University of Tokushima, Shomachi, Tokushima, 770-8505, Japan

Introduction

Native chemical ligation (NCL) which features the use of peptide thioesters is the most practical fragment condensation method for the synthesis of proteins [1]. However, preparation of peptide thioesters using Fmoc SPPS has encountered some problems including decomposition of thioesters during peptide chain elongation. Previously, we reported that *N*-sulfanylethylanilide (SEAlide) peptides **1** as a peptide thioester equivalent could be synthesized using *N*-Fmoc amino acyl *N*-sulfanylethyl anilide linkers **2** by Fmoc SPPS [2]. Although requisite amino acyl linkers **2** have been prepared using Fmoc amino acyl chlorides resulting from treatment of Fmoc amino acids with SOCl$_2$, such treatment induces the loss of acid-labile protections such as *tert*-butyl group. In this paper, we report an efficient introduction method of Fmoc amino acid derivatives to the *N*-sulfanylethyl aniline linker **3**. In addition, the synthesis of a human GM2 activator protein (GM2AP) analog, which is an essential glycoprotein co-factor for degradation of ganglioside GM2 by β-hexosaminidase A (HexA) [3], using the SEAlide peptides **1** is also described.

Results and Discussion

Tesser, et al. reported the condensation of protected amino acids with an aniline derivative through the use of POCl$_3$ as a mild coupling agent [4]. Condensation between Fmoc amino acids and *N*-sulfanylethyl aniline linker **3** has been examined on the basis of the use of POCl$_3$. After preliminary experiments, we found that reaction protocol based on addition of an Fmoc amino acid activated by POCl$_3$ in the presence of Et$_3$N to preformed sodium anilide **4** showed wide applicability to coupling of acid-labile Fmoc amino acid derivatives to linker **3** (Figure 1, Table 1) [5,6]. The application of resulting linkers **2** to Fmoc chemistry afforded various SEAlide peptide **1**. The ligation of SEAlide peptides **1** with *N*-terminal Cys peptide yielded NCL products within 24-72 h at a low level amount of epimerization (H-**VQGS**-**Xaa-CFGRK**-NH$_2$: **Xaa** = Ser, 6.7%; **Xaa** = Ala, 1.0%; **Xaa** = His, 2.1%).

Next, we challenged the synthesis of a GM2AP analog, a glycoprotein consisting of 162 amino acid residues, using SEAlide peptides **1**. Our strategy for preparation of a GM2AP analog is shown in Figure 2. We utilized the synthetic approach, featuring the substitution of cysteine for glycosylated asparagine, to provide a glycosylation site by an S-alkylation protocol. The GM2AP analog was synthesized by the convergent synthesis using five peptide fragments. Here, sulfanyl groups on Cys and SEAlide moiety unrelated to NCL were protected by Acm group. After NCL, S-monoglycosylation at the ligation site followed by removal of Acm groups furnished the N-half segment as SEAlide peptide. One-pot/sequential ligation using SEAlide peptide followed by opening of 1,3-thiazolidine afforded the C-half segment in high chemoselective manner. Finally, the convergent assembly of the N-half (SEAlide peptide) and C-half (*N*-terminal cysteinyl peptide) segments followed by folding yielded the monoglycosylated GM2AP analog [7].

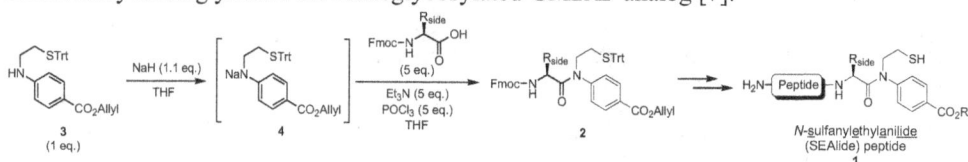

Fig. 1. Preparation of Fmoc amino acid-incorporated aniline linkers 2.

Table 1. Summary of condensation of Fmoc amino acids with aniline linker.

Fmoc amino acid	Isolated yield (%)	Racemization (%)	Fmoc amino acid	Isolated yield (%)	Racemization (%)
Asp(Ot-Bu)	66		Val	87	
Asn(Trt)	89		Met	80	
Thr(t-Bu)	84		Ile	87	
Ser(t-Bu) (r.t.)	90	1.8	Leu	88	
Ser(t-Bu) (4	91	0.4	Tyr(t-Bu)	86	
Glu(Ot-Bu)	72		Phe	88	
Gln(Trt)	93		His(τ-Trt)	41	23
Pro	39		His(π-MBom)	88	0.4
Gly	65		Lys(Boc)	80	
Ala	67		Arg(Pbf)	95	
Cys(Trt)	72	Not detected	Trp	92	

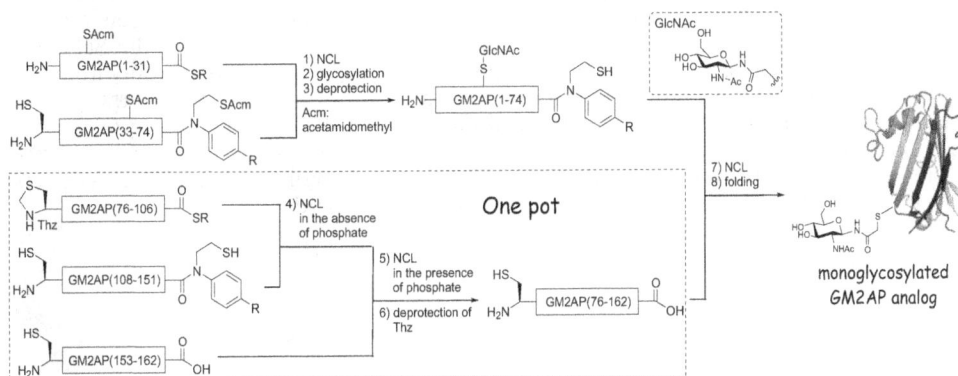

Fig. 2. Our synthetic strategy for monoglycosylated GM2 activator protein analog.

In conclusion, we developed the efficient synthetic protocol for the preparation of the Fmoc amino acid-incorporated N-sulfanylethyl-anilide linkers **2**. Application of the SEAlide peptide **1** to the synthesis of GM2AP analog was achieved. Development of a practical synthetic route for the bioactive GM2AP analog enables medicinal chemistry-based evaluation of glycoproteins as protein therapeutics. Such research is under progress in our laboratory and the results are presented in due course.

References

1. Dawson, P.E., Muir, T.W., Clark-Lewis, I., Kent, S.B.H. *Science* **266**, 776-779 (1994).
2. Tsuda, S., Shigenaga, A., Bando, K., Otaka, A. *Org. Lett.* **11**, 823-826 (2009); Sato, K., Shigenaga, A., Tsuji, K., Tsuda, S., Sumikawa, Y., Sakamoto, K., Otaka, A. *ChemBioChem* **12**, 1840-1844 (2011); Otaka, A., Sato, K., Ding, H., Shigenaga, A. *The Chemical Record* **12**, 479-490 (2012).
3. Kolter, T., Sandhoff, K. *Ann. Rev. Cell. Dev. Biol.* **21**, 81-103 (2005).
4. Rijker, D.T.S., Adams, H.P.H.M., Hemker, H.C., Tesser, G.I. *Tetrahedron* **51**, 11235-11250 (1995).
5. Sakamoto, K., Sato, K., Shigenaga, A., Tsuji, K., Tsuda, S., Hibino, H., Nishiuchi, Y., Otaka, A. *J. Org. Chem.* **77**, 6948-6958 (2012).
6. Hibino, H., Nishiuchi, Y. *Tetrahedron Lett.* **52**, 4947-4949 (2011).
7. Sato, K., Shigenaga, A., Kitakaze, K., Sakamoto, K., Tsuji, D., Itoh, K., Otaka, A. *Angew. Chem. Int. Ed.*, in press (doi: 10.1002/anie.201303390).

Proceedings of the 23rd American Peptide Symposium
Michal Lebl (Editor)
American Peptide Society, 2013

Synthesis of Photoswitchable Homodimeric Polypeptides: Towards Biological Applications

Silvia Sonzini[1], Frank Biedermann[2], and Oren A. Scherman[1]

[1]*Melville Laboratory for Polymer Synthesis, Department of Chemistry, University of Cambridge, Cambridge, CB2 1EW, UK;* [2]*School of Engineering and Science, Jacobs University Bremen, Campus Ring 1, D-28759, Bremen, Germany*

Introduction

Homodimeric polypeptides have been largely investigated on account of their enhanced properties, such as gene delivery and selective targeting, as compared to the unmodified peptides [1,2]. Dimer formation is generally achieved through cysteine bridges derived from monomer oxidation, which can be very problematic in the presence of more than one cysteine residue in a sequence. Herein, we report the synthesis of photoswitchable homodimeric polypeptides by the incorporation of an anthracene moiety. Photochemical reactions have found widespread use in the biological sciences since they are characterized by spatial and temporal control, and a short reaction time. Classic [4+4]-ene photodimerization reactions such as that exhibited by anthracene dimerization, have been successfully applied to sensing applications, biomolecule-ligations and the formation of hydrogels [3,4]. Experimental work on the rate enhancement of the [4+4]-photodimerization of anthracene through non covalent host-guest complex formation in the presence of the macrocyclic host cucurbit[8]uril (CB[8]) [5] has already been carried out in the Scherman group. CB[8], which can accommodate two guests simultaneously in its large hydrophobic cavity, brings together two anthracene moieties in a face-to-face π−π-stacked arrangement, which results in a much faster photochemical dimerization than in the absence of the host molecule. The use of a supramolecular approach will allow us to obtain a stable photoswicthable homodimer, covalently bound at a precise point, avoiding any synthetic doubt raised by the presence of more cysteine residues in a peptide sequence.

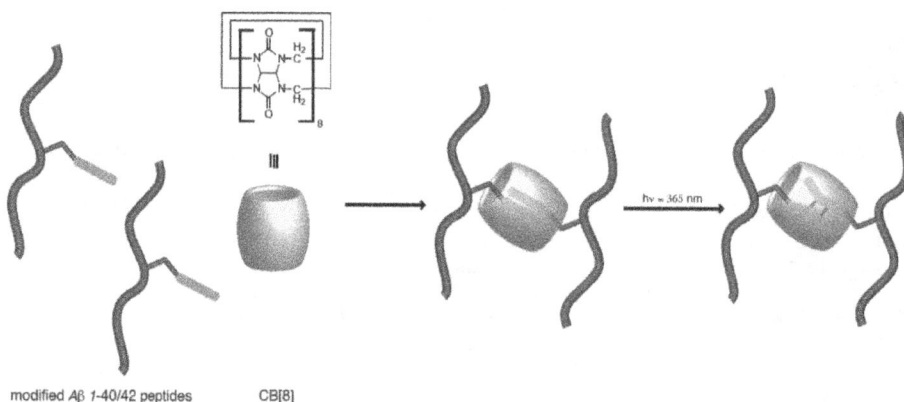

Fig. 1. Scheme of the dimerization strategy using CB[8]. The orange rectangle represents an anthracene moiety.

Results and discussion

We prepared an anthracene guest moiety for CB[8], named AntA, that can readily be inserted along a peptide sequence by solid phase peptide synthesis (SPPS). Up to now, AntA has been inserted in short peptide sequences as a proof of concept, in order to check its stability under SPPS conditions and to test its ability to bind inside the host molecule CB[8]. These short sequences include H-KGG-NH$_2$, H-OrnGG-NH$_2$, H-GGKGG-NH$_2$, H-GGOrnGG-NH$_2$. All of them were synthesized on solid support by standard Fmoc based protocols and Lysine (Lys) and Ornithine (Orn) side chains were protected with a 4-methyltrityl (Mtt) group, which allows the deprotection of the e-amine group without detaching the peptide from the resin. The deprotected sequences were coupled overnight with AntA using HATU as coupling reagent and DIPEA as base; for each sequence, the reaction reached completion after a single coupling step using three equivalents of the AntA moiety. All peptides were then checked by ESI-MS and HPLC for purity. Both the ternary complexes formed from the dimers of H-KGG-NH$_2$ and H-OrnGG-NH$_2$ have been checked by UV-visible and fluorescence spectroscopies, as well as isothermal titration calorimetry (ITC) (Figure 2); both of the sequences present absorbance and emission quenching upon CB[8] addition, and, importantly, the ITC data displays a strong binding with CB[8] with a 2:1 stoichiometry of peptide to macrocycle. Moreover the formation of the dimer upon light irradiation has been detected by UV-Vis. These preliminary but very promising results are highly suggestive that photoswitchable homoternary complexes will also form on longer and more complex sequences leading to a new approach towards peptide homodimerizations.

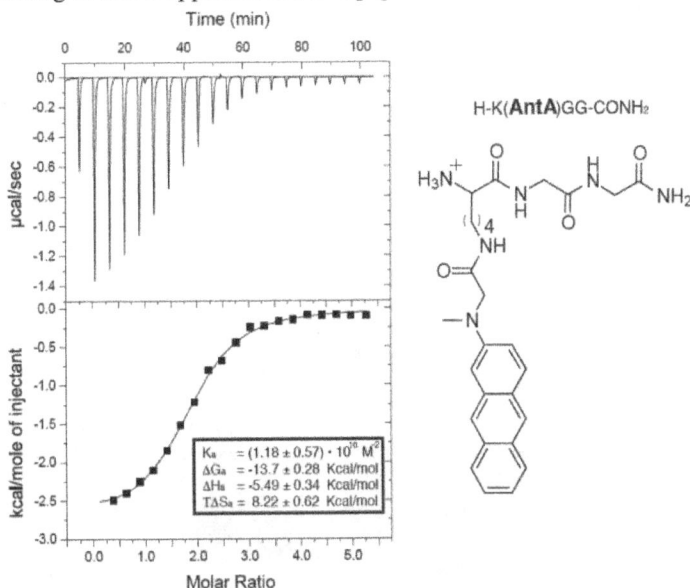

Fig. 2. ITC of 1 mM solution of H-K(AntA)GG-NH$_2$ in 10 mM phosphate buffer into 50 mM solution of CB[8] in water.

Acknowledgments

We thank the EPSRC, which supports our research. Moreover, we thank Trinity Hall College and American Peptide Society (APS), which allowed our participation at the symposium.

References

1. Kim, B.K., Lee, T.J., et al. *Bioorg. Med. Chem. Lett.* **22**, 5415-5418 (2012).
2. Aggarwal, S., Denmeade S.R., et al. *Cancer Research* **66**, 9171-9177 (2006).
3. Trenor, S.R., Shultz, A.R., Love, B.J., Long, T.E. *Chem. Rev.* **104**, 3059-3078 (2004).
4. Yamada, S., Kawamura, C. *Org. Lett.* **14**, 1572-1575 (2012).
5. Florea, M., Nau, W.M. *Angew. Chem. Int. Ed.* **50**, 9338-9342 (2011).

Proceedings of the 23rd American Peptide Symposium
Michal Lebl (Editor)
American Peptide Society, 2013

Synthesis and Analysis of a Lipid-PEG-Octreotate Conjugate for Targeted Delivery of Therapeutic Agents in Liposomes

Zhiyong Tao*, Daniel R Studelska, James R Wheatley, Eric A Burge, Miranda Steele, Debbie Brame, Claire Brook, Alex Micka, Christie Belles, and Todd Osiek

Dept. of Peptide Chemistry, Mallinckrodt, 3600 North Second Street, St. Louis, MO, 63147, U.S.A.

Introduction

Tyr[3] Octreotate binds the SSTR2 receptor, over expressed in some neuroendocrine cancers. DSPE-PEG$_{5000}$-Tyr[3] Octreotate conjugation enables the decoration of liposomes with this targeting moiety. Tyr[3] Octreotate has an *N*-terminal and lysine NH$_2$ available for solution phase conjugation with DSPE-PEG$_{5000}$. The NH$_2$ on lysine is important for the binding of the peptide to the SSTR2 receptor and needs to be protected during conjugation. We describe the solid phase synthesis of Cyclic [2,7]-Tyr[3]-Lys(ivDde)-Octreotate, its solution phase conjugation with DSPE-PEG$_{5000}$ and the removal of the ivDde protecting group from the lysine (Figure 1). The analytical monitoring of the conjugate during solution phase conjugation and ivDde removal by hydrazine is presented. An assay to determine *N*-terminal or side chain functionalization of Tyr[3] Octreotate by the DSPE-PEG$_{5000}$ is shown in Figures 2-7.

Synthesis

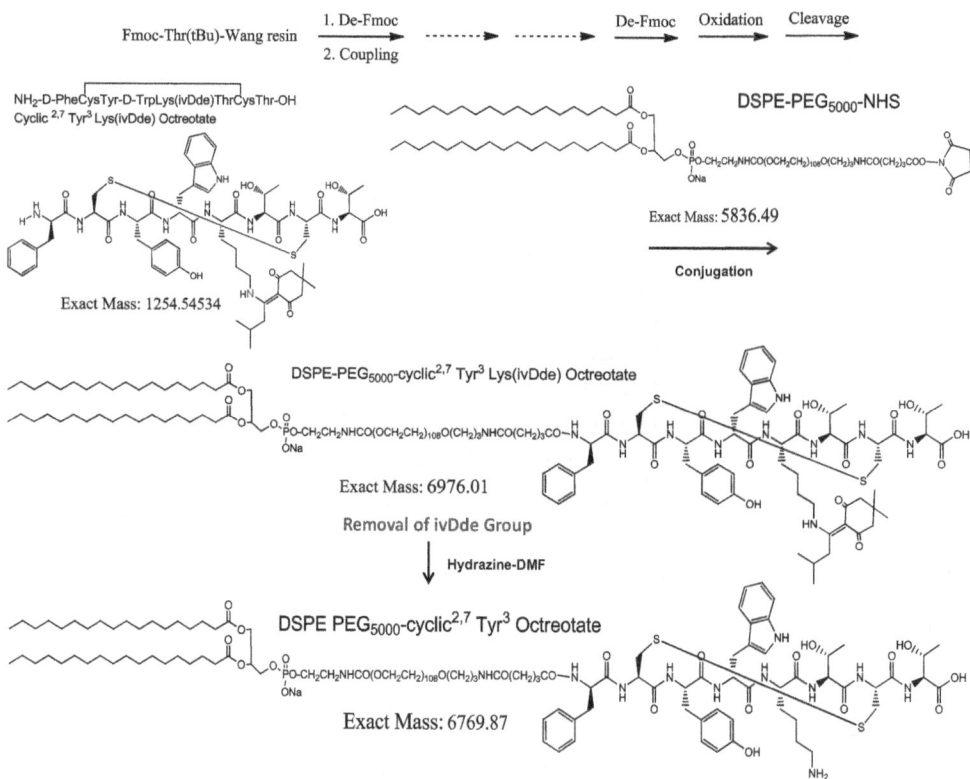

Fig. 1. Solid phase peptide synthesis of Cyclic[2,7] Tyr[3] Lys(ivDde) Octreotate; solution phase conjugation with DSPE-PEG$_{5000}$-NHS and removal of ivDde protecting group.

Analysis and Assay for NH₂ Coupling

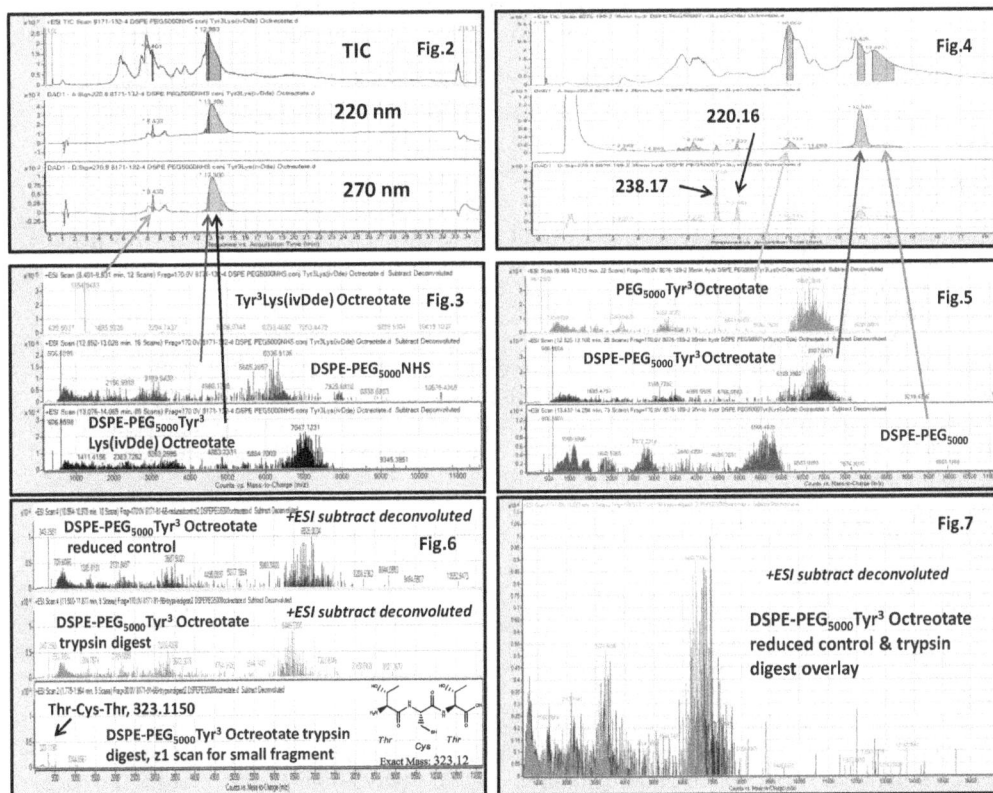

Figs. 2 & 3. LC-MS of Crude DSPE-PEG$_{5000}$-cyclic^{2,7} Tyr^3 Lys(ivDde) Octreotate; Figs. 4 & 5. LC-MS of Crude DSPE-PEG$_{5000}$-cyclic^{2,7} Tyr^3 Octreotate after 35 min 1.0% hydrazine to remove ivDde; Figs. 6 & 7. ESI$^+$ subtract deconvoluted DSPE-PEG$_{5000}$Tyr^3 Octreotate reduced control & trypsin digest.

Summary

DSPE-PEG$_{5000}$-Octreotide conjugates cannot be eluted from C18 columns with acetonitrile. They require at least the eluant strength of IPA/MeOH. There are many minor impurities in commercial DSPE-PEG$_{5000}$NHS preparations readily detected in positive mode by the Agilent LC-TOF we employed. Details were not shown for clarity. The major DSPE-PEG$_{5000}$NHS peak is consumed during solution phase conjugation. Removal of ivDde from the lysine of DSPE-PEG$_{5000}$-Tyr^3-Lys(ivDde)-Octreotide by 1% hydrazine/DMF requires nucleophilic attack; unchecked, the deprotection proceeds to cleave lipid from the desired product. Protection of Tyr^3 Octreotide lysine by ivDde efficiently directs conjugation of DSPE-PEG$_{5000}$NHS to the *N*-terminus of the peptide, as confirmed by a trypsin digestion assay.

References

1. Zhang, J., et al. *Molecular Pharmaceuticals* **7**, 1159-1168 (2010).
2. Simpure Iikka, European Patent Application EP1 738 770 A1 (2007).

Proceedings of the 23rd American Peptide Symposium
Michal Lebl (Editor)
American Peptide Society, 2013

Solvent-Free Synthesis of Bhc-Cl and Application to Cysteine Protection in Solid Phase Peptide Synthesis

Mohammad M. Mahmoodi, Tomohiro Kubo, Daniel Abate-Pella, Jane E. Wissinger, and Mark D. Distefano

Department of Chemistry, University of Minnesota, Minneapolis, MN, 55455, U.S.A.

Introduction

Photoremovable protecting groups are useful for a wide range of applications in peptide chemistry. Recently we have begun to explore the use of the brominated hydroxycoumarin (Bhc) group for thiol protection, a group that can be removed by both one and two photon excitation. Here, we first report a new method for the synthesis of Bhc-Cl using solvent free conditions. Next, we describe using Bhc-Cl to protect thiol functionality in a peptidomimetic farnesyl transferase inhibitor (FTI). Then, we describe the synthesis of Fmoc-Cys(Bhc)-OH and its incorporation into several peptides based on the *C*-terminus of the K-Ras protein, KKKSKTKCVIM. Finally, we report on the results of the photolysis reactions. Interestingly, Bhc removal proceeded smoothly in one case while in several others, a rearranged product was observed.

Results and Discussion

Photoremovable protecting groups (caging groups) have become a useful tool in addressing a wide range of issues in biology, because the light induced release of bioactive molecules inside living systems is orthogonal to other triggers. By adding a caging group to a molecule of interest, it can be rendered inactive to form a species called a "caged compound". Upon irradiation, the active molecule is released (uncaged) at only the time and position where the light is irradiated [1].

Cysteines play important roles in peptide and protein chemistry, such as protein folding, formation of disulfide bridges, and various protein modifications. One important biological process that involves cysteine modification is protein prenylation. In this process, enzymes called protein prenyltransferases catalyze the attachment of either a farnesyl (C_{15}) or geranylgeranyl (C_{20}) isoprenoid to conserved cysteine residues near the *C*-termini of specific proteins. Protein prenylation has been shown to be important in the cellular signal transduction processes and is implicated in numerous diseases including multiple types of cancers. Development of photocleavable thiol-protected inhibitors and peptides, which are only sensitive to light for activation, would be very useful tools to study the role that prenylation plays in diseases, as well as be useful for studying other cysteine related biological processes [2].

Recently, our lab has begun to explore the use of the brominated hydroxycoumarin (Bhc) group for thiol protection. Bhc is an efficient caging group with high one- and two-photon sensitivity which has been used for the protection of various functionalities. Reported procedures for synthesis of Bhc [3] involve using concentrated acids as a reaction solvent, which is not environmentally friendly. To avoid using these strong acids, we developed a straightforward solvent-free procedure for the synthesis of Bhc. In this method, 4-bromoresorcinol and ethyl chloroacetoacetate were ground together for 15 min and left undisturbed overnight at room temperature. With the addition of water, followed by filtration, Bhc was successfully obtained as a white solid in 30% yield (Figure 1A). However, due to the heterogeneous nature of this solvent-free reaction, the yield is highly variable. Hence, efforts to improve this process are currently underway.

Once synthesized the Bhc was used to produce a caged FTI. Initially, the photolysis of the caged FTI in buffered aqueous solution (Figure 1B) was studied by following the disappearance of the starting material upon photolysis and the appearance of the FTI. Further analysis by LC-MS also confirmed the formation of free FTI upon photolysis [4]. Being able to release FTI through irradiation, we next examined the effect of uncaging the FTI on cellular properties. Ciras-3 cells, a fibroblast cell line, which had been previously engineered to constitutively express an oncogenic ras gene that resulted in an aberrant smaller, rounded cell morphology were used to study FTI uncaging. This cell type was chosen because treatment with free FTI results in their conversion from the small, circular morphology to a larger and more spread phenotype. Cells treated with the Bhc-FTI showed no change in morphology. However, upon irradiation cells incubated with Bhc-FTI caused the cells to take on the characteristic spread appearance. This result showed that the oncogenic ras has been suppressed by the photo-released of the FTI (Figure 1C).

After successful use of Bhc for thiol protection and UV deprotection of an FTI, we decided to use it for development of caged cysteine peptides. We synthesized Bhc protected Fmoc-Cys-OH, through alkylation of Fmoc-Cys-OMe using MOM-Bhc-Cl, followed by methyl ester hydrolysis with $(CH_3)_3SnOH$ (Figure 1D). This Fmoc-Cys(Bhc)-OH was incorporated into a K-Ras sequence peptide through solid phase peptide synthesis. The final sequence was 5-Fam-KKKSKTKC(Bhc)VIM. A buffered aqueous solution of the purified caged peptide was irradiated with UV light and the photolysis reaction was monitored by LC-MS. Interestingly, the LC-MS data showed that photolysis resulted in isomerization of the peptide ($[M+3H^+]$=635.2 m/z calc., 635.3 m/z obs.), and not uncaging. Analysis of the MS/MS fragmentation pattern indicated that upon photolysis the Bhc group remains attached to the cysteine residue and does not migrate to any neighboring residue (Figure 1E). Based upon this data, we proposed that photolysis leads to formation of the isomerized product shown in Figure 1F, but this has yet to be confirmed. In summary, we have successfully used a green synthetic method to synthesis Bhc and then used this Bhc as a thiol-caging group for an FTI. However, Bhc removal in caged cysteine peptides led to the formation of a rearranged product. This suggests Bhc photo-deprotection is highly sensitive to its chemical environment. Future work will be directed toward isolating and confirming the photo rearranged product and also exploring other caging groups for thiol protection.

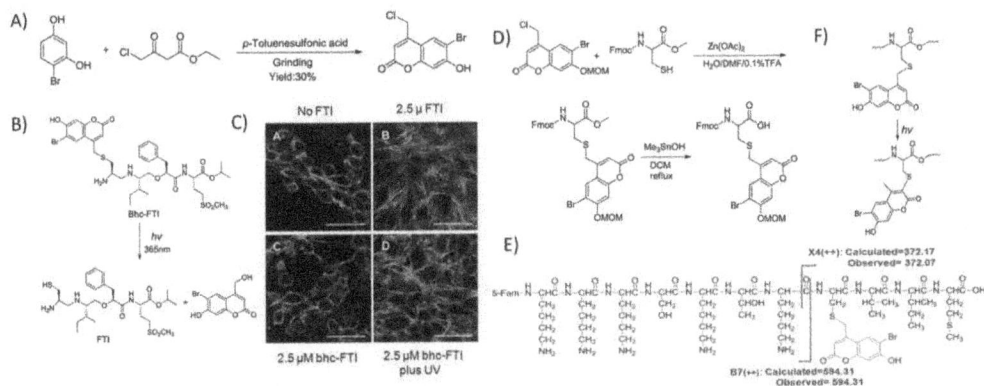

Fig. 1. A) Solvent-free synthesis of Bhc. B) Photolysis of reaction of Bhc-FTI which produces free FTI. C) Morphology of Ciras-3 cells treated with FTI and Bhc-FTI [4]. D) Synthesis of Bhc protected Fmoc-Cys-OH. E) Two key MS/MS fragments showing rearrangements occurs on cysteine. F) Hypothesized photoisomerization reaction Fmoc-Cys(Bhc)-OH.

Acknowledgments

We thank the LeClaire-Dow Instrumentation Facility for ESI-MS and NMR measurements. This research was supported by the National Institutes of Health (GM058842 and GM084152, MDD).

References

1. Ellis-Davies, G.C.R. *Nat. Methods* **4**, 619-628 (2007).
2. Gelb, M.H., et al. *Nat. Chem. Biol.* **2**, 518-528 (2006).
3. Furuta, T., et al. *Proc. Nat. Acad. Sci. USA* **96**, 1193-1200 (1999).
4. Abate-Pella, D., et al. *Chembiochem* **13**, 1009-1016 (2012).

Author index

Keyword index

www.ingramcontent.com/pod-product-compliance
Lightning Source LLC
Chambersburg PA
CBHW051206200326
41519CB00025B/7024